STP 1282

Sampling Environmental Media

James Howard Morgan, Editor

ASTM Publication Code Number (PCN):
04-012820-56

ASTM
100 Barr Harbor Drive
West Conshohocken, PA 19428-2959
Printed in the U.S.A.

Library of Congress Cataloging-in-Publication Data

Sampling environmental media / James Howard Morgan, editor.
   (STP ; 1282)
   "ASTM publication code number (PCN) : 04-012820-56"
   Includes bibliographical references and index.
   ISBN 0-8031-2043-5
   1. Pollution--Management--Congresses. 2. Environmental sampling--Congresses. I. Morgan, James Howard, 1950- . II. Series: ASTM
TD193-S26 1996
628.5'028'7--dc20
96-27603
CIP

Photocopy Rights

Peer Review Policy

Each paper published in this volume was evaluated by three peer reviewers. The authors addressed all of the reviewers' comments to the satisfaction of both the technical editor(s) and the ASTM Committee on Publications.

To make technical information available as quickly as possible, the peer-reviewed papers in this publication were prepared "camera-ready" as submitted by the authors.

The quality of the papers in this publication reflects not only the obvious efforts of the authors and the technical editor(s), but also the work of these peer reviewers. The ASTM Committee on Publications acknowledges with appreciation their dedication and contribution to time and effort on behalf of ASTM.

Printed in Philadelphia, PA
September 1996

Foreword

This publication, *Sampling Environmental Media,* contains papers presented at the symposium of the same name, held on 5–7 April 1995. The symposium was sponsored by ASTM Committee D-34 on Waste Management. James Howard Morgan of The MITRE Corporation at Brooks Air Force Base, Texas presided as symposium chairman and is editor of the resulting publication.

Contents

Overview

In order to protect human health and ecology, measurements and samples of the earth's many different media are required to characterize and remediate pollution in our environment. Samples and measurements of the environment's condition are organized to depict conceptual site models (CSMs), representative of actual conditions by environmental professionals, who use them as decision-making tools. The systems for collecting, preserving, analyzing, and applying the information obtained from samples representative of various environmental media are often not comparable or well understood. Frequently, sampling error distorts or adversely impacts the conceptual models and the consequent decisions affecting pollution management.

The Symposium for Sampling Environmental Media was sponsored by the ASTM Committee D-34 for the purpose of encouraging the exchange of knowledge about environmental sampling. Sampling equipment, techniques, and systems were examined during the symposium to determine their representativeness with regard to a true picture of environmental conditions and the critical elements to successfully apply and use those sampling tools. Samples of all the earth's materials and media are collected to characterize real-world environmental conditions. ASTM environmental committees have traditionally organized themselves by materials and media association, (for example, the Committee D-18 on Soil and Rock or Committee D-34 on Waste Management). It was the intention of the organizers that this symposium would address issues requiring integration of resources and expertise from across all of the ASTM environmental committees. Session chairpersons, presenters, and authors of papers presented at the symposium represented the membership of ASTM Committees D-18, D-19, D-22, D-34, E-47, E-50, and E-51. Topics presented at the symposium required integrated analysis from the disciplines: chemistry, geology, engineering, biology, and risk assessment, as well as an understanding of technical challenges when sampling environmental media in air, soil and rock, soil gas, surface water, wastewater, groundwater, or solid waste. Thirty-one peer reviewed papers are collected in this volume. Support and cooperation from each of the ASTM main committee's writing environmental standards has brought both the symposium and this volume to fruition.

Papers in this publication are organized according to their associated sessions at the symposium. Individual sessions were presented on the following topics:

- Sampling Systems
- Worker Safety and Risk Characterization
- Direct Push Sampling
- Sampling Media
- Sampling Subsurface Media
- Sampling Strategies
- Soil and Soil Gas Sampling
- Innovative Measurements
- Quality Assurance/Quality Control

Readers of this ASTM publication will find it to be an informative and useful reference on many topical environmental sampling issues. ASTM STP 1282 focuses on sampling issues related to the construction or analysis of CSMs. This volume may also serve as a resource guide for identifying ASTM standards related to environmental management and environmental sampling. Included with A. Ivan Johnson's paper entitled "The Accelerated Development of Standards for Environmental Data Collection," is an appendix listing all the environmental standards developed by ASTM. Use of these standards have assisted in improving sample comparability across the environmental management profession.

A number of important themes are consistently woven throughout the papers included within. Among them are: (1) faster, cheaper, better; (2) practical, common-sense approaches; (3) applications for unique, imaginative and innovative science; and (4) integrated environmental systems management. Each of the themes reflect current issues and concerns facing the environmental industry. Many of the papers address real solutions to problems that challenge the application of these themes when constructing a CSM. For example, from Colorado Springs, Colorado, Susan Soloyanis' paper entitled, "A Common Sense Strategy to Expedite Hazardous Waste Site Cleanup" incorporates elements of all four themes and provides a practical guide for achieving cost-effective and timely clean-up remedies that are protective of human health and the environment. From Birmingham, England, P. D. Hedges' paper discussing "Airborne Remote Sensing as a Tool for Monitoring Landfill Sites Within an Urban Environment" describes a practical use of very advanced remote sensing tools and further advances the theme "applications of unique, imaginative and innovative science." One pragmatic site characterization sampling tool discussed frequently at the symposium was the use of a cone penetrometer to characterize contaminants in soil, soil gas, and groundwater. Several papers presenting unique and imaginative methods for applying this sampling tool are:

- Methods of Determining In-Situ Oxygen Profiles in the Vadose Zone of Granular Soils
- The Multiport Sampler: An Innovative Sampling Technology
- Detailed Characterization of a Technical Impracticability Zone Using Drive Point Profiling
- Research and Standardization Needs for Direct Push Technology Applied to Site Characterization

Other papers of special note are:

- Estimation of Volatile Organic Compound Contamination in the Vadose Zone: A Case Study Using Soil Gas and Methanol Preserved Soil Sample Results
- Utilization of Soil Gas Monitoring to Determine the Feasibility and Effectiveness of In-Situ Bioventing in Hydrocarbon Contaminated Soils
- Innovations to the CERCLA Remedial Investigation Process at Closure Bases

The collection of samples and data representative of a media's real environmental condition is the most fundamental challenge to construction of a realistic CSM. Collection of representative samples and development of realistic CSMs are the environmental industries' foundation for effective human health and ecological risk management. Each year numerous advances occur with respect to collecting representative samples of environmental media. With each advance, a new technical issue or applied integration problem also occurs. Consequently, every environmental professional is challenged to seek a broader environmental data base and future symposia focusing on advances of the environmental sampling techniques and methods will be needed. A second symposium for sampling environmental media is being planned by ASTM's environmental committees for the spring of 1997.

James Howard Morgan

The MITRE Corporation,
Symposium Chairman and Editor.

Acknowledgments

The editor gratefully acknowledges the voluntary support of the numerous individuals who served as session chairpersons and topical peer reviewers. Their contributions made the symposium presentation and this publication possible. Special acknowledgment is extended to Mark Marcus, Ph.D., who served as mentor, counselor, and co-editor throughout the development of STP 1282.

Session Chairpersons for the Symposium for Sampling Environmental Media

Mark Marcus, Ph.D.
Recra Environmental, Inc.
Amherst, New York

Fred T. Price, Ph.D.
The MITRE Corporation
McLean, Virginia

Gwen Eklund
Radian Corporation
Austin, Texas

Thomas Doane, Ph.D.
Battelle Memorial Institute
San Antonio, Texas

Richard Lewis
Groundwater Technology, Inc.
Norwood, Massachusetts

Gomes Ganapathi, Ph.D.
Bechtel—Oak Ridge Corporate Center
Oak Ridge, Tennessee

Scott Macrae
Waste Policy Institute
San Antonio, Texas

Richard Brown, Ph.D.
Horne Engineering and Environmental
    Services
Alexandria, Virginia

Mitzi Miller
Environmental Quality Management
Knoxville, Tennessee

Brian Anderson
SCA Business Services
Sauk City, Wisconsin

James Mickam
O'Brien and Gere Engineers, Inc.
Syracuse, New York

A. Ivan Johnson[1]

THE ACCELERATED DEVELOPMENT OF STANDARDS FOR ENVIRONMENTAL DATA COLLECTION

REFERENCE: Johnson, A. Ivan, ''The Accelerated Development of Standards for Environmental Data Collection,'' Sampling Environmental Media, ASTM STP 1282, James Howard Morgan, Ed., American Society for Testing and Materials, 1996.

ABSTRACT: There has been an increasing need for uniformly high quality basic data for making decisions related to the regulation of wastes and the protection of the environment, especially the quantity and quality of the nation's water resources. Hundreds of Federal and non-Federal agencies have recognized the need for high-quality standards that provide greater comparability, compatibility, and usability of all types of environmental data. The wide use of computerized data banks, also has accelerated the need for stored data of known quality.

Federal co-operation in the development of standard methods was initiated in the early 1970's through the U.S. Geological Survey's Office of Water Data Coordination. As a result of a co-ordinated effort of Federal agencies, with review by a group of non-Federal representatives, a "National Handbook of Recommended Methods for Water Data Acquisition" was produced. This handbook contained many ASTM existing standards and made recommendations for many others that could be developed by a variety of ASTM committees.

As the needs increased for high-quality data, and more regulatory and quality assurance requirements were developed, the need for consensus-type standards also increased. As a result, during the past several years joint efforts by the EPA, U.S. Geological Survey, and U.S. Navy have resulted in a funding process whereby highly accelerated development has been made possible by ASTM for a wide variety of standard methods, practices, guides, and terminology. The process of developing such standards in relation to a variety of water resources, waste management, and environmental problems are described.

[1] President, A. Ivan Johnson, Inc., 7474 Upham Court, Arvada, CO 80003.

KEYWORDS: contamination, environmental assessment, ground

water, open-channel flow, petroleum releases, rock, soil,
storage tanks, vadose zone, water, waste management

INTRODUCTION

In the United States over 30 Federal Agencies, hundreds
of State agencies,and thousands of local agencies, universities, and private organizations are involved in collecting
and disseminating water and environmental data. In the last
two decades, hundreds of Federal and non-Federal agencies
have had increasing needs for uniformly high-quality data
for making decisions related to development, conservation,
management, and regulation of the quantity and quality of
the Nation's Water resources and for protection of the
environment.
The search for appropriate data to solve rapidly increasing water problems led to an increased demand for
water-resources data in the early 1970's. The search
brought to light numerous deficiencies in the data base, one
of which was the difficulty of ascertaining the quality of
existing data. Another deficiency was the lack of comparability and compatibility of data collected by different
organizations, or in some cases even by different parts of
the same organization. Many of these problems were caused
by differences in the methods used for acquiring, handling,
and archiving the data.
The decades of the 70's and 80's saw a huge prolifera-
tion of computerized data banks, most of which had no indication of the quality of the stored data or the method by
which it had been collected. During the same decades many
regulations were developed by EPA and other regulatory
agencies that required legally-binding enforcement actions.
These regulations initially impacted surface water quantity
and quality, and later impacted ground water quantity and
quality.

EARLY EFFORTS TO DEVELOP RELIABLE WATER DATA

In an effort to meet the increasing demands for efficient and economical water data, the Office of Management
and Budget (OMB) issued Circular A-67 in 1964. This Circular prescribed guidelines for coordinating water quality and
quantity data-acquisition activities in the Nation's
streams, lakes, estuaries, reservoirs, and ground waters.
Responsibility for implementing Circular A-67 was assigned
to the U.S. Geological Survey (USGS), the agency that ac-
quires about 70 percent of the water data at Federal level.
To carry out the Circular's mission, the USGS established the Office of Water Data Coordination (OWDC). The
major functions assigned to this office were (1) design a
national network for acquiring water data, (2) coordinate
the national network and specialized water-data acquisition,

(3) maintain a central catalog of information on water data, and (4) develop a Federal plan to acquire needed water data. This author was detailed to Washington, D.C. in 1964 for several months staff study to assist in organizing the Office of Water Data Coordination, and in 1971 was transferred to that office and served there until 1979.

To provide advice and counsel in implementing Circular A-67, the Secretary of the Interior established a Federal and a non-Federal advisory committee--the Interagency Advisory Committee on Water Data and the Advisory Committee on Water Data for Public Use, respectively. The Federal committee consists of representatives and alternates from about 30 Federal agencies, and the non-Federal committee consists of about 30 representatives of national, State, and Regional organizations, universities, consulting firms, and technical and public-interest societies. Members of the non-Federal Advisory Committee include representatives from organizations such as ASTM and the American Society of Civil Engineers, American Water Resources Association, Association of American State Geologists, Association of Western State Engineers, National Water Well Association, and the Universities Council on Water Resources. Both committees have a number of subcommittees and ad hoc working groups to address a variety of special problems (Figure 1).

In developing early plans to implement OMB Circular A-67, the need for standardizing data-collection methods and data handling and dissemination received attention by two working groups of the Federal advisory committee. Following two years of interagency efforts, these two groups produced significant recommendations that were approved not only by the Federal advisory committee but also by the non-Federal advisory committee. The recommendations of the data-handling working group provided the design characteristics for the National Water Data Exchange (NAWDEX)--a national confederation of water-oriented organizations working together to improve access, through a computerized index, to water data that was collected by approximately 30 Federal agencies and hundreds of state and local organizations.

The Federal Working Group on Designation of Standards for Water-data Acquisition recommended an interagency activity to develop a "National Handbook of Recommended Methods for Water-Data Acquisition." During the 1970's, a coordinated move in the direction of standardization of water data acquisition was carried out by the Office of Water Data Coordination through its Federal interagency committee. With this author serving as Methods Coordinator, approximately 200 technical personnel from 26 Federal agencies were assigned to 12 working groups to produce the "National Handbook" (Figure 1). The Working Groups' products in turn were coordinated with the Non-Federal Advisory Committee on Water Data for Public Use, all members of which had the opportunity to review chapters of the "National Handbook" as they were developed. This particular coordination activity went a long way towards improving the comparability, compatibility, and usability of water data.

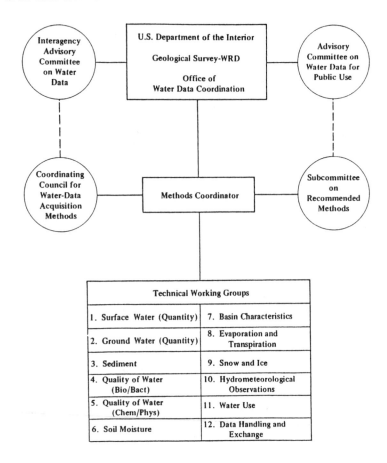

FIG. 1--Organizational Relationships in Developing
the "National Handbook of Recommended
Methods for Water-Data Acquisition."

Beginning in 1976, efforts were made to coordinate this
interagency methods activity with standards activities of
ASTM committees, whereby certain ASTM committee standards
would be considered for adoption in the National Handbook.
In turn, methods from the manuals of various agencies were
made available to work into ASTM standard format and then
move through the ASTM balloting and approving process to
become the widely accepted voluntary consensus standards for
which ASTM is well known. However, the standards development
was deficient in standards for environmental investigations-
-mainly because most such methods had never been put into a
printed form that had been developed by the broad consensus
process. Following contact between the Methods Coordinator
and the ASTM committees, the committees saw the need and
organized new subcommittees appropriate for processing

standards related to the needs of the specific chapters of
the National Handbook. The interchange of resource materials
and ASTM standards related to water investigations continued
throughout development of the "National Handbook."

IMPACT OF NEW REGULATIONS

During the past two decades Congress tackled the increasing environmental contamination problems by enactment
of Federal Laws, such as the Resource Conservation and
Recovery Act (RCRA) and the Comprehensive Environmental
Response, Compensation, and Liability Act (CERCLA, or
"Superfund"). Subsequently EPA developed regulations to
implement these laws. The regulations for RCRA require
establishment of ground water or unsaturated zone monitoring
systems, or both, at all hazardous waste treatment, storage,
and disposal facilities. In addition, the regulations for
CERCLA, covering cleanup of "uncontrolled" hazardous waste
sites, require installation of a system to monitor the
extent of environmental contamination and progress of the
cleanup. Ground water and unsaturated zone monitoring also
is done in connection with other regulatory programs, as
well as for many non-regulatory purposes.

The interest in use of standards development organizations by Federal agencies became even stronger in the 1970's
and resulted in the 1980 release of OMB Circular A-119. The
Circular prescribed that whenever possible Federal agencies
should use public sector standards employing a wide consensus process rather than developing standards in-house. A
1993 update of the OMB Circular prescribed even stronger
standards development guidelines for Federal agencies.

THE NEED FOR STANDARD METHODS

In the past few years, the variety of regulatory and
quality assurance requirements have accelerated the need for
more environmental standards. Especially needed are standards developed, reviewed, and approved by the consensus of
a broad interdisciplinary group of experts that represent
industry, academia, consultants, and equipment
manufacturers, as well as government at all levels. Practitioners of environmental investigation and monitoring, as
well as data collecting and regulatory agencies, have realized that establishing credibility for environmental investigations required improvement of the state of the art
through development of useful and widely accepted consensus
guidelines and standards. However, the idea of developing
standards has been especially controversial for the very
interdisciplinary sciences with which environmental investigators have to relate.

Participants in the new ASTM Committee activities have
found that such standards development may be difficult but
is not impossible. One must simply remember that the basic
ASTM definition of a "standard" is "a rule for an orderly

approach to a specific activity, formulated and applied for
the benefit, and with the cooperation, of all concerned."
The two key points in this definition are that it is "an
orderly approach" and that it is "applied for the benefit,
and with the cooperation, of all concerned."

A SPECIFIC NEED FOR GROUND WATER STANDARDS

As was mentioned earlier in this paper, the development
of reliable ground water and other water standards initially
was explored as early as 1976 within ASTM. Standards that
could be used for laboratory tests related to some quantitative aspects of ground water investigations had been developed by ASTM Committee D18 on Soil and Rock, whose scope
includes "fluids contained therein." In the mid 1980's a
section on ground water monitoring standards was organized
within Subcommittee D18.01 on Surface and Subsurface Characterization. Standards related to chemical, physical, and
biological quality aspects of water in general had been
developed by Committee D19 on Water. That committee also
decided to take on development of fluvial sediment and
stream flow measurement standards. Committee D22 on Atmospheric Sampling and Analysis added a meteorological subcommittee to develop standards related to hydrometeorological
methods. Committee D34 on Waste Management organized a
Subcommittee on sampling and monitoring waste products.
To provide a bank of resource material on new methods
that possibly could lead to the development of the needed
new standards on ground water field methods, ASTM Committees
D18 and D19 organized a "Symposium on Field Methods for
Ground Water Contamination and Their Standardization" held
in February 1986 in Cocoa Beach, Florida. Most of the accepted papers related to the quantitative aspects of ground
water and vadose zone investigations, subjects for which
there were existing sections and expertise within Committee
D18. Selected papers from that symposium were published in
ASTM Special Technical Publication (STP) 963, edited by A.G.
Collins and A.I. Johnson, and published in 1988. This symposium provided additional momentum and resource documents for
development of additional ground water and vadose zone
methods in Committee D-18.

DEVELOPING NEEDED GROUND WATER STANDARDS

Following the Cocoa Beach symposium, the Ground Water
Section in D18.01 was reorganized into Subcommittee D18.21
on Ground Water Monitoring, with 10 sections. To develop
more resource material to initiate development of standards,
an ASTM-EPA-USGS sponsored "Symposium on Standards Development for Ground Water and Vadose Zone Monitoring" was con-
vened in January 1988 during the ASTM D18 meeting in Albu-
querque, New Mexico. Selected papers from that symposium
are published in ASTM STP-1053 "Ground Water and Vadose Zone
Monitoring," edited by David M. Nielsen and A. Ivan Johnson,

and published in 1990.

Subsequently, Subcommittee D18.21 on Ground Water Monitoring changed its name to "Ground Water and Vadose Zone Investigations" to more properly indicate its broad subsurface coverage and interest in all types of investigations in the saturated and unsaturated zones. The Subcommittee was subdivided into Sections organized to address a variety of narrower subject areas. These subjects included Surface and Borehole Geophysics; Vadose Zone Monitoring; Well Drilling and Soil Sampling; Hydrogeologic Parameters; Well Design and Construction; Well Maintenance, Rehabilitation, and Abandonment; Ground Water Sample Collection and Handling; Design and Analysis of Hydrogeologic Data Systems; Monitoring in Karst and Fractured Rock Terrains; and Ground Water Modeling. With this organizational structure in place, and a membership of over 300 engineers, hydrologists, geologists, soil scientists and other ground water specialists, it became possible to launch a concentrated effort to use the full ASTM consensus process to develop standards needed to insure the collection of high-quality ground water data that are comparable, compatible, and usable no matter where or by whom collected.

FEDERAL AGENCY PARTNERSHIP

In late 1988, discussions between this author, as representative of ASTM, and members of the Environmental Monitoring Systems Laboratory (EMSL) of the Environmental Protection Agency (EPA), Las Vegas, Nevada, and of the Office of Water Data Coordination (OWDC), of the U.S. Geological Survey (USGS), Reston, Virginia, resulted in a cooperative agreement between these organizations and ASTM to accelerate development of ground water and vadose zone monitoring standards in six high-priority subject areas. Under this agreement expenses were provided for six 3-person task groups to hold three to four extra 2-day Task Group meetings per year. In addition to the three specialists from the non-Federal sector, at least one EPA and USGS specialist was assigned to each task group. The first funded task groups were Borehole Geophysics; Surface Geophysics; Vadose Zone Monitoring; Well Drilling and Soil Sampling; Determination of Hydrogeologic Parameters (aquifer field testing); Monitoring Well Design and Construction; and Monitoring Well Maintenance, Rehabilitation, and Abandonment. As a result of that agreement, over 20 detailed standards were produced for subcommittee or committee ballot by the end of 1989.

In early 1990, discussions between this author and a staff member of the U.S. Navy Engineering Facilities Command in Port Hueneme, California resulted in the Navy adding funding to the cooperative agreement and four new subject area task groups were added -- Ground Water Sample Collection (which included representation from Committees D19 on Water and D34 on Waste Management, as well as Committee

D18); Design and Analysis of Hydrogeologic Data Systems; Special Problems of Monitoring Karst and Fractured Rocks; and Ground Water Modeling. The USGS also started funding the development of standards on surface water hydraulic measurements and fluvial sediment sampling. In early 1993, EPA and USGS started additional funding of ground water related task groups for four existing sections in Subcommittee D18.01 on Surface and Subsurface Characterization. These new task groups were on Site Characterization for Environmental Purposes; Geostatistics; Site Characterization for Septic Systems; and GIS and Mapping. Also in 1993, an office of the EPA, Washington, D.C. added funding for four task groups of Committee D-34, primarily related to sampling of waste and a standard on development of Data Quality Objectives (DQO's). Four Task Groups on Underground Storage Tanks in Committee E50 on Environmental Assessment were added by Las Vegas EPA Laboratory funding in 1994.

DEVELOPMENT OF STANDARD TERMINOLOGY

Another important result of the program has been development of standard terminology related to the subjects of the accelerated standards. For example, the definitions for ground water terms were taken for balloting from a ground water glossary prepared by a Federal interagency working group impaneled by the USGS Office of Water Data Coordination and its Federal Advisory Committee. Over 200 ground water terms and definitions have been balloted to assist in having a common language for communication among ground water, vadose zone, and related environmental specialists.

SUMMARY

The completed ground water and vadose zone standards, as well as the standard terminology, of Committee D18 on Soil and Rock are published in the Annual Book of ASTM Standards, volumes 04.08 and 04.09. In addition the funded standards approved as of early 1994 were available in a special separate publication entitled "ASTM Standards on Ground Water and Vadose Zone Investigations," which in 1995 became available in an expanded second edition. The other three funded Committees also are rapidly developing standard guides, practices, methods, and terminology related to their respective specialties. Standards of these latter committees are being published in the Annual Book of ASTM Standards, volumes 11.01 and 11.02 for Committee D19 on Water and volume 11.04 for Committees D34 on Waste Management and E50 on Environmental Assessment.

By late 1995, the dedicated work of the Federally-funded task groups have produced nearly 200 standards or drafts related to ground water, waste management, surface water hydraulics, and environmental assessment. Of these, over 100 drafts have been approved as fully accepted ASTM consensus standards and the others are in some form of

subcommittee, committee, or Society ballot. In the Appendix
of this Special Technical Publication, Tables 1 through 4
list the standards under the jurisdiction of the four ASTM
Committees having Federally funded task groups--D18 on Soil
and Rock, D19 on Water, D34 on Waste Management, and E50 on
Environmental Assessment.

      To date, ASTM Committee D18 on Soil and Rock has com-
pleted 66 standards and 68 more are being developed. Committee D19 on Water has completed 23 standards, D34 on Waste
Management has completed 13 standards and is developing an
additional 10, and E50 on Environmental Assessment has
completed 5 standards with 7 more in development. Most of
the completed standards are available in the "ASTM Compi-
lation", the Annual Book of Standards, or as individual ASTM
separates. A number of these standards are very detailed,
extensive productions, but all represent a extensive efforts
by members of the task groups. These efforts include considerable volunteer time between Task Group meetings in
addition to funded time at the Task Group meetings. The
success of this accelerated standards development program is
due to the dedicated task group members, the foresight of
the three Federal agencies funding the task groups, and the
extensive assistance and co-operation of the ASTM staff.

      Information concerning the standards or dates when any
of the listed Committees meet can be obtained by contacting
the office of Robert Morgan, Director of Standards Development, at ASTM headquarters (Fax 215/299-2630; Telephone
215/299-5505) or by contacting this author, who serves as
Task Group Co-ordinator, 7474 Upham Court, Arvada, CO 80003
(Fax and Telephone: 303/425-5610).

APPENDIX

by A. Ivan Johnson
President, A. Ivan Johnson, Inc.
7474 Upham Court, Arvada, CO 80003

In this appendix are lists of standards containing useful information on terminology and methodology related to the primary subjects presented in this ASTM Special Technical Publication, namely the sampling, monitoring, and investigation of environmental media. The standards listed are only those that have been completed to date as a result of Federal funding of the appropriate Task Groups. Additional information on availability of these standards or about any of the many other available standards related to the same general categories may be obtained by contacting ASTM Customer Service by mail at 100 Barr Harbor Drive, West Conshohocken, PA 19428-2959, by telephone (610) 832-9585, or by fax (610) 832-9555.

TABLE 1 -- **Standards completed as a result of Federal funding of task groups of ASTM Committee D18 on Soil and Rock (and fluids contained therein)**

D420-93	Guide to Site Characterization for Engineering Design and Construction Purposes
D653-90	Standard Terminology Relating to Soil, Rock, and Contained Fluids (includes nearly 200 terms related to ground water added since 1990)
D3404-91	Guide for Measuring Matric Potential in the Vadose Zone Using Tensiometers
D4043-91	Guide for Selection of Aquifer-Test Field and Analytical Procedures in Determination of Hydraulic Properties by Well Techniques
D4044-91	Test Method (Field Procedure) for Instantaneous Change in Head (Slug Tests) for Determining Hydraulic Properties of Aquifers.
D4050-91	Test Method (Field Procedure) for Withdrawal and Injection Well Tests for Determining Hydraulic Properties of Aquifer Systems
D4104-91	Test Method (Analytical Procedure) for Determining Transmissivity of Confined Nonleaky Aquifers by Overdamped Well Response to Instantaneous Change in Head (Slug Test)

D4105-91   Test Method (Analytical Procedure) Determining
Transmissivity and Storativity of Nonleaky
Confined Aquifers by the Modified Theis
Nonequilibrium Method

D4106-91   Test Method (Analytical Procedure) for Determining
Transmissivity and Storativity of Confined
Nonleaky Aquifers by the Theis Nonequilibrium
Method

D4630-91   Test Method for Determining Transmissivity and
Storativity of Low Permeability Rocks by In Situ
Measurements Using the Constant Head Injection
Test

D4631-95   Test Method for Determining Transmissivity and
Storativity of Low Permeability Rocks by In-Situ
Measurements Using the Pressure Pulse Technique

D4696-92   Guide for Pore-Liquid Sampling from the Vadose
Zone

D4700-91   Guide for Soil Sampling from the Vadose Zone

D4750-93   Test Method for Determining Subsurface Liquid
Levels in a Borehole or Monitoring Well
(Observation Well)

D5092-95   Practice for Design and Installation of Ground
Water Monitoring Wells in Aquifers

D5126-90   Guide for Comparison of Field Methods for
Determining Hydraulic Conductivity in the Vadose
Zone

D5254-92   Practice for Minimum Set of Data Elements to
Identify a Ground-Water Site

D5269-92   Test Method (Analytical Procedure) for Determining
Transmissivity of Nonleaky Confined Aquifers by
the Theis Recovery Method

D5270-92   Test Method for (Analytical Procedure) for
Determining Transmissivity and Storage Coefficient
of Bounded, Nonleaky, Confined Aquifers

D5299-92   Guide for Decommissioning of Ground Water Wells,
Vadose Zone Monitoring Devices, Boreholes, and
Other Devices for Environmental Activities

D5314-92   Guide for Soil Gas Monitoring in the Vadose Zone

D5408-93   Guide for the Set of Data Elements to Describe a
            Ground-Water Site; Part 1 - Additional
            Identification Descriptors

D5409-93   Guide for the Set of Data Elements to Describe a
            Ground-Water Site; Part 2 - Physical Descriptors

D5410-93   Guide for the Set of Data Elements to Describe a
            Ground-Water Site; Part 3 - Usage Descriptors

D5447-93   Guide for Application of a Ground-Water Flow Model
            to a Site-Specific Problem

D5472-93   Test Method for Determining Specific Capacity and
            Estimating Transmissivity at the Control Well

D5473-93   Test Method (Analytical Procedure) for Analyzing
            the Effects of Partial Penetration of Control Well
            and Determining the Horizontal and Vertical
            Hydraulic Conductivity in a Nonleaky Confined
            Aquifer

D5474-93   Guide for Selection of Data Elements for Ground-
            Water Investigations

D5490-93   Guide for Comparing Ground-Water Flow Model
            Simulations to Site-Specific Information

D5518-94   Guide for Acquisition of File Aerial Photography
            and Imagery for Establishing Historic Site-Use and
            Surficial Conditions

D5521-94   Guide for Development of Ground-Water Monitoring
            Wells in Granular Aquifers

D5549-94   Guide for Reporting Geostatistical Site
            Investigations

D5609-94   Guide for Defining Boundary Conditions in Ground-
            Water Flow Modeling

D5610-94   Guide for Defining Initial Conditions in Ground-
            Water Flow Modeling

D5611-94   Guide for Conducting a Sensitivity Analysis for a
            Ground-Water Flow Model Application

D5714-95   Specification for Digital Geospatial Metadata

D5716-95   Test Method to Measure the Rate of Well Discharge
            by Circular Orifice Weir

D5717-95   Guide for the Design of Ground-Water Monitoring
            Systems in Karst and Fractured-Rock Aquifers

D5718-95   Guide for Documenting a Ground-Water Flow Model Application

D5719-95   Guide to Simulation of Subsurface Air Flow Using Ground-Water Flow Modeling Codes

D5730-95   Guide to Site Characterization for Environmental Purposes, with Emphasis on Soil, Rock, the Vadose Zone, and Ground Water

D5737-95   Guide to Methods for Measuring Well Discharge

D5738-95   Guide for Displaying the Results of Chemical Analyses of Ground Water for Major Ions and Trace Elements--Diagrams for Single Analyses

D5753-95   Guide for Planning and Conducting Borehole Geophysical Investigations

D5754-95   Guide for Displaying the Results of Chemical Analyses of Ground Water for Major Ions and Trace Elements -- Trilinear and other Multi-coordinate Diagrams

D5777-95   Guide for Using the Seismic Refraction Method for Subsurface Investigation

D5781-95   Guide for the Use of Dual-Wall Reverse-Circulation Drilling for Geoenvironmental Exploration and the Installation of Subsurface Water-Quality Monitoring Devices

D5782-95   Guide for Use of Direct Air-Rotary Drilling for Geoenvironmental Exploration and the Installation of Subsurface Water-Quality Monitoring Devices

D5783-95   Guide for the Use of Direct Rotary Drilling with Water-Based Drilling Fluid for Geoenvironmental Exploration and the Installation of Subsurface Water-Quality Monitoring Devices

D5784-95   Guide for the Use of Hollow-Stem Augers for Geoenvironmental Exploration and the Installation of Subsurface Water-Quality Monitoring Devices

D5785-95   Test Method (Analytical Procedure) for Determining Transmissivity of Confined Nonleaky Aquifers by Under-damped Well Response to Instantaneous Change in Head (Slug Test)

D5786-95   Practice (Field Procedure) for Constant Drawdown Tests in Flowing Wells for Determining Hydraulic Properties of Aquifer Systems

D5787-95   Practice for Monitoring Well Protection

D5809-95   Guide for Use of Fluid Flow and Contaminant
           Transport Modeling in the Risk-Based Corrective
           Action Process

D5850-95   Test Method (Analytical Procedure) for Determining
           Transmissivity, Storage Coefficient and Anisotropy
           Ratio from a Network of Partially Penetrating
           Wells

D5855-95   Test Method (Analytical Procedure) for Determining
           Transmissivity and Storage Coefficient of a
           Confined Nonleaky or Leaky Aquifer by the Constant
           Drawdown Method in a Flowing well

D5872-95   Guide for Use of Casing Advancement Drilling
           Methods for Geoenvironmental Exploration and the
           Installation of Subsurface Water-Quality
           Monitoring Devices

D5875-95   Guide for Use of Cable-Tool Drilling and Sampling
           Methods for Geoenvironmental Exploration and the
           Installation of Subsurface Water-Quality
           Monitoring Devices

D5876-95   Guide for Use of Direct Rotary Wireline Casing
           Advancement Drilling Methods for Geoenvironmental
           Exploration and the Installation of Subsurface
           Water-Quality Monitoring Devices

D5877-95   Guide for Displaying the Results of Chemical
           Analyses of Ground Water for Major Ions and Trace
           Elements--Diagrams Based on Data Analytical
           Calculations

D5879-95   Practice for Surface Site Characterization for On-
           site Septic Systems

D5880-95   Guide for Subsurface Flow and Transport Modeling

D5881-95   Test Method (Analytical Procedure) for Determining
           Transmissivity of Confined Nonleaky Aquifers by
           Critically Damped Well Response to Instantaneous
           Change in Head (Slug Test)

D5903-95   Guide for Planning and Preparing for a Ground-
           Water Sampling Event

D5911-95   Practice for Minimum Set of Data Elements to
           Identify a Soil Sampling Site

TABLE 2 --Standards completed as a result of Federal funding of task groups of ASTM Committee D19 on Water

---

D1941-91   Test Method for Open Channel Flow Measurements of Water with the Parshall Flume

D3858-95   Test Method for Open-Channel Flow Measurement of Water by Velocity-Area Method

D3975-93   Practice for Development and Use (Preparation) of Samples for Collaborative Testing of Methods for Analysis of Sediments

D4408-84   Practice for Open Channel Flow Measurement by Acoustic Means

D4409-95   Test Method for Velocity Measurement of Water in Open Channels with Rotating Element Current Meters

D4411-93   Guide for Sampling Fluvial Sediment in Motion

D4822-94   Guide for Selection of Methods of Particle Size Analysis of Fluvial Sediments (manual methods)

D5089-95   Test Method for Velocity Measurements of Water in Open Channels with Electromagnetic Current Meters

D5129-95   Test Method for Open Channel Flow Measurement of Water Indirectly by Using Width Contractions

D5130-95   Test Method for Open-Channel Flow Measurement of Water Indirectly by Slope-Area Method

D5242-92   Test Method for Open-Channel Flow Measurement of Water with Thin-Plate Weirs

D5243-92   Test Method for Open-Channel Flow Measurement of Water Indirectly at Culverts

D5387-93   Guide for Elements of a Complete Data Set for Non-Cohesive Sediments

D5388-93   Test Method for Measurement of Discharge by Step-Backwater Method

D5389-93   Test Method for Open Channel Flow Measurement by Acoustic Velocity Meter Systems

D5390-93   Test Method for Open-Channel Flow Measurement of Water With Palmer-Bowlus Flumes

D5413-93   Test Methods for Measurement of Water Levels in Open-Water Bodies

D5541-94   Practice for Developing a Stage-Discharge Relation for Open Channel Flow

D5613-94   Test Method for Open-Channel Measurement of Time of Travel Using Dye Tracers

D5614-94   Test Method for Open Channel Flow Measurement of Water with Broad-Crested Weirs

D5640-95   Guide for Selection of Weirs and Flumes for Open Channel Flow Measurement of Water

D5674-95   Guide for Operation of a Gaging Station

TABLE 3 -- Standards completed as a result of Federal funding of task groups of ASTM Committee D34 on Waste Management
_____

D4687-95   Guide for General Planning of Waste Sampling

D5358-93   Practice for Sampling with a Dipper or Pond Sampler

D5451-93   Practice for Sampling Using a Trier Sampler

D5495-94   Practice for Sampling with a Composite Liquid Waste Sampler (COLIWASA)

D5633-94   Practice for Sampling with a Scoop

D5658-95   Practice for Sampling Unconsolidated Waste from Trucks

D5679-95A   Practice for Sampling Consolidated Solids in Drums or Similar Containers

D5680-95A   Practice for Sampling Unconsolidated Solids in Drums or Similar Containers

D5743-95   Practice for Sampling Single or Multilayered Liquids, with or without Solids, in Drums or Similar Containers

D5745-95   Guide for Developing and Implimenting Short Term Measures or Early Actions for Site Remediation

D5746-95   Classification of Environmental Condition of Property Area Types

D5792-95   Practice for Generation of Environmental Data Related to Waste Management Activities: Development of Data Quality Objectives

PS37-95   Practice for Conducting Environmental Baseline Surveys

TABLE 4 -- Standards completed as a result of Federal funding of task groups of Committee E50 on Environmental Assessment

---

ES 40-94   Practice for Assessment of Buried Steel Tanks Prior to the Addition of Cathodic Protection

E1430-91   Guide for Using Release Detection Devices with Underground Storage Tanks

E1526-93   Guide for Evaluating the Performance of Release Detection Systems for Underground Storage Tank Systems

E1599-94   Guide for Corrective Action for Petroleum Releases

E1739-95   Guide for Risk-Based Corrective Action Applied at Petroleum Release Sites

Sampling Systems

Susan C. Soloyanis[1]

A COMMON SENSE SAMPLING STRATEGY TO EXPEDITE HAZARDOUS WASTE SITE CLEANUP

REFERENCE: Soloyanis, S. C., "**A Common Sense Sampling Strategy to Expedite Hazardous Waste Site Cleanup,**" Sampling Environmental Media, ASTM STP 1282, James Howard Morgan, Ed., American Society for Testing and Materials, 1996.

ABSTRACT: A common criticism of the remedial investigation/feasibility study process at Comprehensive Environmental Response, Compensation and Liability Act (CERCLA) sites is that too much time and money are spent studying the problem before remediation can be started. A factor leading to this perception may be that the timing, type, and spacing of soil and groundwater sampling at these sites are inappropriate. The purpose of this paper is to propose a 3-step sampling strategy for quickly obtaining sufficient information to make remediation decisions and thus reducing overall characterization costs. The significant element in the strategy is that most of the expensive, time-consuming sampling and analyses using permanent monitoring devices are reserved for the end of the 3-step process as confirmation of the conceptual site model developed as a result of the investigation.

Step 1 is to identify the presence and source of contamination. Sample collection methods for this step should be nonintrusive or minimally intrusive; few permanent monitoring devices should be installed; and field screening analytical methods should be emphasized. The resulting data are used to assess the potential for risk to human health and the environment and identify potential pathways and barriers to migration that are the focus of subsequent investigation.

Step 2 is to delineate the contaminant source and determine contaminant fate and transport. Many samples may be collected at discrete intervals (both lateral and vertical),

---

[1]Principal Scientist, The MITRE Corporation, Brooks Air Force Base, San Antonio, Texas 78235-

The views expressed are those of the author and do not reflect the official policy or position of the U.S. Air Force, Department of Defense, or the U.S. Government.

using temporary devices. The resulting geochemical and geotechnical data are used to determine feasibility of remediation and identify presumptive remedies.

Step 3 is to quantify risk and determine cleanup goals. Permanent monitoring devices should be installed if long-term monitoring will be necessary. The resulting data are used to document the presence of contaminants above background or regulatory criteria.

KEYWORDS: field screening, sampling, site characterization

INTRODUCTION

This paper calls for flexible, timely, and cost-effective soil and groundwater sampling in order to expedite hazardous waste site cleanup. The theme is not new. The U.S. Environmental Protection Agency has issued guidance for accelerated cleanup of both private and federal Superfund sites. Most people involved in hazardous waste site cleanup realize that sampling, analysis, and data validation costs can be as much as fifty percent of the total investigation costs [1].

This paper is a reminder to use common sense when designing the sampling and analysis strategy for soil and groundwater during a remedial investigation/feasibility study (RI/FS) at any hazardous waste site. The strategy should allow investigators to sort out *obviously clean* from *obviously contaminated* sites and also to focus the appropriate time and money on complicated sites where the answers are not so obvious. This common sense approach should be used throughout the Comprehensive Environmental Response, Compensation and Liability Act (CERCLA) RI/FS process to consolidate the preliminary assessment, site investigation, remedial investigation, and feasibility study sampling efforts.

This paper describes a 3-step process to streamline the soil and groundwater sampling effort, use data in the field, and use new and innovative technologies to expedite investigations leading to hazardous waste site cleanup.

The overall purpose of an investigation should guide all aspects of the sampling strategy. The purpose of an RI/FS is to identify the source and extent of contamination, the pathways of possible migration or releases to the environment, and the extent of potential human or other environmental exposure to contamination [2] and to evaluate the remedial alternatives available to manage that risk. Only information "*sufficient to support an informed risk management decision regarding which remedy appears to be most appropriate for a given site*" [3] is needed.

Two examples are summarized below to illustrate different scales of investigation: a federal facility, with many sites, on the National Priorities List (NPL) and a dense, nonaqueous phase liquid (DNAPL) contaminant site.

Example NPL Federal Facility

Forty-three sites have been identified for investigation under the Installation Restoration Program for this facility. In ten years, approximately 500 monitor wells and 4,000 soil borings have been installed, and 400,000 sample analyses have been performed at a total cost of $43 million. It has been determined as a result of this work that 25 sites require no further response actions, 11 sites require minimal source removal, and 7 sites will require some actual remediation.

The investigations at these sites were conducted in stages, in an iterative manner, as a discovery process. This has been the conventional approach for site characterization. In retrospect, it can be assumed that many wells were used to plume chase and many borings were unnecessary. Initial investigations designed to scope future work should not have used time-consuming and expensive fixed-base laboratory analyses with detailed data validation.

Example DNAPL Site

An example of a statement commonly found in RI reports is "This site is well characterized; we have put in 25 monitor wells and 50 soil borings." Monitor wells are the standard installations used to characterize groundwater at hazardous waste sites, but plume chasing and DNAPL hunting using monitor wells can be inefficient and expensive. Soil borings are the standard installations used to characterize the unsaturated zone at hazardous waste sites. The number of wells and borings is not as important as having wells and borings in the right locations, both laterally or vertically. A well-characterized hazardous waste site is one for which there is a detailed understanding of both contaminant distribution and the hydraulic conductivity controlling that distribution. Soil borings and monitor wells may not be the appropriate devices to characterize a DNAPL site. A discussion of these devices and a recommendation for the use of state-of-the-art sampling systems are given in the following paragraphs.

Conventional soil borings--Sampling devices (Shelby tube, split spoon, piston corer) are incorporated in drilling or drive point systems. Sample collection is not always continuous, but should be because the presence of significant permeability interfaces may be overlooked if cores are not collected continuously from ground surface to total depth of the boring. Auger driven borings may generate significant quantities of waste that must be managed.

Monitor wells--Monitor wells can be expensive and time-consuming to install. They are permanent installations that require maintenance, may leak, and must ultimately be decommissioned. The high per-well cost for installation can limit the ability to gather sufficient data to understand contaminant transport in three dimensions [4]. Because monitor wells are usually installed in phases--separated by several seasons, sometimes years--dynamic groundwater conditions may change sufficiently to preclude meaningful data integration from one phase of investigation to the next.

Conventional monitor wells (wells with single, filterpacked, screened intervals that have been sealed from the top of the filterpack to the surface) yield depth integrated values (average values that are a combination of the individual values of all of the hydraulic units exposed in the screened interval) for chemistry and conductivity and may yield samples that underestimate contaminant solute concentration by an order of magnitude.

Contaminant concentrations *in situ* will be underestimated if the vertical sampling interval in a well is greater than either the thickness of a contaminant plume or the thickness of low permeability layers [5]. Thin contaminant plumes can result when there is transport from a source in the unsaturated zone to the saturated zone or when the source itself has an irregular distribution, such as occurs with DNAPL. Thin, low permeability zones are underrepresented because water flows more readily from surrounding high permeability layers, so that the resulting sample is diluted.

State-of-the-art sampling systems--It is now possible to obtain a detailed chemical profile and a profile of hydraulic conductivity using drive point profiling or some conventional drilling methods supplemented by a variety of samplers and instruments. These sampling systems can be used to expedite the characterization of less complicated sites and adequately characterize DNAPL sites. For example, methods that incorporate a clear liner in boring devices enable rapid use of visual information derived from soil cores. In areas of known stratigraphy, cone penetrometer technology can be used to gather physical property information that can be correlated to existing borings. Drive point sampling systems can be used to collect soil gases, soils, and groundwater. The availability of these evolving technologies should enable site investigators to use common sense more often.

A common sense strategy that uses state of the art sampling systems, rather than monitor wells, in the early stages of investigation -- in order to reduce time and cost of investigation while gathering data appropriate to the identified problem -- is discussed below.

SOIL AND GROUNDWATER SAMPLING STRATEGY

A common sense sampling strategy should use a phased approach as described below in Steps 1 through 3. The use of a phased approach to sampling, for data quality objectives, and to develop a conceptual site model is not new.

For the purpose of this paper, it is assumed that validated data are "better" (and more expensive) than data with less documentation. The data validation requirements for each step are determined by the data quality objectives of each step. The sampling and analysis plan to implement the strategy should identify data quality objectives by addressing the following questions:

- What data are required to achieve the purpose of the step?

- How will the data be used?

- How "good" do the data need to be for this use?

All available information should be used. The sampling effort should be streamlined so that data are used in the field to plan the next sampling step. The conceptual site model should be continuously revised and refined so that the final product is not simply information pieced together in a report, but rather data integrated into a coherent representation of the site. The term "conceptual site model" is used in this paper in the classical sense of a source-pathway-receptor characterization of a hazardous waste site. The purpose of the conceptual site model is to consolidate site information into a set of facts, assumptions, and concepts that can be evaluated in a regional context. The draft ASTM standard, Standard Guide for Developing Conceptual Site Models for Contaminated Sites (being developed by Subcommittee E47.13), should be consulted for a detailed discussion of the elements of a conceptual site model and the methodology for developing that model.

Step 1

This purpose of Step 1 is to determine the presence and source(s) of contamination. The elements of the step are described below.

Gather information--Start with a table or list of what is known about the site. The draft ASTM standard, Standard Guide to Site Characterization for Environmental Purposes with Emphasis on Soil, Rock, the Vadose Zone, and Ground Water (being developed by Subcommittee D18.01), provides a comprehensive list of sources of site information as well as methods of investigation. Site history and the rationale for suspecting contamination at the site will provide clues for developing a list of chemicals of concern for the site. Historical maps and aerial photography can be useful in determining site history when no written records exist. A map and a cross section based on a site visit and existing maps and surveys should be constructed. They can be schematic--together they are simply a three-dimensional representation of what is presumed to be present. All of this information should be used as the first version of the conceptual site model.

Establish points of control--Points of control can be existing boring logs, natural or manmade exposures of subsurface layers, or simply published descriptions of the subsurface geology in the area. Nonintrusive methods such as aerial photograph interpretation and minimally intrusive methods such as geophysical surveys [6] can be used to correlate among the points of control. Areas for future investigation using intrusive methods are identified using all the information gathered to this point.

Sample using temporary devices--The object of this step is to gather information without establishing permanent monitoring points. Sample locations must be adequately documented so that the information can be incorporated into the database for the site. Grab samples of soil and water collected using temporary devices are all that may be necessary. Geophysical surveys can be used to detect actual source objects, such as tanks or drums, in the subsurface. The purpose of Step 1 is only to determine whether there is a source of contamination and to estimate its nature and distribution. Known or suspected sources can be investigated using best professional judgment to bias sampling points. Unknown sources can be investigated using a systematic grid. Each sampling plan should be adapted during the sampling effort to incorporate information as it is gathered.

Analyze quickly--Analytical methods for soils, soil gases, and water that require only hand-held equipment or portable laboratory equipment are preferred. The object is to analyze many samples rapidly so that the data are available for decisionmaking within hours. Hand-held or portable field equipment includes immunoassay kits (for PCBs, petroleum products, pesticides, and explosives), specific ion electrodes, chemical detector tubes and papers, and organic vapor analyzers. An example of rapid groundwater contaminant plume mapping is given by Pitkin [7], who used a photoionization detector to evaluate groundwater sample head space contaminant concentrations. Mobile laboratories can perform analyses using gas chromatographs, mass spectrometers, X-ray fluorescence spectrometers, atomic absorption, and inductively coupled plasma. The choice of analytical method depends largely on its detection limits. Some percentage of samples may be analyzed more rigorously or with additional validation documentation, if necessary, at this stage of the investigation.

Incorporate field data into the conceptual site model--The questions that should be answered at this time are: Is there a source? Where is the source? Is it a continuing source that should be considered for an early removal action? Is the source in the unsaturated zone or in the saturated zone? Has contamination spread from the unsaturated zone to the saturated zone or *vice versa*? The conceptual site model should be revised to account for all observations. The conceptual site model should be used to evaluate the potential for risk to human health and the environment.

Determine heterogeneity--Lateral and vertical variations in hydraulic conductivity (resulting from changes in grain size, sorting, cementation, and fracturing) as well as the presence of conduits such as utility corridors, drains, and sewer systems can control contaminant migration or act as barriers to migration. An estimate of the spacing of these variations and knowledge of the locations of potential conduits is necessary to plan sampling for Step 2.

The result of Step 1 is an assessment of the potential for risk to human health and the environment based on an estimation of the nature and extent of contamination.

Step 2

The purpose of Step 2 is to delineate the contaminant source and determine contaminant fate and transport. The elements of the step are described below.

Sample--Many samples at discrete lateral and vertical intervals may be necessary in order to develop a quantitative three-dimensional picture of the site. These samples need not be taken at permanent monitoring points, but can be obtained using temporary installations as in Step 1. Sampling layout should be biased by the information about contaminant distribution determined in Step 1. The spacing of samples should be at the same scale as the variations in hydraulic conductivity determined in Step 1. The importance of closely spaced vertical sampling to delineation of chemical and hydrological gradients in groundwater has been demonstrated by several researchers. The most widely reported research effort has been conducted by the U.S. Geological Survey at Otis Air Force Base on Cape Cod, Massachusetts [8].

In this step, it is appropriate to use geophysical surveys to map detailed stratigraphy and water table elevation. Soil gas surveys can be used to map contaminant distribution in the unsaturated zone, and grab samples of groundwater can be used to map contaminant distribution in the saturated zone.

Analyze--This step may require the largest number of samples and analyses. Methods that minimize analytical costs should be emphasized. It is appropriate to map contaminant concentrations in orders of magnitude because the purpose of this step is to determine the feasibility of remediation; the choice of remedy can usually be based on contaminant concentrations measured in orders of magnitude.

Incorporate new data into the conceptual site model--This step should result in a three-dimensional picture of the source of contamination and the extent of migration of the contaminant. Rates of migration should be estimated, but multiple sampling iterations (using semipermanent or permanent devices) or a time period of observation may be required in order to calculate these rates. Uncertainties in the conceptual site model are the data gaps that should be filled by the end of this step.

Evaluate the necessity and feasibility of remediation--This will enable the planning of final samples in Step 3. If the source is to be removed, Step 3 sampling may be limited to confirmation sampling. If there are no receptors for the affected media, there may be no need for remediation.

Identify potential remedies--Remedies may include source removal, source control, management of migration, long-term monitoring, and receptor protection. It may be possible to limit additional sampling to the confirmation of removal. Other remedies may require additional sampling in Step 3 for the design of remediation. Certain remedies are considered to be presumptive and can be identified at this stage.

The result of Step 2 is the identification of possible remedial strategies.

Step 3

The purpose of Step 3 is to assess risk and determine cleanup goals. The elements of the step are described below.

Sample--All previously collected data should be used to identify the appropriate locations and depths in specific media for sampling in this step. Sampling can be done using temporary or permanent installations. The number of permanent installations should be minimized and chosen carefully to be useful as long-term monitoring stations, if possible. Sampling in this step will include collection of background data in all affected media for comparison with contaminated media and may include additional data validation documentation.

Sampling for remedial actions should be designed to acquire all the geotechnical and geochemical data necessary to evaluate the potential remedial alternatives identified in Step 2. Examples of these data are soil organic carbon content and oxygen availability for bioremediation and soil moisture, grain size, and air conductivity for soil vapor extraction. These data are not usually collected in Steps 1 and 2 and are critical to the choice of remedial technologies.

Calculate risk--The accumulated data from all steps should be organized to show—for each affected medium—contaminants of concern, contaminant concentrations present, distribution, and comparison to background. This information plus the analytical results from soil and groundwater samples collected in this step are used to calculate risk to human health and the environment. A complete baseline risk assessment will include other media and pathways that are not discussed in this paper.

Determine cleanup goals--These goals can be based on regulatory criteria, risk reduction, or comparison to background concentrations of chemicals of concern.

The result of Step 3 is the quantification of risk presented by soil and groundwater to receptors and determination of the necessity for cleanup. This information should be used by the remedial project managers to manage the risk presented by the site.

SOIL AND GROUNDWATER SAMPLING EQUIPMENT AND METHODS

Choice of sampling equipment is limited only by the investigator's imagination and data quality objectives. Many new technologies are enabling faster and more comprehensive acquisition of data. Sampling equipment can be, but is not always, medium specific.

There are many methods of acquiring samples at hazardous waste sites. There are ASTM standards for monitor well installation (D5092, Standard Practice for Design and Installation of Ground Water Monitoring Wells in Aquifers) and sampling for groundwater (D4448, Standard Guide for Sampling Groundwater Monitoring Wells), soils (D1452, Standard Practice for Soil Investigation and Sampling by Auger Borings; D4700, Standard Guide for Soil Sampling from the Vadose Zone; and others), soil gas (D5314, Standard Guide for Soil Gas Monitoring in the Vadose Zone), soil pore liquids (D4696, Standard Guide for Pore-Liquid Sampling from the Vadose Zone), sediments (D4823, Guide for Core Sampling Submerged Unconsolidated Sediments and D4411, Guide for Sampling Fluvial Sediments in Motion), and surface water (D5358, Standard Practice for Sampling with a Dipper or Pound Sampler).

Soil, soil gas, and groundwater sampling devices for use in conventional well drilling applications are readily available and include screened hollow stem augers and samplers that can be lowered inside drillstem. A technology particularly applicable to Steps 1 and 2, drive point profiling, is discussed below.

Drive Point Profiling

Drive point devices provide a temporary installation for sampling discrete vertical intervals. They are usually efficient, rapid, and cost-effective for sampling soils, soil gases, and groundwater. They generate very little investigation-derived waste. The systems are variable and can be mounted on a platform, a truck, or a rig or can be as simple as a hammer-driven rod. Many systems are used in combination with or include a mobile chemical analytical laboratory so that a chemical profile can be obtained in real time.

Data collected during driving and at the discrete sampling points can be used to develop a detailed stratigraphy and profile of hydraulic properties in addition to the chemical profile.

Waterloo profiler--An example is the Waterloo vertical drive point profiler, developed at the University of Waterloo Centre for Groundwater Research [9]. This system was developed as a groundwater quality profiler. It is installed using hand-operated equipment, can acquire multiple samples at numerous depths in a single hole without coming out of the borehole between samples, and can be used to grout the borehole as the equipment is retrieved. The sampling mechanism is a smooth, ported stainless steel tip that is flush-mounted on drill rod. The sampling interval can be developed, screen size in the ports can be varied, and the device has been used to depths greater than 20meters.

Potential for cross contamination--It has been suggested that dragdown of contaminated media occurs more readily with drive point devices than with conventional well drilling apparatus. Sampling results from several studies [7,10] indicate that sharp interfaces in dissolved contaminant plumes can be seen and sampling results using drive point methods can be duplicated with bundled piezometers. The possibility exists that dragdown may occur below pooled DNAPL.

RESULTS

Using this common sense approach, it is possible to make remediation decisions more quickly and less expensively. The time and money spent studying the problem will have been spent to develop a complete conceptual site model that is accurate in three dimensions. More data, at an appropriate scale and spacing, will have been collected. Investigators can realize savings of as much as 50-60 percent in costs and perhaps more than that in terms of time.

CONCLUSIONS

More data can be collected and used and the time for characterization of hazardous waste sites can be shortened if common sense is used to develop a sampling strategy that matches decision making information needs with sampling requirements. Validated analytical results are not always necessary. Validated, quantitative analytical results may be required for the calculation of risk, determination of regulatory exceedances, and to design remediation, but not every sample taken at a site should be obtained from a permanent monitoring device or subjected to rigorous analytical protocols.

Using common sense, characterization costs in terms of time, number of monitor wells, disposal of investigation-derived waste, and laboratory analyses can be reduced. Long-term monitoring costs can also be reduced because there will be fewer wells, fewer samples, fewer analyses, less documentation, less maintenance, and fewer abandonments associated with a site. The tools and the common sense approach, with the associated savings in time and resources, are available now. They only await implementation.

ACKNOWLEDGMENTS

I thank my colleagues at MITRE, the Air Force Center for Environmental Excellence, and on the Pease Disposal Management Team for urging me to remind them and others to use common sense.

REFERENCES

[1]   Zemo, D. A., Pierce, Y. G., and Gallinatti, J. D., "Cone Penetrometer Testing and Discrete-Depth Ground Water Sampling Techniques: A Cost-Effective Method of Site Characterization in a Multiple-Aquifer Setting", Ground Water Monitoring and Remediation, Fall 1994, pp. 176-182.

[2]   Arbuckle, J. G., et al., Environmental Law Handbook, Tenth Edition, Government Institutes, Inc., 1989.

[3]   U.S. Environmental Protection Agency, Guidance for Conducting Remedial Investigations and Feasibility Studies under CERCLA, Office of Emergency and Remedial Response Directive 9355.3-01, EPA/540/G-89/004, 1988.

[4]   Cherry, J. A., "Ground Water Monitoring: Some Current Deficiencies and Alternative Approached" Chapter 13 in Hazardous Waste Site Investigations, Gammadge and Berven (eds), Lewis Publishers, 1993.

[5]   Johnson, R. L., "Sampling for Volatile Organics in Ground Water: Are We Losing It?" Chapter 16 in Hazardous Waste Site Investigations, Gammadge and Berven (eds), Lewis Publishers, 1993.

[6]   Benson, R. C., Glaccum, R. A., and Noel, M. R., Geophysical Techniques for Sensing Buried Wastes and Waste Migration, EPA/600/7-84/064, 1984.

[7]   Pitkin, S. E., "A Point Sample Profiling Approach to the Investigation of Groundwater Contamination," Unpublished Master of Science thesis, University of Waterloo, Waterloo, Ontario, Canada, 1994, 242p.

[8]   Smith, R. L., Harvey, R. W., and LeBlanc, D. R., "Importance of closely spaced vertical sampling in delineating chemical and microbiological gradients in groundwater studies," Journal of Contaminant Hydrology, vol. 7, pp.285-300, 1991.

[9]   Broholm, M. M., Ingleton, R. A., and Cherry, J. A., "The Waterloo Drive-Point Profiler - Detailed Profiling of VOC Plumes in Groundwater Aquifers," Transport and Reactive Processes in Aquifers, Dracos and Stauffer (eds), Balkema, Rotterdam, 1994, pp.95-100.

[10]   Pitkin, S. E., Ingleton, R. A., and Cherry, J. A., "Use of a Drive Point Sampling Device for Detailed Characterization of a PCE Plume in a Sand Aquifer at a Dry Cleaning Facility," National Ground Water Association Eighth National Outdoor Action Conference on Aquifer Restoration, Ground Water Monitoring and Geophysical Methods, Minneapolis, 1994.

Eric Y. Chai[1]

A SYSTEMATIC APPROACH TO REPRESENTATIVE SAMPLING IN THE ENVIRONMENT

REFERENCE: Chai, E. Y., 'A Systematic Approach to Representative Sampling in the Environment,' Sampling Environmental Media, ASTM STP 1282, James Howard Morgan, Ed., American Society for Testing and Materials, 1996.

ABSTRACT: In environmental sampling, samples are taken for the purpose of solving a problem in a target population. The problem may not be correctly solved if bias is present in the samples, in the measurement of the samples, and/or in the use of the sample data to make conclusions about the population. This paper discusses a systematic approach which considers all these potential sources of bias when attempting to obtain a representative set of samples for making inference about the population.

KEYWORDS: Representative sampling, sampling bias, measurement bias, statistical bias, systematic approach

If the target population is totally homogeneous, then one single sample, no matter how small in weight or volume or where it is taken, would be completely representative of the population with respect to the population parameter(s) or characteristic(s) of interest. Precisely because typical populations are inherently heterogeneous to one degree or another that the question of a representative sample assumes importance.

A sample or a set of samples is often taken from a target population because there is interest in some parameter of that population and yet the population cannot be observed in its entirety. Therefore, the first principle in taking a set of samples is how representative the samples are of the population. If the samples are not representative, then what is observed in the samples would not be what is actually contained in the population. Thus, the information derived from the samples would not provide a complete or correct picture of the target population. If this lack of valid information in the samples were known to the sampler, the sampler could either simply view the sampling as a wasted effort, or make appropriate adjustment so that the information can be useful. As often is the case,

[1]Staff Research Mathematician, Statistics Department, Shell Development Company. P. O. Box 1380, Houston, TX 77251-1380.

this deficiency in the samples is generally not known to the sampler. Thus, when the sample data are used to make conclusions about the population, the conclusions would likely be off the mark, and any decision based on these sample data may be in error. In environmental sampling, such errors can be very costly.

A systematic approach to representative sampling is therefore needed.

REPRESENTATIVE SAMPLE

A representative set of samples is taken to mean a set of physical sample(s), when taken from a defined population, that collectively mirrors or reflects some properties or characteristics of interest in that population. For this paper, the abbreviated term of "a representative sample" will be used in its place.

"A representative set of samples" is different in notation from "a set of representative samples" or "a representative sample", where a sample is an individual physical sample. "A representative set of samples" is a set of samples which is collectively representative of the population, though the physical samples may or may not be representative individually. "A representative sample" is a physical sample which by itself is representative of the population. This happens either when the population is totally homogeneous (or nearly so) or when sampling is done in such a manner as to make it possible, such as a composite of several individual samples suitably taken from the population.

Representative sampling is used to mean a process by which a set of samples is obtained from a target population to collectively mirror or reflect certain properties of the population. The word reflection implies not only closeness of the physical/chemical properties of the samples to those of the population as a whole. It also implies closeness of the sample data to those which would have been obtained by measuring the entire population and closeness of the inferred value (or estimate using the sample data) to the true population value.

A SYSTEMATIC APPROACH

A systematic approach is one which first defines the desired end result and then designs a process by which such a result can be obtained. The process also describes the essential elements and their relationships in the process.

In environmental sampling, the desired end result is a set of sample data which can be used to make valid inference about the population. Essential elements of a process to obtain this result include clearly defined sampling objective(s), a well defined population, the use of correct sampling procedures and equipment, and preservation of sample integrity before measurement. When these elements are managed properly, sampling bias is minimized and a representative sample is the likely result.

However, these samples will need to be measured to obtain the data and the data will need to be analyzed for making inference about the population. Thus, the total process also includes two additional elements in terms of a measurement process and a statistical procedure which are free or nearly free of bias.

The objective of a systematic approach is to minimize these potential sources of bias in the entire process. This approach is described in a schematic diagram (Fig. 1).

ELEMENTS IN THE SYSTEMATIC APPROACH

1.   Sampling Objective

Sampling objective defines the problem to be solved. When this is clearly stated, it provides assurance that the right problem will be solved.

2.   Population

Given the sampling objective, the population can be viewed as follows:

Target versus sampled population--The target population is the real population of interest. The sampled population is the actual population sampled, which could be simply a part of the target population accessible to sampling. When the target and sampled populations are not the same, there is a potential for biased results. The first effort in sampling is, therefore, to make sure that the sampled population is identical to the target population.

The target population is most easily defined by its spatial and/or temporal boundaries. If there is difficulty in sampling from the entire population, either these difficulties need to be overcome or the concerned parties need to agree on the portion to be sampled (sampled population). When these two populations are not the same, the potential impact, or lack thereof, on the sampling results need to be stated.

Spatial/temporal boundaries--The population can often be defined by the spatial and/or temporal boundaries. The spatial boundaries define the horizontal as well as vertical limits of the areas to be sampled. The temporal boundaries define the time

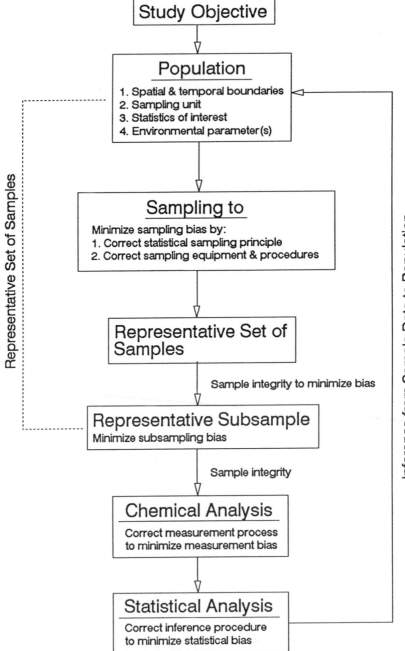

FIG.1 -- Schematic of Representative Sampling

interval of interest when making the inference from sample data to the population. For example, if the interest is in the estimation of the average concentration of a continuous process such as a stream over, say, a 1-year period, then the temporal boundary is a specified calendar year. Sampling can then take place over the year or over only part of the year, depending on the behavior of the process. Using concentration distribution of a contaminant as an example, if the concentration is random and steady over time, then sampling of only a portion of the 1-year interval would still lead to a representative sample. Only when there is systematic patterns to the concentration distribution over time, such as seasonality, then sampling over the entire 1-year interval may be necessary.

Sometimes spatial or temporal data have autocorrelations, in that samples closer in space or time are more likely to have similar results (e.g., concentration) than those which are further apart. One way to minimize the effect of autocorrelation is to take samples suitably far apart in space or time. The variogram method can be used to determine this distance [1].

Sampling unit--This is the weight or volume of a physical sample to be taken from the population. The size of the sampling unit needs to be specified beforehand, based on consideration of laboratory needs, nature and extent of heterogeneity in the population, and the sampling objective. Pitard [1] has done extensive work in this area, in particular the relationship between the variance of "fundamental error" and sample weight, where sample weight is determined as a function of particle size and shape and so forth. (Fundamental error is the part due to the inherent heterogeneity in the sampling material when the sampling operation is carried out correctly). In general, for a given particle size (diameter), the larger the sample weight is, the smaller the variance of the fundamental error is. It is to be noted that, unless the sampling unit is large enough to capture the largest particle in the sampling materials, sampling bias is likely to occur, leading to an unrepresentative sample of the population.

Environmental parameter(s) of interest--The population may contain various constituents. These constituents need to be identified for sampling purposes. Often only a subset of the constituents drives the decision-making process. If so, the subset needs to be identified.

Statistics of interest--This is the statistical parameter(s) of interest, which could be something such as the mean or variance of concentrations of constituents in the population. Identification of these statistical parameters beforehand is important, because sampling operations such as physical modification of the sampling materials may render estimation of some of these parameters impossible. For example, compositing would render estimation of the variance of the individual samples impossible. Estimation of this variance often

is a goal in a pilot study. The variance can be used to
determine the number of samples needed in the formal sampling
plan. On the other hand, if composite samples are the ultimate
sampling units of interest in the sampling plan, then the
variance of the composite samples is the one of interest.
Compositing of the individual samples would, in this case, do no
harm.

Often it makes sense to divide the original population into
separate units, where the decision is to be made for each unit
based on the sample data from that unit. When this is the case,
each unit is the population for sampling and decision-making
purposes. An example is an area highly stratified either by
contaminant concentrations or by contaminant types. If the
objective is to estimate the mean concentration for each stratum,
then each stratum is a population for sampling purposes. In such
cases, sampling design such as simple random sampling may be
used. On the other hand, if the objective is in estimating the
mean concentration over all the strata, then the entire
population is the decision unit. In such cases, sampling design
such as stratified random sampling may be used.

3. Sampling Bias

The first step in a sampling operation is the taking of a
set of physical samples. At this stage, proper sampling protocol
needs to be followed to minimize the sampling bias.

A representative sample is obtained when the expected
sampling bias is zero or nearly so.

Bias in a set of samples, when estimating the mean
concentration of a population, is:

$$b_1 = \overline{x} - \mu \tag{1}$$

where

b_1 = sampling bias from a single set of samples

$\overline{x}$ = sample mean concentration

μ = true mean concentration of the population.

Due to $\overline{x}$ being subject to variability, b_1 may not be zero.
But if we should repeat taking many sets of the same number of
samples, the average or expected value of $\overline{x}$ may or may not
deviate from the true value. Thus, the expected bias of $\overline{x}$ is:

$$b_s = E(b_1) = E(\overline{x}) - \mu \tag{2}$$

where

b_s = expected or average sampling bias when many sets of samples are taken by the same sampling protocol

$E(\bar{x})$ = expected or average of the sample means of many sets of the same number of samples.

The difference between Equations (1) and (2) is subtle but fundamental. Suppose an estimate of the population parameter is unbiased. Since the estimate is subject to random variation, the random error may cause the estimate to deviate from the true value on <u>any</u> occasion (therefore the bias in Equation (1) may not be zero), but the random errors do balance out on the average, leading to a zero bias in Equation (2).

When b_s in Equation (2) is zero or nearly so, a representative set of samples is said to have been obtained.

The sampling bias of interest in environmental sampling is typically the quantity b_s in Equation (2), not b_1 in Equation (1).
This definition of representativeness is different from Pitard's [2], where he defines the degree of representativeness as the sum of the squared bias (b_s^2) and the sampling variance.

This definition has one undesirable feature in that a set of samples will be considered unrepresentative to some degree even though the bias is zero. The degree of unrepresentativeness in this case is a function of how large the sampling variance is, which is in turn a function of the population heterogeneity and cannot be controlled in the original population (may be controlled during sampling through methods such as compositing).

The use of Equation (2) to define bias, and thereby the degree of representativeness, is that b_s in Equation (2) is a pure measure of the expected degree of representativeness of a set of samples, for a given population with an inherent degree of heterogeneity. Judgment on how close the mean of a single set of samples is to the true value certainly will depend on the size of the sampling variance. But this issue of the size of sampling variance can be easily handled by increasing or decreasing the number of samples to achieve the desired closeness of the sample mean to the true value. This issue is therefore better addressed elsewhere by methods such as statistical confidence limits as a function of the number of samples.

If subsampling is to take place either in the field or in the laboratory before measurement, the original sample can be construed to be the target population and proper sampling protocol needs to be followed so that the sampling bias is minimized at this stage of operation.

Sampling bias in Equation (2) cannot be measured because the true value μ is typically not known. Efforts to ensure a certain degree of sample representativeness, however, can be made by two means:

a. Use of proper statistical sampling design--Proper statistical sampling design, such as simple random sampling or stratified random sampling, would typically improve precision in estimation as well as achieve equal probability of selecting the units in the populations. Unequal probabilistic sampling could easily introduce sampling bias.

b. Use of proper sampling procedures and equipment--This applies to all phases of the sampling operation (when physically taking the samples). When proper protocol is not followed, this area has great potential for sampling bias. More details are given by Pitard [2].

The above discussions center on the population mean concentration. For population parameters other than the mean concentration, similar concepts apply.

4. Sample Integrity

Once the physical samples have been taken, integrity of the samples before measurement is the next important issue. Proper procedures in container use, transportation, storage, sample preparation and such are needed to prevent alteration of the samples to the extent that the content of the samples is no longer similar to that of the sampled population, thereby introducing bias. Since these elements are not part of the measurement process, so the bias from these sources are viewed as part of the sampling bias.

5. Measurement Bias

Measurement bias can occur when there is systematic error(s) in the measurement device or process, such as instrument drift or systematic human error. It is defined in a manner similar to the sampling bias in Equation (2). In the example of mean concentration, the measurement bias is:

$$b_m = E(\,\overline{X}\,) - \mu \qquad\qquad (3)$$

where b_m = expected or average measurement bias when many samples are measured by the same measurement process.

Often bias in the measurement process can be reasonably estimated in a laboratory testing setting when the true value is a known quantity. Laboratory samples spiked with known quantities of a chemical or contaminants or certified reference standard can often be used for this purpose. Minimization of or adjustment for such estimable bias in the measurement process is essential in order to obtain data which are unbiased. When estimation of measurement bias is not possible, care in measurement protocol and training are probably the only recourse.

6. Statistical Bias

After the sample data have been obtained, the data need to be used to estimate the population parameter(s).

A statistical bias is the difference between the expected value of a statistical estimator and the true value of the population parameter. The phrase "expected value of an estimator" implies that the estimator as an estimate of the population parameter is subject to variation (or, in statistical terminology, has a sampling distribution), but in the long run the mean of such estimates will converge to an expected value. If this expected value is equal to the population parameter, then the estimator is said to be unbiased.

In general, a statistical bias is:

$$b = E(\bar{t}) - (t) \tag{4}$$

where $\bar{t}$ = a statistical estimator of the population parameter t
       t = a known population parameter once the population
           distribution is specified or assumed
$E(\bar{t})$ = expected (or average) value of the estimator
           when the same sampling procedure is repeated
           many times.

In the example of mean concentration, the sample mean $\bar{x}$ is the estimator of the population parameter mean μ, and the bias in the statistical estimator is $[E(\bar{x}) - \mu]$. If the population distribution is specified, then whether $\bar{x}$ is an unbiased estimator or not can be mathematically determined. For example, when the population is normally distributed, $\bar{x}$ is an unbiased estimator of μ since $E(\bar{x}) = \mu$.

Equation (4) implies that the statistical bias can arise from three different situations:

a. Selection bias--This bias arises when the samples are not taken according to a statistical sampling design, as discussed in Section 2, since the sample data so obtained will give a biased value to the estimator. An extreme case of selection bias occurs when a certain portion of the population has zero probability of selection, leading to the sampled population being different from the target population.

b. Statistical bias in estimator--Not all statistical estimators are unbiased. For example, the sample variance below is not an unbiased estimator of the population variance σ^2, since:

$$E[\Sigma_{i=1,n}(x_i - \overline{x})^2/n] = [(n-1)/n]\sigma^2 \qquad (5)$$

On the other hand, the statistic $\Sigma_{i=1,n}(x_i - \overline{x})^2/(n-1)$ is an unbiased estimator since:

$$E[\Sigma_{i=1,n}(x_i - \overline{x})^2/(n-1)] = \sigma^2 \qquad (6)$$

Some estimators may be biased when the number of samples i small, but can tend to be unbiased when the number of samples is large. The estimator in Equation (5) is the case since, when n is very large, the ratio $(n-1)/n$ tends to 1.

c. Bias in distribution assumption--Because the estimator in Equation (4) has a sampling distribution, the expected value of the estimator will not be equal to the true value unless the sampling distribution is correctly specified. A simple example is the assumption of normal distribution when the real underlyir distribution is lognormal distribution. And a statistical test under misspecified distribution is a biased test.

Distribution assumption can be examined and tested by methods such as probability plotting, Wilk-Shapiro test or chi-squared test [3].

In many ways, Equation (4) is where the three sources of bias (sampling, measurement and statistical) come together.

In the first place, if the physical samples are not representative of the population, then the estimate $\overline{t}$ would hav been obtained from a set of biased samples and it cannot be expected to be equal to the true population value t, leading to the need to minimize the sampling bias. In the second place, even when the physical samples are unbiased, if the measurement process is not unbiased, the sample data from this measurement process cannot be expected to produce an estimate $\overline{t}$ equal to t, leading to the need to minimize the measurement bias. In the third place, if the statistical estimator is biased, then its expected value is by definition not equal to the true value, leading to the need to minimize the statistical bias.

PHYSICAL MODIFICATION AND REPRESENTATIVENESS

Often the samples are physically modified before or during the measurement process. Modifications such as crushing and grindir maintain the original size of the sampling unit, while compositing changes the size of the sampling unit.

Physical modification may or may not affect sample representativeness of the population, depending on factors such as whether modification affects the results of the measurement and the sampling objective.

Result of measurement--If the modified samples produce a set of measured results different from that of unmodified samples, then a judgment needs to be made regarding which set of results is consistent with the sampling objective. If the unmodified results meet the need of the sampling objective while the modified results do not, then physical modification is not advised, unless proper adjustment can be made to account for the difference.

An example could be made in measuring the concentration of a contaminant in the soil. Suppose that a certain portion d of the total concentration x is bonded to the soil matrix such that, in the unmodified form, the measurable concentration is (x-d). On the other hand, the measurable concentration is x when the samples are modified by some means. Then the question is which measurement, x or (x-d), is consistent with the defined problem. Oftentimes concentration x is of interest. On occasion when the bonded portion d is construed to be safe, then the quantity of interest could be (x-d).

Sampling objective--Often the sampling objective is to estimate the sampling variance. This is often the case in a pilot study. The estimated variance is then used to determine the number of samples needed in a subsequent sampling plan. When this is the case, physical modification of the sampling materials may render estimation of the variance impossible. If this is the case, physical modification should not take place. A particular example is sample compositing. If the purpose of sampling is to obtain an estimate of the variance of the individuals samples, then compositing will make it impossible.

If the final sampling plan calls for the taking of composite samples, then it is perfectly alright to take composite samples in the pilot study.

CONCLUSION

The suggested approach to representative sampling is a systems approach to problem solving by ensuring that no important elements in the entire process are neglected in the prevention and minimization of bias. When this is done, the inference from the sample data to the population will have a high degree of validity.

Some bias, such as the sampling bias as a result of using

improper sampling equipment, often cannot be measured.   But effort still needs to be undertaken to assure a high degree of representativeness in the samples.

ACKNOWLEDGMENT: Many ideas in this paper were developed during work with an ASTM task group (D34.02.08).   I am grateful to the task group members for their ideas and discussions.   Thanks also go to the referees of the paper for the suggested changes.

REFERENCES

[1]   Journel, A. G. and Huijbregts, Ch. J., _Mining Geostatistics_, Academic Press, New York, 1978.

[2]   Pitard, F. F., _Pierre Gy's Sampling Theory and Sampling Practice_, Volumes I and II, CRC Press, Boca Raton, Florida, 1989.

[3]   Hahn, G. J. and Shapiro, S. S., _Statistical Models in Engineering_, John Wiley & Sons, New York, 1968.

Worker Safety and Risk Characterization

T.J. Lyons,[1] Joseph P. Knapik,[2]

WORKER EXPOSURE DURING ENVIRONMENTAL SAMPLING: THE HAZARDS INVOLVED

REFERENCE: Lyons, T. J. and Knapik, J. P. "Worker Exposure During
Environmental Sampling: The Hazards Involved," Sampling Environmental
Media, ASTM STP 1282, James Howard Morgan, Ed., American Society for
Testing and Materials, 1996.

ABSTRACT: As a result of the expansion of environmental site assessment
(ESA) practice, there has developed a potential for field investigators
to put themselves at risk from the materials being sampled or the site
being investigated. The authors propose the development of a supporting
standard to define the measures necessary to recognize, evaluate and
control hazards which environmental professionals may encounter during
investigative activities. Existing standards addressing environmental
investigations and remediation work peripherally cite the importance of
safe work practices while mandating procedures which may result in
unnecessary or unknowing exposures. Federal and State regulations
define what work is to be completed, and what goals are to be met but do
not provide specific information on hazards to the workers involved.
The authors illustrate this need through citation of various ASTM
Standards, presentation of case studies and a recommended outline for
development of a standard site safety and health plan.

KEYWORDS: action level, environmental site assessment, health and
safety audit, MSDS (Materials Safety Data Sheet), personal protective
equipment (PPE), site safety and health plan

BACKGROUND

When Apollo 11 returned from its successful landing upon the Moons
surface, instead of a warm embrace and handshake with President Nixon,
they were immediately installed into the "Mobile Quarantine Facility"
and flown to Houston, Texas where they remained for 16 days. The
reason? The risk of team members infecting Earth as a result of their
extensive "field trip". Environmental professionals routinely complete
the same scenario when investigating properties during environmental
site assessments (ESA), sub-surface geophysical investigations and
material sampling.

[1]Associated Safety Professional, Lebanon Valley Environmental Health and
Safety, West Lebanon, NY 12195
[2]Certified Industrial Hygienist, Ames Plating Inc. 24 Montville St.
Chicopee, MA 01201

Several American Society of Testing and Materials (ASTM) standards
identify a potential for hazards during field operations. A review of
the latest edition of the 1994 Annual Book of ASTM Standards Section
11.04 Pesticides; Resource Recovery; Hazardous Substances and Oil Spill
Responses; Waste Management; Biological Effects discuss several
standards where risk is considered. Some of the standards identified
were:

D4448	Guide for Sampling Groundwater Monitoring Wells
D4687	Guide for General Planning of Waste Sampling
E1391	Guide for Collection, Storage, Characterization, and Manipulation of Sediments for Toxicological Testing
E1527	Practice for Environmental Site Assessments: Phase I Environmental Site Assessment Process
E1528	Practice for Environmental Site Assessments: Transaction Screen Process
E1599	Guide for Corrective Action for Petroleum Release

In most cases, the Scope for the ASTM standard, clearly states "This
standard does not purport to address all the safety problems associated
with its use. It is the responsibility of the user of this standard to
establish appropriate safety and health practices and determine the
applicability of regulatory limitations prior to use - ASTM E1527.

CURRENT STANDARDS FOR SITE ASSESSMENTS

A benchmark in the practice of ESA's has been the publication of ASTM
E1527. This standard and a related document ASTM E1528 have been widely
accepted by industry.

Though the safety issue is covered under a standard disclaimer, there
are several areas within the document that could result in worker
exposure. As an example, ASTM E1527 8.4.2.5 states: "Strong, pungent,
or noxious odors shall be described in the report and their sources
shall be identified in the report to the extent visually or physically
observed or identified from interviews or records review." This section
indicates that the investigator should make an assessment as to whether
an odor is strong, pungent or noxious and further to identify its
source. The risk of exposure to an unknown chemical or mixture is
obvious.

A more practical approach is referenced in ASTM E1528 9.5.1.2 and
states: "Hazardous substances or petroleum products may often be
unmarked. "The preparer should never open any unmarked container at the
facility because they may contain explosives or acids." This warning
indicates the concern for the investigators safety by the developers of
the standard.

The authors advocate the creation of a standard to define the manner by
which written safety and health procedures will be used in the conduct
of practice consistent with ASTM methodologies.

CASE HISTORIES

The authors have completed numerous site investigations ranging from complex industrial sites to small commercial properties. In performing ESA's, in particular the site reconnaissance, there is an allure for some, to open every drum and crawl in every crawlspace. A detective's mindset is necessary, tempered with forethought, to prevent unnecessary risk.

Perhaps there is a primal need to find bad things. Generations of "war stories" make for interesting water cooler symposia and the more colorful (or dangerous) the tale, the more respect one expects to find. However, as both authors can attest, the implications to such hazards can and do result in injury and needless exposure.

The authors have detailed several instances where hazards to the assessors was noted and could easily have been avoided.

Case Study #1 - Textile Mill Site Investigation - Western Massachusetts

Every environmental professional has at least one site where the lessons were learned from mistakes made. In this case, a project was awarded to a large consulting firm to complete a site assessment of an abandoned mill complex that previously made wool textiles in the late 1800's and electrical capacitors in the mid 1900's.

In addition to the obvious hazards of such site, there was a lack of information offered by the owner; who was the client. In this case, it became apparent to the investigators that the client was hiding several site conditions.

After completing a two day inspection of the building interior the author, with an investigator-in-training, toured the exterior of the facility. We observed a brand-new sheet of plywood screwed to the exterior wall at the basement level. This covering was removed revealing a door. A quick survey from the exterior revealed an empty room and the potential for a confined space entry. The author entered the room and after three steps retreated. A grease-like material was discovered across the concrete floor which made walking difficult and caused concern.

The author immediately removed his boots, and went home in his socks. With a new pair of boots, the author revisited the site with appropriate Personal Protective Equipment (PPE) (Fig. 1). Subsequent analysis of the material from his boots identified it as Polychlorinated Biphenyl (PCB). The highest concentration (180,000 parts per million) ever sampled in that particular laboratory.

Owners have an interesting approach to site assessment. They will try their best to show off the clean parts of their property while glossing over the contaminated. Having the owner along during the reconnaissance of the property is the owners right, but keep in mind they do not have the same level of training as the site investigator.

During the walk through phase of this particular project, the client approached an unmarked drum and indicated that it was empty. To prove it, he kicked the drum. The drum was not empty. This became readily apparent when the bottom of the drum split along the seam and a white powder spilled across the floor.

A hasty retreat from the room followed. Leaving the owner to clean his boots, the author asked the owner to assume all the containers were full unless the tops were off. Anecdotal information gathered as a result of the investigation indicated the contents of the drum was virgin asbestos insulating material used for boiler repairs.

The above case highlights the potential hazards to the investigator and to the investigators co-workers and family. With reference to the PCB issue, an untrained investigator may have simply scraped this Vaseline-like material off his shoes and subsequently contaminated his vehicle, his office flooring and surfaces throughout his residence.

The potential for the investigator to contaminate others *must* be a fundamental consideration when developing risk based protection during site assessment.

The drum release is more apparent. Training must be conducted to help employees avoid such a situation. Drum investigations, with the exception of confined space hazards may pose the highest level of risk during site assessments.

As in the first case, a risk classification of the site should have been completed prior to any entry. The history of the site was readily available and would have highlighted the potential hazards encountered long before exposure could develop.

Case Study #2 Burned out plastic extrusion plant - Arson Data Collection

The scope of services for this site assessment required a visual assessment and collection of environmental samples to be analyzed for petroleum fuel residuals from the burned out remains of a 50,000 square foot plastics extrusion plant. This included air samples for residual volatiles and/or accelerants.

The focus of the investigation was to develop a sampling and analysis plan to identify the residual presence of fuel elements suspected of being used as an accelerant. The fire had resulted in substantial structural damage and destruction of interior features.

FIG. 1--Decontamination of soil sampling equipment.

FIGURE 2

Within the space and overhead, not visible from the exterior, were several "bucket" di-electric capacitors. Refer to Fig. 2.

FIGURE 3

The bottoms of the units had been removed and the material spilled to the floor. Missing from the scene was one of the buckets. This was later found being used as a waste basket.

Safety and health assessment (SHA) and controls were followed by the investigative team that included the project leader, a Certified Industrial Hygienist, an environmental chemist, a chemical engineer and an industrial hygiene technician. The highest priority was given to safe entry to the site.

A structural engineer conducted an investigation and reported periodic entry was allowable if the structure remained undisturbed. A site-specific safety and health plan (SHP) was drafted and the team was briefed by the project leader. The team mobilized at the site with Level C PPE and field decontamination equipment. A photoionization detector (PID) and multiple gas meter were used for initial atmosphere characterization. The structure was open to the elements and above grade with a noticeable styrene odor. Entry was made by a three person team with the chemist coordinating sample preparation and collection at the control point. A cellular phone, programmed to 911 was within reach of the chemist. The entry team maintained line of sight and voice communications with each other and the chemist.

While not a true ESA, this investigation presented some of the hazards common to damaged or abandoned facilities.

Case Study #3 ESA and short term abatement measures at an abandoned manufacturing facility.

The scope of services was to assume environmental assessment responsibilities, regulatory interfacing and remediation project management on behalf of the facility owner replacing a consultant with five years experience at the site. The facility, closed since 1986, consisted of several buildings on a 25 acre site. Manufacturing activities included foundry operations, machine work, casting and warehousing. Some facility structures dated to the 1880's with Manhattan Project activities conducted in one building (1942 to 1946).

Project activities commenced in 1988. State environmental agency directives included updated assessment of releases of oil or hazardous materials (OHM) and control of an imminent asbestos hazard under the United States Environmental Protection Administration (USEPA) National Emissions Standard for Hazardous Air Pollutants (NESHAPs). The site was listed as a confirmed waste disposal site triggering the requirements of the Occupational Safety and Health Administration (OSHA) Hazardous Waste Operations and Emergency Response regulation (HAZWOPER, 29 CFR 1910.120). In 1990, a comprehensive site specific safety and health plan was drafted using historical information, existing assessment records and reports relating to the Manhattan Project activities. Personnel working on the project site were required to have site specific HAZWOPER training.

In 1992, project operations included a facility wide asbestos abatement project. The SSHP and training requirements were updated to include remedial contract operations. Additionally, several facility structures had roof failure and edifice crumbling. A licensed structural engineer was retained to provide an opinion on facility integrity during remediation operations. The engineer determined work would be unsafe in several locations even though visual conditions appeared satisfactory. The remediation project did not address Manhattan Project contamination.

In 1994, Manhattan Project decontamination commenced under separate contract. A key part of this contract was to update the radiological survey data using current technology. There was a concern that the extent of contamination was more extensive than previously reported. This contingency was not addressed in previous safety and health planning by the consultant. As with case study #2, this example shows the hazards of investigating uncontrolled sites and clearly illustrates the need to carefully plan the health and safety aspects of site entry and assessment.

RISK BASED ANALYSIS TO SITE ASSESSMENTS

What the above case histories highlight is the potential for exposure. With the wealth of information now available, a more complete picture of a site can be developed from the office. The following resources at a minimum should be consulted prior to any sampling or assessment action:

Local Resources

The town historian, or "local history" section of the library will give indications of previous uses of the property and what compounds or materials may have been used or produced at the site. A textile mill will have consumed copious amounts of chlorinated hydrocarbons. A rayon mill will have used sodium hydroxide and lead shielding. A metal finishing plant will have solvents, acids, thinners, lead pigments and lacquers etc.

Under the Superfund reauthorization, local communities are mandated to list those businesses where hazardous or regulated materials are stored. These material summaries are found in most town or city clerks office, the fire department or the local emergency management coordinators office. This information is updated annually under the EPA's Emergency Planning Community Right-to-know Act (EPCRA) program and available to anyone.

Federal Sources

Electronic databases are available that list just about any spill, release, leak, and hazardous/regulated material storage within one mile of the target site. These databases compile information from numerous state and federal sources and will report in detail what may be found at the site to be investigated. Databases include listing of underground storage tanks, reported spills and releases and inventory locations of permitted waste generators.

On-Site Information

One of the best, and chronologically last resources that can be used is
the hazardous materials inventory at the site. Under the requirements
of the OSHA Hazard Communications Act (HAZCOMM) a listing of materials
and a materials safety data sheet (MSDS) is to be available for both
employees and any other interested or affected party. These documents
will define what materials are in drums, tanks and vaults at the site
and can be used as an indicator of what waste is being produced from the
operation and what PPE may be required.

DISCUSSION OF IMPACT ON PHASE I & PHASE II ACTIVITIES [1] [2] [3]

Environmental site assessments are focused on identifying conditions,
events or activities which have or may have resulted in release of oil
or hazardous materials to the environment. These investigations by
definition and design are preliminary in scope and not intended to
define the nature and extent of OHM contamination or indicate
appropriate remedial measures.

The individuals conducting these investigations include the "user" and
the "environmental professional" (E1527 & E1528). Neither definition
nor the authors' experiences indicates these individuals consistently
have knowledge or training in industrial hygiene or health and safety
and health assessment.

The investigators need to understand and respect that their work may
involve exposures to physical and chemical hazards. The degree of
definition and control of these hazards is largely related to the
current status of the site. An operating manufacturing facility is
likely to have hazard management programs in place. An abandoned
facility, conversely, may not have been closed in an environmentally
diligent manner and thus pose extensive hazards to the environmental
professional.

A significant implication of the ESA process is discovery of conditions
where OHM was actually released to the environment or where existing
conditions indicate the release of OHM is likely or imminent. These
conditions frequently necessitate notification of local, state or
federal authorities. Subsequent investigation may result in
"uncontrolled hazardous waste site" designation (depending on regulatory
nomenclature). Where further investigative activities, mandated or
voluntary corrective actions commence at these sites, the Hazardous
Waste Operations and Emergency Response (HAZWOPER) regulation applies.
The HAZWOPER regulation's (29 CFR 1910.120) paragraph (b) Safety and
Health Programs mandate thorough assessment, analysis and management or
control of site hazards and proposed activities before site operations
may commence. Failure to plan for hazards may result in unnecessary
risk to workers. The conditions which may exist or evolve at such a site
may result in the classification of the site as an uncontrolled
hazardous waste site.

In an ESA, however, the assessor may be discovering the conditions and hazards which ultimately will result in HAZWOPER status but without benefit of the assessment and control measures necessary to protect themselves from site hazards.

There are three major site categories the assessor may encounter: active, abandoned and undeveloped. Hazards or conditions which may be encountered during the ESA process can also be reduced to three categories with respect to the site categories listed: (1) hazards common to all three, (2) hazards which may be common to some but not all and (3) site specific hazards.

Characterizing these hazards and the measures necessary to control them for the purposes of conducting an ESA poses a challenge. Time constraints, resource allocation (low bids), expertise, personnel limitations and site access are factors which affect safety and health planning. The short duration, high intensity nature of ESA's requires planning that incorporates a high degree of flexibility.

Worker protection is an employer's first obligation. It has been stated that the health and safety of all response and cleanup personnel depend on effective evaluation of potential exposures and expert determination of the appropriate means to protect against those exposures. The following outlines an approach to safety and health planning using standard procedures, site specific hazard and task management measures. Development of a cost effective, useful plan requires the involvement of a CIH, an occupational physician, environmental professionals, management and project personnel.

Proposed Outline
ESA STANDARD OPERATING PROCEDURES FOR
SAFETY AND HEALTH MANAGEMENT

1.0   Purpose: Protection of worker safety and health during the conduct of ESA's and environmental sampling.

2.0   Written Program: standard company policy

2.1   Description of activities which necessitate the plan; use the ESA or environmental sampling scopes of services.

2.2   Organizational plan: Key personnel and responsibilities from principal to environmental technician.

2.3   Training: List minimum levels of training personnel receive including training in HAZWOPER, asbestos assessment, ESA techniques, confined space entry, respiratory protection and topics specific to the work proposed in section 2.1.

2.4   Medical surveillance: Identifies medical surveillance requirements specific to anticipated or documented hazards and use of required PPE.

2.5   Personnel Monitoring: Highly dependent on the nature of the anticipated contaminants and may be driven by activities like noise monitoring due to drilling operations or photoionization detector monitoring for work around OHM contamination. Monitoring must be designed to assess potential exposures to specific contaminants.

2.6   Activity Hazard Analyses (AHA): The AHA should be developed
          for each of the tasks and activities performed. The AHA
          should present specific tasks, hazards of the task, control
          measures and equipment necessary to perform the task.
          Examples include well bailing or unmarked drum sampling.
2.7   Personal Protective Equipment: PPE is necessary where a
          hazard cannot be controlled. PPE for ESA personnel may
          include hardhats, safety glasses, protective gloves,
          respirators, outer garments and footwear. PPE selections
          must be specific to hazards and personnel must receive
          documented training.
          2.7.1   Levels of PPE: Tiered levels of protection
                    (LOP) may be necessary (HAZWOPER uses LOP's A to
                    D for management of the most severe to least
                    severe hazards).
          2.7.2   LOP Upgrade/Downgrade: The means by which LOP's
                    may be changed based on site-specific
                    conditions.
2.8   Site Control Measures: These can be integrated with
          site-specific information to show where specific activities
          will occur and for a means of presenting standard orders for
          site operations.
2.9   Decontamination Plan: Provide a decontamination plan if
          contact with OHM is likely.
          2.9.1   Sequencing of decontamination operations
          2.9.2   Inventory of decontamination equipment
2.10   Hazards: This section should include information on hazards
          to which personnel may be exposed and associated health
          effects.
          2.10.1   Chemicals: Hydrocarbons, chlorinated solvents,
                    heavy metals, pesticides and other OHM that may
                    be encountered. Include MSDS information and
                    sampling methods.
          2.10.2   Physical: Cold, heat, energy, compressed gases,
                    pipes, excavations, confined spaces.
          2.10.3   Biological: Insects, spiders, snakes, plants.
          2.10.4   Extreme Weather: Wind, lightning, snow, ice,
                    temperature extremes, rain and fire hazard.
2.11   Emergencies: Include first aid, accident/injury reporting,
          communication with emergency services, extreme weather and
          responsibilities.
2.12   Appendices: Include the organization's existing written
          programs, historical monitoring results, medical
          surveillance records and training records.

3.0   Active Sites: Consult existing programs and activities
          before conducting the site investigation. This information
          may pinpoint specific hazards that would likely be
          encountered. The following lists resources which should be
          considered in developing site-specific procedures.
3.1   Site orientation programs, key site manager, occupants
3.2   Chemical and hazardous waste storage management plans
3.3   Permit-required confined space programs
3.4   Hazard communication program

3.5   Hearing conservation program
3.6   General safety and health program
3.7   Lockout/tagout program (control of hazardous energy)
3.8   Previous ESA information, environmental sampling data and
            related historical records
3.9   Site layouts, drawings or plans

4.0   Inactive Sites: A site may be inactive due to bankruptcy,
            obsolescence, loss of chemical storage or occupancy permits
            or other mechanism of condemnation. The extent of hazardous
            materials closures or hazardous facility conditions will
            vary. The following lists resources which should be
            considered in developing site-specific procedures:
4.1   Items listed in section 3.0 may be available to varying
            degrees. They may not represent existing conditions.
4.2   Determine the reason for the inactivity or abandonment.
            Were permits revoked? Was it bankruptcy?
4.3   Review state and federal databases for operating and closure
            files relating to the site.
4.4   Modify standard programs (section 2.0) to reflect what may
            be known of the site. This can include site-specific
            confined space entry to investigate subgrade storage spaces
            and similar measures.
4.5   Determine from local sources if there are hazards relating
            to site inactivity including structural integrity, fire
            hazard, unauthorized trespass or human occupation, vermin
            and similar conditions.
4.6   Conduct perimeter site reconnaissance where potential risks
            are identified or if little is known of the site.

5.0   Undeveloped Sites: This category may include urban,
            suburban or rural environments ranging from open city lots
            to large tracts of wooded land.
5.1   Historic research may identify unknown prior uses or
            activities. Topographical and aerial maps and similar
            references may provide data about site access and
            conditions.
5.2   Determine from local authorities or reference sources if
            there are hazards relating to the site including water
            hazards, rockfall, forest fire, unauthorized trespass or
            human occupation, vermin and similar conditions. Site
            conditions may warrant additional contingencies and
            personnel.
5.3   Access to and communication with emergency services may need
            additional planning to address unique conditions (islands,
            undeveloped roads, extreme distances, adverse weather).
            5.3.1   Communications: Cellular phones, two way
                        radios, ham radios or land line.
            5.3.2   Determine emergency service response times to
                        and from the site.
            5.3.3   Supplemental first aid and emergency gear to
                        address foreseeable emergencies.

5.4   Additional equipment may be necessary ranging from all terrain vehicles to specialized PPE (snowshoes, hipwaders, personal flotation devices, boats). These contingencies need to be planned and personnel must be trained on the operation of each.

5.5   Staff Reporting: Staff should notify supervisors of specific activities planned at the site, start and finish times and deviations from anticipated plans and activities. Supervisors must provide for communication during site activities.

6.0   Site-Specific Information: Site-specific information should include the findings from sections 3.0, 4.0 or 5.0 in addition to the following. Include emergency phone numbers, planned confined space entries, PPE and monitoring for specific hazards and similar "fine tuning". This information should be attached to the standard site safety and health plan as a separate section and organized on a standard form.

6.1   Scope of services: This may be a proposal, work order or other document which defines the project activities.

6.2   Historical Findings and Interview: These records can define prior site uses and identify hazards. Conduct historical investigations and interviews in advance of the site visit.

6.3   Site-specific modifications should be developed from section 2.0 for at least the following:

6.3.1   Organization: personnel assigned to the project.

6.3.2   Medical Surveillance: provision for specific monitoring or follow-up examinations.

6.3.3   AHAs: Updated AHAs.

6.3.4   Personnel Monitoring: Monitoring methods, criteria and action levels for likely contaminants.

6.3.5   PPE: PPE anticipated and criteria for use.

6.3.6   Site Control: Drawings, maps, access routes, hazard areas, meeting areas, equipment storage and site specific information.

6.3.7   Decontamination: Depends on AHA's, hazards, and PPE selections.

6.3.8   Emergencies: Telephone numbers, communication, emergency services access, etc.

6.3.9   Training: Site specific training and briefings.

The recommended outline is a guide. Federal, state and local regulations, reference publications and competent professionals will also need to be consulted.

SUMMARY

Common sense is the best approach to protecting yourself and field operatives from the hazards of ESA activities. Of prime consideration is the ability to recognize and control hazards.

Assembly of a comprehensive plan may not seem feasible, however, given the realities of the ESA and environmental services industry it is vital. The production of the standard site safety and health plan (see outline section 1.0-5.0)is manageable and cost effective. As with any written program, the plan must be periodically reviewed and updated.

The development of a standard to protect workers during investigations and activities consistent with ASTM methods is recommended. Though current standards make some reference to safe work practices, their purposes are directed to assessment, data collection or remediation. The proposed standard would encompass activities defined under ASTM methods and represent a minimum duty of care to protect environmental professionals.

REFERENCES

[1] George and Florence Clayton, <u>Patty's Industrial Hygiene + Toxicology</u> 4th edition Volume 1 Part B, John Wiley and Sons 1991, New York, NY.

[2] <u>Health and Safety Plan Users Guide USEPA</u>, 9258.8-01 EPA 540-c-93-002 7/93. Washington, DC. 1993

[3] NIOSH-OSHA-USCG-EPA <u>Occupational Safety and Health Guidance Manual for Hazardous Waste Site Activities</u> US Dept. of Health and Human Services. NIOSH Publication no. 85-115. Arlington, MA. 1985

Peter D. Hedges,[1] Robert J. Ellis,[2] and John Elgy[1]

AIRBORNE REMOTE SENSING AS A TOOL FOR MONITORING LANDFILL
SITES WITHIN AN URBAN ENVIRONMENT

REFERENCE: Hedges, P. D., Ellis, R. J., and Elgy, J., ''Airborne
Remote Sensing as a Tool for Monitoring Landfill Sites Within
an Urban Environment,'' Sampling Environmental Media, ASTM STP 1282,
James Howard Morgan, Ed., American Society for Testing and Materials,
1996.

ABSTRACT: The use of remote sensing imagery to facilitate the
management of landfill sites in urban areas has, to date, been primarily
confined to the use of conventional aerial photography. The potential of
airborne video thermography, and the high resolution and the range of
spectral bands offered by the Airborne Thematic Mapper (ATM) for
monitoring landfill sites are under evaluation, with the ultimate aim of
establishing an economic methodology that can be employed for rapidly
surveying landfill sites within a large urban area. Detailed
investigations of one urban landfill site in the West Midlands of the UK
have been undertaken using ATM imagery and aerial photography obtained
in 1992, and ATM imagery and thermography acquired in 1994. This work
has enabled recommendations to be made regarding the techniques employed
in monitoring by remote sensing which revolve around the use of the
thermal infra-red band.

KEYWORDS: remote sensing, landfill, waste disposal, monitoring, gas
migration, leachate, thermography, airborne thematic mapper, thermal
infra-red.

HISTORY OF TIPPING IN THE BLACK COUNTRY

As with many of the older industrial conurbations of the United
Kingdom, the Black Country area in the West Midlands (Fig. 1) has a long
history of using worked out quarries for tipping both industrial and
domestic waste. By the mid-16th century the Black Country was an
established centre of iron production and metal working. However,

---

[1]Lecturer, Department of Civil Engineering, Aston University,
Aston Triangle, Birmingham, B4 7ET, UK.

[2]Research Student, Department of Civil Engineering, Aston
University, Aston Triangle, Birmingham, B4 7ET, UK.

it was not until the main thrust of the industrial revolution from the
1760s onwards that the

> "roaring and blazing furnaces, the smoke of which
> blackened the country as far as the eye could see"
>   (James Nasmyth, 1830, as quoted by Skipp [1])

gave this 300 square kilometres or so of countryside its name.

Iron workings were originally attracted to the area since coal,
ironstone, fireclay and limestone were all readily available, as were
streams offering good sources of water power. Later in the mid-
nineteenth century, smelting processes designed to use the local coal,
the harnessing of steam power and a good transport network, provided
first by canals and later by the railways, established the area as an
international centre of metal working.

The needs of the metalwork industry in turn encouraged the
development of sister industries - initially to service their
requirements, but later many becoming established enterprises in their
own right. Amongst these manufacturing interests were: the production of
bricks, tiles and earthenware (for which the local marls were
extensively quarried); glassmaking; and, chemical industries.

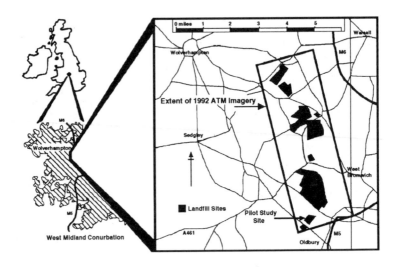

Fig. 1--The Location of the Black Country showing the extent of the
1992 ATM imagery and major landfill sites

Throughout these times in the Black Country, interwoven into the
grim landscape alongside the industrial production, quarrying and
transport networks, were the villages and dwellings of the workforce,
together with the everyday services necessary for their survival.

Although a degree of dereliction persists from the region's
decline in the early decades of the twentieth century, the tradition of
metalworking and the legacy of the industrial revolution have persisted
in the Black Country up to the present day. However, the current phase

of redevelopment, from a heavy manufacturing base towards light industry, is often hampered by the unknown nature of the ground under consideration.

As can only be expected from an era where industrial production was the raison d'être, and there was little or no awareness of environmental considerations, indiscriminate dumping of industrial and domestic waste into the convenient holes left by mining and quarrying was the norm. This practice continued from the earliest days of metalworking, circa 1650, up to the 1970s when strict controlling legislation was enacted. Much of the land within the Black Country is now made ground, and the exact location of tips is often lost. Furthermore, since historical records prior to 1974 are poor, even when the location of a tip is known the nature of the fill is often unknown.

REMOTE SENSING AND LANDFILL SITES

In large industrial conurbations with high development densities and often difficult access, the techniques of earth resources remote sensing have much to offer the environmental manager. In particular, the acquisition of vertical images of large areas at a low cost, often provides an opportunity to filter the data and enable scarce resources to be focused on potential problem areas.

Experience in the use of remote sensing for locating and monitoring landfill sites has to date been largely restricted to the use of conventional aerial photography [2], [3], [4], [5] and [6]. Sangrey and Philipson [7] and Haynes et. al. [8] explored the use of alternative remote sensing techniques in landfill management, but other than the work of Jones [9] little research has been undertaken in recent years. Currently, although aerial thermography is seen as promising in certain quarters [10], the regulatory agencies in the UK are not yet convinced of its value.

The low resolution currently offered by commercial satellite imagery is usually inadequate for distinguishing individual features within the urban environment. On the other hand, digital imagery acquired from aircraft offers rapid coverage of an area at a high resolution in a format suitable for computer processing.

Research with the overall aim of developing an economic methodology, using remote sensing techniques, for rapidly surveying and monitoring landfill sites within an urban area was initiated in 1992.

THE INTERACTION OF ELECTROMAGNETIC RADIATION WITH FEATURES ON THE EARTH'S SURFACE

Reflected Electromagnetic Radiation

When incident electromagnetic radiation (E_I) strikes a surface it is either reflected (E_R), absorbed (E_A) or transmitted (E_T). The proportions of each will depend upon the interaction of the radiation with the specific surface under consideration, and this in turn will vary from wavelength (λ) to wavelength: thus

$$E_R(\lambda) = E_I(\lambda) - [E_A(\lambda) + E_T(\lambda)] \qquad (1)$$

Furthermore, the way in which the radiation is reflected is dictated by the relative size of the elements forming the feature's surface roughness when compared with the wavelength of the incident energy interacting with it. Thus the reflectance characteristics of individual features on the earth's surface will vary according to both the wavelength of the incident radiation and the nature of the surface itself. The amount of radiation reflected by an object and recorded by a remote sensing sensor is called the spectral reflectance, and is defined as:

$$\rho_\lambda = \frac{E_R\,(\lambda)}{E_I\,(\lambda)} = \frac{\text{energy of wavelength } \lambda \text{ reflected from object}}{\text{energy of wavelength } \lambda \text{ incident upon object}} \times 100 \qquad (2)$$

Fig. 2 illustrates the variation in reflectance with wavelength for soil, water and vegetation within the visible (0.4 - 0.7 µm), near infra-red (0.7 - 1.3 µm) and mid-infra-red (1.3 - 3 µm) regions of the spectrum.

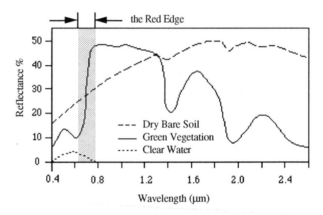

FIG 2--Idealised reflectance curves for vegetation, soil and water.

<u>Emitted Electromagnetic Radiation</u>

In the thermal portion of the spectrum, bodies will emit varying amounts of radiation depending upon both their temperature, their thermal properties and on how closely they resemble black bodies. Thus different bodies will also display different emittance characteristics in the thermal infra-red region of the electromagnetic spectrum.

<u>Image Interpretation</u>

Since the earth's surface features exhibit different spectral characteristics, the varying reflectances and emittances recorded by remote sensing sensors can be used to differentiate between objects (see Fig. 2). This can be achieved either through visual inspection and interpretation of images, or through the use of classification methods applied to digital imagery. In the latter, the information held in a number of wavebands is combined and simplified using statistical techniques to produce a single image in which each individual pixel is assigned to a particular land cover category. For a more detailed description of classification techniques the reader is referred to standard remote sensing texts: e.g. [11] and [12].

THE AIRBORNE THEMATIC MAPPER

The Daedalus AADS 1268 Airborne Thematic Mapper (ATM) is of particular interest to the monitoring of landfill sites, as it offers both a wide number of spectral bands and high resolution. The ATM is a Multispectral Linescanner, which, when mounted in an aircraft, records the electromagnetic radiation reflected from or emitted by objects on the ground surface. The system operates by virtue of an oscillating mirror scanning a swathe of consecutive lines normal to the direction of flight, and immediately beneath the aircraft as it progresses forward (Fig. 3). The information is recorded on magnetic tape, and after processing, when the scan lines are laid side by side a continuous image of the ground along the flight path is built up.

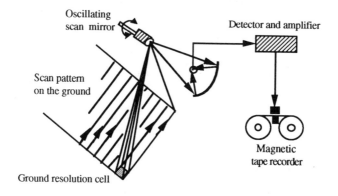

FIG. 3--Schematic diagram of the ATM and its operation.

TABLE 1--ATM bands.

ATM Band	Band Width (µm)	Description
1	0.42 - 0.45	violet
2	0.45 - 0.52	blue
3	0.52 - 0.60	green 1
4	0.60 - 0.625	green 2
5	0.63 - 0.69	red
6	0.695 - 0.75	red/infra-red edge
7	0.76 - 0.90	near infra-red 1
8	0.91 - 1.10	near infra-red 2
9	1.55 - 1.75	mid infra-red 1
10	2.08 - 2.35	mid infra-red 2
11	8.50 - 13.00	thermal infra-red

The ATM splits the incoming electromagnetic radiation into its component wavelengths, and records the information from 11 discrete bands, as shown in Table 1. The information from each of the 11 bands is recorded consecutively for each individual scan line, before progressing forward to the next line - a process referred to as 'band-interleaved by line' (BIL).

To enable the data to be processed by a computer, the original continuous incoming trace is broken into short segments, each of which is assigned a number (the Digital Number) on a scale of 0 to 255 - the Grey Scale. The Digital Number (DN) represents the intensity of electromagnetic radiation recorded from a small area of ground - called a pixel, when displayed on a VDU.

ACQUISITION AND MANAGEMENT OF THE ATM DATA

In August 1992, as part of the Natural Environment Research Council's Airborne Remote Sensing Campaign, the 3 by 9 km area of the Black Country shown in Fig. 1 was flown in four passes at a height of 800 m, giving a pixel resolution of approximately 1.5m. This flight was made around mid-day and both ATM data and colour aerial photography were acquired. Subsequently, in April 1994 a second set of ATM imagery was obtained for the same area, but this time the flight was made at dawn - thus only the thermal infra-red band is of use.

The first problem encountered in managing the ATM data was its sheer volume - for the first flight it was supplied on 9 magnetic tapes. The area covered during each flight was approximately 27 km^2, and the 11 bands at a 1.5 m resolution yielded some 350 Megabytes of data.

In BIL format, the data comes as a continuous stream in the form:

(line 1/band 1),(line 1/band 2) ... (line 1/band 9),(line 1/band 10), (line 1/band 11),(line 2/band 1),(line 2/band 2),(line 2/band 3) .. etc.

Normally the number of bytes in each line is the same, so separation of the interleaved data is achieved by taking a prescribed amount of information and placing it sequentially in 11 discrete files - one for each of the bands. Once appropriate code has been written for disentangling the data, providing the data is not corrupted in any way, the procedure is relatively straightforward. However, it is extremely time consuming.

On viewing the 1992 daytime ATM imagery it became clear that some degree of atmospheric interference had been experienced, and pixel intensity values varied across scan lines - for most bands the intensity being greater towards the ends of the lines than in the centre (Fig. 4). The imagery was corrected by applying the following procedure:

i)   calculate the mean intensity values for columns of pixels along the line of flight;
ii)   derive an algorithm for normalising these values;
iii) rectify the value of each individual pixel by application of the algorithm.

No problems with atmospheric interference were experienced with the 1994 ATM data due to it being a dawn flight with only the thermal band being of interest.

To date no practical technique has been developed for successfully geocorrecting ATM imagery. During the scanning process variations in the aircraft flight pattern due to pitch, roll, yaw, atmospheric

disturbances etc., result in the displacement of individual scan lines
adjacent to each other. Thus although the features and their locations
within the imagery are readily recognisable, it cannot be used directly
as a map.

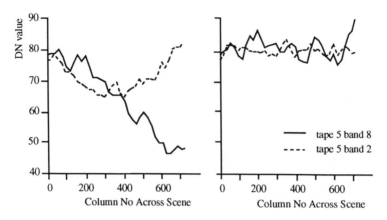

Fig. 4a Before correction   Fig. 4b After correction

FIG. 4--Atmospheric interference of the 1992 ATM data illustrated
through the variation in mean pixel intensity value across a
scene.

THE AERIAL THERMOGRAPHIC SURVEY

The thermographic survey of the pilot study site, and an adjacent
active landfill, was undertaken using a modified Agema 450 Thermovision
infra-red thermal imager containing only one detector. The recording was
made from a helicopter flying at heights of 150 m and 450 m with the
door on the viewing side removed. The imager was mounted in parallel
with a conventional camcorder on a specially constructed facility and
directed at the ground by the operator at an angle of approximately 45°.
The infra-red survey results are recorded on VHS video tape either
in conventional black and white or using the ironbow colour scale,
whereby an internal program automatically assigns a colour to represent
the sensed thermal intensity level. During the flight the operator can
view the imagery in real time, thus enabling a more detailed record of
areas of particular interest to be obtained.
The consultants collaborating in the study selected this
particular imager, which senses thermal infra-red within the 2 to 5 μm
waveband, on the grounds that it is inexpensive and provides rapid
results of an adequate quality. They also argue that viewing at an
oblique angle from a helicopter provides the operator with the
flexibility to vary height, hover, etc., in order to investigate
locations of particular interest in detail - a facility which is not
possible with a similar sensor mounted vertically in a fixed wing
aircraft flying in essentially straight lines.
The VHS tape of the imagery is available immediately the flight is
completed and no preprocessing is required. The natural colour recording

from the camcorder, when viewed at the same time as the thermal imagery, facilitates in the identification of the location under consideration.

Under normal circumstances a thermographic survey would be made at approximately sunrise. However, due to management difficulties this was not possible and the flight was undertaken at mid-afternoon on a dull overcast day in late November. Thus, although not ideal, the effects of solar warming were minimised.

When viewed in normal mode on a VDU, the imagery is adequate for qualitative interpretation. However, frozen scenes and those 'grabbed' for digital processing display considerable noise. It is understood [13] that this can be overcome by either employing a similar camera using more than one detector, or a system sensing in the 8 to 14 μm waveband. Both of these options are more expensive than the system used, and the latter is more unwieldy in that it requires cooling by liquid nitrogen.

IDENTIFICATION OF LANDFILL SITES

Historically due to the nature of ad hoc urban development, where outmoded buildings are pulled down, the land reclaimed and further development takes place, many tip sites have been covered and built upon. The remote sensing techniques under consideration simply record the spectral characteristics of the surface features and do not probe beneath the immediate cover. Thus, unless there is some surface indicator of buried features, historical redeveloped landfill sites cannot be located by these methods.

A characteristic of more recent landfill sites in urban areas is that on completion they are either reclaimed for formal recreational purposes or are provided with a soil cap and left derelict. Thus in attempting to locate old landfill sites, the main clue available to remote sensing technicians is that they will be relatively large open spaces within the urban setting. Derelict land will be characterised by scrub vegetation and patches of bare ground, and sites reclaimed for recreation tend to have uniform swards of grass as their distinguishing feature.

ATM Imagery Or Aerial Photography for Site Identification ?

Fig. 5 shows an area of the 1992 ATM imagery containing a landfill site which has been classified using the Maximum Likelihood Classifier. The image has been produced by consideration of Bands 5, 7, 10 and 11, together with an index (Normalised Difference Vegetation Index), created by performing arithmetic on those bands which show the best response to vegetation health. The classification procedure is semi-automatic, in that the operator interactively identifies areas of known land cover (training areas). The image processor subsequently uses the pixel statistics from these training areas to classify the remainder of the image.

To date it has not been possible to use straight forward classification techniques alone to identify large open spaces. The cut grass swards of playing fields, and the poor quality vegetation and bare ground of derelict land can be satisfactorily identified. However, it is not possible to employ these simple classes on their own to differentiate between open spaces and areas of urban development. Since housing and industrial developments contain a patchwork of vegetation (both grass and other forms of cultivation) and soil and building materials (which often have similar spectral characteristics to bare ground), these too exhibit the same classes. Thus in Fig 5, although

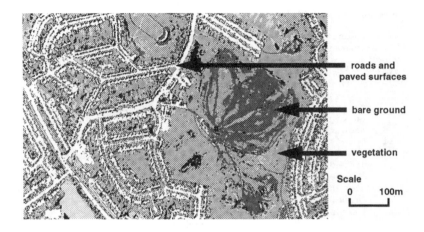

FIG. 5--Section of a Classified ATM Image Showing a Landfill Site

relatively large areas of a single class can be seen, these classes can
also be found within the rapidly changing land cover classifications of
the neighbouring housing developments.

As a consequence of the above, a second stage of identification
for likely open spaces is required. This is achieved simply by
identifying areas of a single class over a given threshold size - in the
case of the Black Country study 5 ha has been selected.

Clearly at the current state of research, the visual scanning of
either natural colour (a composite of the red, green and blue bands) or
false colour red (usually a composite of green, red and near-infra-red
bands) photographic imagery is as effective in identifying potential
landfill sites as the more complex semi-automatic process of
classification described above. Thus, ATM imagery appears to offer no
advantages over classical aerial photography either in terms of
economics or effectiveness.

However, the potential of applying texture analysis (studying the
rate of variation of either pixel DN value or land class) to the ATM
data is currently under investigation, and indications are that this may
prove a useful tool in differentiating between large areas of a single
class, and urban areas where the land cover changes rapidly over a short
distance, such as in housing developments.

Video Imagery for Site Identification

As discussed below, thermography is well suited to particular
aspects of the investigation of individual landfill sites, or series of
sites. However, the rapid survey of a large urban area for landfill site
identification requires either a mosaic of high resolution small scale
single frames, digital imagery, or multispectral data - none of which
can currently be provided adequately by videography. Furthermore, it is
often surprisingly difficult to locate individual scenes from moving
video coverage of a large area which has been taken from the air.

MONITORING LANDFILL SITES

In monitoring completed landfill sites, the responsible authority is concerned with the integrity and safety of the site. Interest is thus focused upon landfill gas migration and leachate emergence. In monitoring landfill sites using remote sensing for gas and leachate there are three factors which the operator can utilise:

i   vegetation stress /die back: landfill gas can be injurious to vegetation - thus areas of dead or sickly vegetation might be indicative of high levels of gas concentration;
ii   colour: soluble ferrous ions dissolved in percolating water will oxidise to red-brown ferric iron and stain the ground where leachate emerges at the surface; dark wet patches of ground may also be due to leachate;
iii elevated temperatures: where gas or leachate emerge may be warmer than the surrounding ground, due to the gas or liquor retaining heat acquired within the landfill as a result of the elevated temperatures within the waste caused by bacteriological decay and exothermic reactions.

ATM Imagery

When Jones [2] utilised ATM data to investigate gas migration in a rural landfill site, since the site had a uniform cover, she was able to identify areas of vegetation stress using the red edge (the rapid change in reflectance between the red and the near infra-red which occurs in the region of 0.6 to 0.8 μm - see Fig 2). The urban sites investigated to date all possess non-homogeneous cover, and it has therefore not been possible to use vegetation stress as an indicator of gas presence.

Although dark wet patches of ground have been identified on the imagery, none of these have proved to be leachate seeps - indicating the unreliability of this approach. However, during periods of extended dry weather it may prove sound. In addition, since the natural colouring of the soil over much of the study area is an orange brown, colour has also proved an inadequate indicator of leachate and no seeps have been located using this method.

Despite the setbacks with the above two approaches, some success has been achieved with the use of temperature as an indicator of gas emergence.

When used alone, the 1992 daylight thermal infra-red band has enabled zones where significant levels of gas have been venting to be identified. At these locations the elevated ground temperature results in higher levels of emittance than in the immediate vicinity. This is recorded on ATM Band 11 as relatively high DN values. Using a technique referred to as density slicing, the operator instructs the image processor to divide a histogram of pixel frequency against DN value into specified intervals (Fig 6a). Each of these intervals is subsequently displayed as a different colour on the image.

For the site shown in Fig 6b, the locations identified are settlement cracks which are actively venting gas. Since the 1992 imagery was acquired in the middle of the day, features such as bare ground and the roofs of buildings had been heated by the sun. These are the other objects displayed in black - also at an elevated temperature relative to their surroundings. At a second landfill site, trenches known to be venting gas were barely discernible.

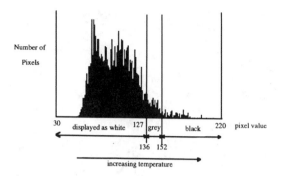

Fig. 6a   Histogram of pixel frequency against DN.

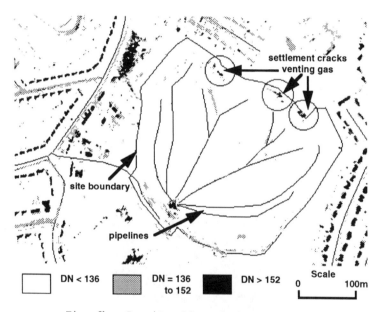

Fig. 6b   Density slice of thermal band.

FIG. 6--Density slice of thermal band showing active venting of gas.

Time constraints have not yet permitted the 1994 thermal imagery to be analysed in detail. However, the flight was made at dawn to take advantage of the enhanced differences in temperature which would be expected from the cooling of the land overnight. Initial indications are that temperature differences are indeed more noticeable and the imagery is sharper (Fig 7) than the equivalent daytime thermal ATM Band.

Following the dawn flight a spike survey of gas levels was undertaken along predetermined transects across the site. However, no correlation was found between the gas measurements and the recorded ATM

FIG. 7--Extract from ATM Band 11 showing pilot study site:
black areas indicate cold, white warm.

spectral emittance levels. It is hoped that a second dawn flight will be
made in 1995 during which gas levels will again be monitored.

Thermographic Imagery

    In the discussions above concerning the acquisition of the
thermographic data, the flexibility of the helicopter mounted thermal
imager system was highlighted, in relation both to the ability to dwell
on sites of interest, and the manoeuvrability and readily variable
altitude of viewing. In addition this system enables real time viewing
and a rapid response to be made in the event of areas of concern being
identified.

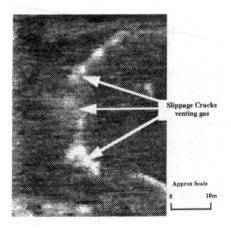

FIG. 8--Digitised thermography image showing gas venting from slippage
cracks in north east corner of pilot site.

At the pilot site, the elevated ground temperatures caused by gas venting from settlement cracks around the periphery of the landfill were readily identified (Fig. 8), both in real time and from the moving video recording. Other 'hot spots' proved to be boreholes and pipelines from the active venting system and bare patches of ground.

Although no leachate emergence is evident at the site, Titman [10] has reported examples of seeps being located using this technique. He

TABLE 2--Comparison of remote sensing techniques employed during study

Feature of the Survey	Aerial Photography	ATM Imagery	Thermography
Sensor	Wild RC-8 Survey Camera	Daedalus 1268 ATM	Agema 459 Thermovision thermal imager
Platform	<- Piper Chieftain aircraft ->		Squirrel twin engine helicopter
Coverage	3 x 9 km	3 x 9 km	25 ha
Cost of acquisition	<- approx \$8 250 -> (£5 500)		approx \$3 750 (£2 500)
Preprocessing	developing and printing	extensive	none
Resolution / scale	1 : 5 000	1.5 m pixel size	100 pixels per line
Image quality	very good	good	noisy
Identification of sites	rapid visual interpretation	significant image processing required	impractical
Monitoring gas migration: - technique	not applicable	i) vegetation stress identified through multispectral analysis ii) thermal Band 11 employed to detect temperature variation in land cover - dawn flight for best results	temperature variations in land cover detected - dawn flight gives best results
Monitoring gas migration: - response time	not applicable	image analysis results available in: ≈ 5 days Band 11 alone ≈ 7 days all bands	real time or on VHS tape immediately flight completed
Accessibility of system	no problems in commissioning flights	currently only one system in the UK	service offered by several UK consultants

has also used this approach to monitor underground fires, and the use of the imager in association with a lorry mounted hydraulic platform for detailed local investigation is also discussed in his paper.

Due to noise, it has not been possible to undertake any meaningful analysis of single digitised frames, such as Fig. 8, captured from the video imagery. Their visual appearance is also disappointing.

COMPARISON OF VARIOUS REMOTE SENSING TECHNIQUES

Table 2 provides a comparison of the airborne remote sensing techniques employed during the Black Country study. From the Table it can be seen that although ATM imagery provides the greatest flexibility, it is more costly to acquire and process than thermography. In addition the processing and analysis time is significant should a rapid response be required. However, the factor which currently militates most against the adoption of ATM for monitoring landfill sites is the lack of available hardware - only one system is operational in the UK.

CONCLUSIONS

Two sets of ATM imagery and colour air photographs have been acquired for a large area of the Black Country in the West Midlands, together with thermography of two specific landfill sites.

Within this urban setting, research undertaken to date indicates that it is not possible to identify old landfill sites that have been redeveloped using conventional remote sensing techniques. It has been found, however, that aerial photography provides the most straight forward and rapid approach for locating open spaces that are potentially historic landfill sites.

Regarding the monitoring of gas migration, some success has been achieved using both the daylight and the dawn ATM thermal imagery for locating zones where significant quantities of gas are venting at the surface and warming the soil. Dawn imagery is a better data source than its daylight counterpart.

The helicopter mounted thermographic survey, although producing poor quality imagery, provides a cheaper and more flexible alternative to ATM data acquired from a fixed wing aircraft. Sites can be viewed in real time, and imagery is available immediately the flight is completed.

For monitoring landfill sites within a large urban area the use of ATM is recommended. However, where image quality is not important and an immediate result is required in relation to a single site, airborne thermography is the cheapest and simplest technique.

The presence of a 'hot spot' on thermal imagery does not necessarily indicate gas migration - the recorded brightness may be due to other factors. Airborne remote sensing studies must therefore be backed up with ground surveys to determine whether locations identified as potential sources of gas are indeed active.

REFERENCES

[1]   Skipp, V., The Centre of England, Eyre Methuen, 1979.

[2]   Garofalo, D. and Wobber, F.J., "Solid Waste and Remote Sensing", Photogramm. Eng. and Remote Sensing, Vol. 40, No. 1, 1974, pp 45-59.

[3]   Erb, T.L., Philipson, W.R., Teng, W.L. and Liang, T., "Analysis of Landfills with Historic Air Photos", Photogramm. Eng. and Remote Sensing, Vol. 49, No. 9, 1981, pp 1363-1369.

[4]   van Genderen, J.L., van de Griend, J.A. and van Stokkom, H.T.C., "An Operational Remote Sensing Methodology for the Detection, Inventory and Environmental Monitoring of Waste Disposal Sites", Proc. EARSeL ESA Symp. on Remote Sensing Applications for Environmental Studies, Brussels, Belgium, pp 149-156, 1983.

[5]   Curran, P.J., "Small Format, Oblique, Colour Aerial Photography: an Aid to the Location of Methane Gas Seepage", Int. Jnl. Remote Sensing, Vol. 7, No. 4, 1986, pp 477-479.

[6]   Airola, T.M. and Kosson, D.S., "Digital Analysis of Hazardous Waste Site Aerial Photographs", Jnl. Water Pollution Control Federation, Vol. 61, No. 2, 1989, pp 180-183.

[7]   Sangrey, D.A. and Philipson, W.R., "Detecting Landfill Leachate Contamination Using Remote Sensors", EPA Report No EPA-600/4-79-060, Las Vegas, Nevada, USA, 1979.

[8]   Haynes, J.E., Wood, G.M., Lawrence, G. and Bourque, H.H., "Remote Sensing and Waste Management", Proc. 7th Canadian Symp. of Remote Sensing, Winnipeg, Manitoba, 1981, pp 316-322.

[9]   Jones, H.K., "The Investigation of Vegetation Change Using Remote Sensing to Detect and Monitor Migration of Landfill Gas", Ph.D Thesis, Aston University, UK, 1991.

[10]   Titman, D.J., "Aerial Thermography - A Cost Effective Approach to Landfill Site Surveys", Proc 3rd Int Conf on Re-Use of Contaminated Land and Landfills, Brunel University, UK, 1994, pp 269-271.

[11]   Lillesand, T.M. and Kiefer, R.W., Remote Sensing and Image Interpretation, Wiley, 1987.

[12]   Mather, P.M., Computer Processing of Remotely Sensed Images: an Introduction, Wiley, 1993.

[13]   Titman, D.J., Pers. Comm., DTTS, Harpenden, Herts, UK, 1995.

Acknowledgements

The authors would like to acknowledge the help, both financial and in terms of time and data provision, of the West Midlands Hazardous Waste Unit, the Black Country Development Corporation, Aspinwall and Co., TBV Stanger Ltd and the Head of the Department of Civil Engineering, Aston University. Furthermore, the authors acknowledge that without the provision of flying time with the ATM by NERC this project would not have been possible.

Direct Push Sampling

Susan C. Soloyanis,[1] Mark A. McKenzie,[2] Seth E. Pitkin,[3] and Robert A. Ingleton[3]

DETAILED CHARACTERIZATION OF A TECHNICAL IMPRACTICABILITY ZONE USING DRIVE POINT PROFILING

REFERENCE: Soloyanis, S. C., McKenzie, M. A., Pitkin, S. E., and Ingleton, R. A., **"Detailed Characterization of a Technical Impracticability Zone Using Drive Point Profiling,"** Sampling Environmental Media, ASTM STP 1282, James Howard Morgan, Ed., American Society for Testing and Materials, 1996.

ABSTRACT: A team of geologists and engineers conducted an investigation in a dense, non-aqueous phase liquid contaminant (DNAPL) source area at Pease Air Force Base (AFB) Installation Restoration Program Site 32. The investigation was designed to (1) demonstrate that vertical drive point profiling using techniques and equipment developed by the Waterloo Centre for Groundwater Research would work at this hydrogeologically complex site and (2) locate solute concentrations indicative of DNAPL.

The original contaminant source at Site 32 was a 1,200-gallon ($\cong$ 4500 L) underground storage tank with an overflow discharge pipe. The tank held waste solvents from aircraft maintenance in Building 113 at the former Pease AFB. The tank was removed in 1988. The overflow discharge pipe and contaminated soil under and along the pipe were removed in 1990. The results of the Site 32 Source Area Remedial Investigation/Feasibility Study indicate that solvent-related contaminants are present in soil and overburden and in bedrock groundwater and that there is a residual DNAPL source resulting in dissolved-phase groundwater contamination. Efforts to locate DNAPL using monitor wells have been unsuccessful. A limited action remedial alternative based on technical impracticability is proposed for this site.

---

[1]Principal Scientist, The MITRE Corporation, Brooks Air Force Base, San Antonio, TX 78235

[2]Installation Restoration Program manager, Air Force Base Conversion Agency Operating Location A, Pease Air Force Base, NH 03803-0157

[3]Graduate student and research technician, respectively, Waterloo Centre for Groundwater Research, University of Waterloo, Ontario, Canada N2L3G1

The views expressed are those of the authors and do not reflect the official policy or position of the U.S. Air Force, Department of Defense, or the U.S. Government.

A drive point profiler was used to obtain 40 discrete interval groundwater samples from a total of five locations in nine days of field work. The drive point profiler enabled sampling of fine-grained, low permeability units in which monitor wells would not usually be installed. Samples were analyzed in a mobile laboratory for volatile organic compounds: trichloroethene, 1,2-dichloroethene, and vinyl chloride. The investigation produced useful information about the highly variable distribution of these compounds in the overburden units.

KEYWORDS: dense non-aqueous phase liquid, drive point profiling, groundwater samples, site characterization, underground storage tank, volatile organic compounds

---

INTRODUCTION

Installation Restoration Program Site 32 (Figure 1) is located in the industrial area of the former Pease Air Force Base in Portsmouth, New Hampshire. The site has been investigated as an area of suspected contamination and one in which dense, non-aqueous phase liquid (DNAPL) contamination is suspected to occur in the subsurface. The investigation results reported here were obtained as part of ongoing efforts to evaluate and document the technical impracticability of groundwater restoration [1] at this site. This report documents an evolution of the conceptual site model for contaminant distribution.

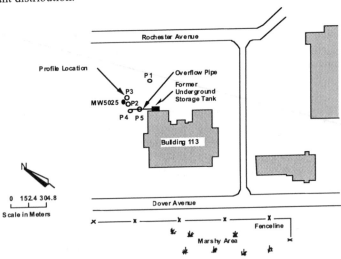

FIG. 1--Site 32 location map. From Pitkin, 1994 [2].

Site History

The original source of contamination at Site 32 was a 1,200 gallon ($\cong$ 4500 L) underground storage tank (UST). Waste trichloroethene (TCE) used in maintaining and cleaning B-47 and B-52 aircraft weapons systems was discharged to the tank through a floor drain in Building 113 from 1956 until 1968. There are no records with which to estimate the volume of wastes disposed to the tank, nor is it possible to determine whether the UST was ever emptied or maintained. An overflow pipe that allowed subsurface discharge from the UST was added at some time during the use of the tank. There is anecdotal evidence that the overflow pipe was extended "whenever it got swampy." After 1968, the use of Building 113 changed from weapons maintenance to avionics maintenance, and it is therefore assumed that very small quantities (less than 1 pint [$\cong$ 30 cc] per year) of solvents were disposed to the UST. In 1977, approximately 1,000 gallons ($\cong$ 3750 L) of TCE were pumped from the UST, and it was then filled with sand. The tank was excavated and removed in 1988. The overflow pipe and 441 tons ($\cong$ 450,000 kg) of associated contaminated soil were removed in 1990. A pilot groundwater treatment plant has been in operation as an interim remedial measure (IRM) since 1991.

Site Description

Five hydrogeologic units (shown in Figure 2) are present at Site 32: upper sand (US), marine clay and silt (MCS), lower sand/glacial till (LS/GT), shallow bedrock, and deep bedrock. All units are hydraulically connected. A seasonally variable water table is present in the uppermost overburden unit, the US, between 0-6 feet ($\cong$ 0-2 m) below ground surface (bgs). The hydraulic units are described below in order of occurrence down from ground surface.

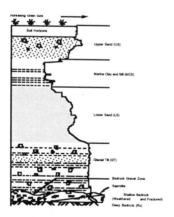

FIG. 2--Generalized stratigraphic column. From Weston, 1992 [3].

Upper sand--This unit consists of poorly sorted, generally fine-grained sand with some silt and gravel and ranges in thickness from 4-10 feet ($\cong$ 1-3 m). Its average hydraulic conductivity is 1×10^{-1} feet/day ($\cong 3.5 \times 10^{-5}$ cm/sec). The US unit has been replaced with artificial fill in areas of building construction and where the UST was removed.

Marine clay and silt--This unit is a glacio-marine clay up to 20 feet ($\cong$ 6 m) thick that varies from plastic clay to silt in composition and behaves as a leaky aquitard. Its average hydraulic conductivity is estimated to range between 10^{-4} and 10^{-3} feet/day ($\cong 10^{-8}$ and 10^{-6} cm/sec). This unit appears to be thicker over bedrock lows and thinner over bedrock highs.

Lower sand/glacial till--This unit consists of fine-grained sand with both silt and gravel layers and lenses. The combined LS/GT is 15-25 feet ($\cong$ 4.6-7.6 m) thick except over bedrock lows (where it can be thicker) and is less transmissive than the underlying shallow bedrock. Its hydraulic conductivity ranges between 4×10^{-2} and 5×10^{-1} feet/day ($\cong 1 \times 10^{-5}$ and 2×10^{-4} cm/sec).

Shallow bedrock--This unit consists of weathered and extensively fractured metamorphic bedrock occurring 10-60 feet ($\cong$ 3-18 m) bgs. This unit has also been described as saprolite and a bedrock gravel zone. The thickness of this transmissive unit varies from 0-15 feet ($\cong$ 0-4.6 m). Its average hydraulic conductivity ranges from 10-40 feet/day ($\cong 10^{-3}$-10^{-2} cm/sec).

Deep bedrock--This unit is competent, fractured, layered biotite quartzite intruded by diabase dikes. Its average hydraulic conductivity ranges from 4×10^{-3} to 3×10^{-1} feet/day ($\cong 10^{-6}$-10^{-3} cm/sec). The bedrock surface topography consists of troughs (lows) and ridges (highs) that resulted from glacial erosion along lines of weakness caused by fractures.

Contaminant source--A waste solvent UST that was formerly located close to Building 113 (Figure 1) is considered the probable source of volatile organic contaminants (VOCs). The tank was buried 2 feet ($\cong$ 0.6 m) into the MCS (11 feet bgs [$\cong$ 3.3 m]). An overflow pipe was buried 4 feet ($\cong$ 1.2 m) bgs and extended approximately 75 feet ($\cong$ 23 m) into a wetland area northwest of the tank.

Site Investigation

The site was identified in a 1983 records search as a potential source for the release of contaminants to the environment. The tank was removed in 1988, at which time the overflow pipe was discovered. An IRM consisting of excavation and removal of the overflow pipe and associated contaminated soil was performed in the fall of 1990 as a source control action. When excavated, the overflow pipe consisted of clay, steel, and iron pipe sections that either were not sealed to each other or had been taped together

(Figure 3). The highest concentrations of TCE detected in soils were in the tank excavation area (169 parts per million [ppm]) and at the top of the MCS unit at the end of the overflow pipe (130 ppm). Throughout this paper, only concentrations of TCE will be cited; its degradation products, 1,2-dichloroethene and vinyl chloride, are also present at the site.

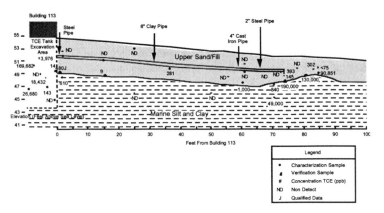

FIG. 3--Tank, pipe, and soil removal diagram. From Weston, 1991 [4].

Remedial investigation--A remedial investigation (RI) was conducted in stages from 1983 through 1992. The Site 32 RI report was issued in 1992 [3]. By this time, a pilot groundwater collection system and treatment plant had also been installed as an IRM for contaminated groundwater. The RI report presented a conceptual site model for the source area (tank excavation and overflow pipe) in which the MCS unit was presumed to have acted as a barrier to downward migration of contaminants (Figure 4). The extent of groundwater contamination was still being evaluated.

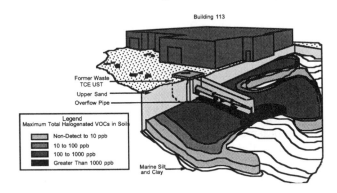

FIG. 4--Conceptual site model in RI report [3].

Feasibility study--A feasibility study (FS) was completed in 1993. The FS [5] concluded that DNAPL may exist or have existed but had not been observed in borings or monitor wells. It was also concluded that contamination appeared to have migrated down to the LS/GT below the MCS at the source area. This conclusion was based on the presence of high concentrations of chlorinated VOCs in groundwater (e.g., 680 ppm TCE in a well screened in the LS). The concentration of TCE remaining in soil at the end of the overflow pipe on top of the MCS was 190 ppm. It was assumed that DNAPL had spread laterally at the MCS surface below the source area and then migrated downward along preferred pathways in the MCS. These pathways could be root zones, fractures, sandy lenses, or artificial penetrations (building footings, borings, or monitor wells).

Technical impracticability evaluation--A report [6] evaluating the technical impracticability of groundwater restoration was issued in draft form in March, 1994. This report documents efforts to locate DNAPL that include the installation of monitor wells screened at the bottom of the LS/GT unit, rotasonic drilling, the use of hydrophobic dye and fluorescence techniques to evaluate cores, and sampling without well development or purging. The highest concentration of TCE in groundwater observed by these efforts was 420 ppm. The report concludes that DNAPL exists as residual saturation in soil pores and bedrock fractures. The report also concludes that dissolved contaminants are traveling in preferential pathways in the shallow bedrock and at the base of the LS/GT unit and that vertical hydraulic gradients--downward at the source, upward downgradient of the source--explain the presence of dissolved contaminants farther away from the source than would be predicted based on average groundwater flow velocities in the LS/GT.

The report evaluates the performance of the IRM implemented to control migration of source area groundwater by pumping from the LS/GT (1991-1992) and the shallow bedrock (1993-1994). Pumping from the LS/GT was inefficient because of low yield (less than 1 gpm [≅ 4 L/min]), and pumping from the shallow bedrock at a rate sufficient for containment may result in undesirable settling of nearby buildings.

The report concludes that limited remedial action is appropriate for the source area at site 32 because the nature of the contaminants present (residual DNAPL) and the complex site hydrogeology combine to make it unlikely that existing technologies could be used to restore the water bearing units to acceptable water quality levels. Management of contaminant migration appears to be feasible.

This investigation--The purpose of this investigation was to characterize the site using a drive point vertical profiling technique. There were two goals: (1) to demonstrate that vertical drive point profiling, using equipment developed at the University of Waterloo [7, 8], would work in the heterogeneous, relatively low permeability units present and (2) to find contaminant solute concentrations indicative of DNAPL. Investigations to date have not answered some questions that are critical to the evaluation of remedial options for the source area and management of migration. These questions include: Is TCE ponded on, and dripping through holes in, the MCS? Is the MCS saturated with TCE throughout its thickness? Are there lenses of DNAPL in the LS/GT?

EXPERIMENTAL METHOD

The experiment tests the concept that contaminant distribution is more variable and solute concentrations higher than previously established using conventional means. Although dissolved concentrations of contaminants indicative of the presence of TCE as DNAPL have been detected in monitor wells at the site (910 ppm; the aqueous solubility of TCE is 1,100 ppm), no DNAPL has been recovered. DNAPL is presumed to be present as residual saturation.

The vertical drive point profiling method demonstrated by Pitkin [8] is a rapid, inexpensive method of obtaining groundwater samples that generates little waste, develops and purges a very small sampling interval (see Figure 5), and can take multiple samples in a single boring without removing equipment from the hole between samples.

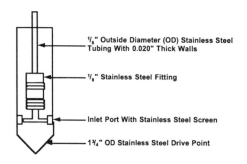

FIG. 5--Drive point profiler tip. From Pitkin et al., 1994 [8].

System Description

The system (see Figure 6) consists of a unique drive point profiler tip (shown in Figure 5) attached to drill rod suspended by a winch in a wheeled scaffolding that can be taken apart for transport. The drill rod is driven with a pneumatic hammer. A sacrificial grouting tip was used for several of the profile points during this investigation and worked successfully, so it is now possible to grout the borings after sampling as the equipment is removed from the hole.

The profiling tip is advanced to the desired sampling depth, and the groundwater sample is collected in sample containers installed in-line so that volatile constituents are not lost to the atmosphere (see Figure 7). There are six screened inlet ports that feed a common reservoir in the profiler tip; stainless steel tubing carries the samples from the ports to the surface apparatus. The volume of water necessary to fill the sampling system is calculated, and appropriate volumes can be purged before a sample is taken. It is also possible to inject water into and withdraw water from the sampling interval in such a way as to develop it before purging and sampling.

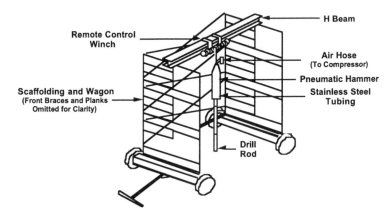

FIG. 6--Field driving configuration. From Pitkin et al., 1994 [8].

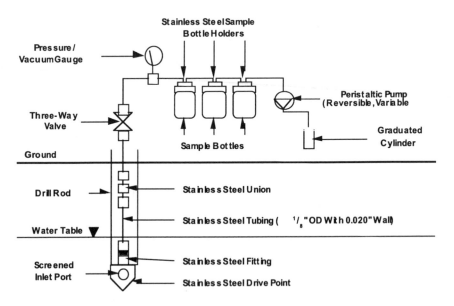

FIG. 7--Sampling configuration. From Pitkin et al., 1994 [8].

Data Collection

Five vertical profiles were sampled at the locations labeled P1 through P5 shown on Figure 1. The results for P1 and P5 will not be discussed. P1 was in a clean area, and P5 sampled only the fill present along the former overflow pipe. The original sampling plan was to grid the site, work from clean to contaminated areas, and take samples every foot from the ground surface to the top of bedrock (approximately 35 feet [$\cong$ 12 m]). This plan was based on rates of sample recovery in more homogeneous, coarser-grained unconsolidated materials than are present at the site. When it was determined there would be insufficient time to sample both clean and contaminated areas, the sampling plan was revised and a transect was established across the end of the former overflow pipe. P2, P3, and P4 were installed in a line perpendicular to and at the farthest extension of the overflow pipe.

It was not possible to acquire samples in most of the MCS unit or in low permeability layers in the LS/GT. When driving through the MCS, sampling ports clogged with clay. The driving procedure was modified so that clean deionized water was being pumped out under pressure as the profiler was driven. This water was recovered, and additional system volumes of groundwater were purged through the system before a sample was taken. Flow rates ranged from 4 minutes per 100 mL in the most permeable sands to 50 minutes per 100 mL in the MCS. Field measurements of specific conductance were used to distinguish between deionized water and groundwater. Samples were analyzed using a gas chromatograph with flame ionization and photoionization detectors in a mobile laboratory. Results were available within hours and were used to evaluate the sampling technique and to plan the next samples. No evidence of dragdown was seen in the samples; decreases in TCE concentrations of at least an order of magnitude between samples only one foot ($\cong$ 0.3 m) apart (at P2 and P3 in the LS/GT) were observed.

RESULTS

The results of this investigation have been previously reported by Pitkin [2]. The results for TCE at P2, P3, and P4 are shown in Figure 8. Variations of two orders of magnitude in solute concentration were seen in the profiles. In general, the discrete samples had solute concentrations that are higher than in the nearby monitor well in the same unit.

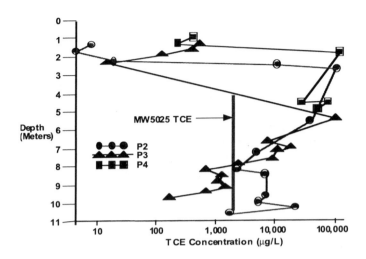

FIG. 8--Overflow pipe transect.

Contaminant Distribution

Stratigraphic units were correlated with nearby wells and borings by depth and by hydraulic conductivity estimated on the basis of sample flow rates. The highest concentrations of TCE sampled were in the MCS: 370 ppm at 9 feet ($\cong$ 3 m) bgs in P2, and 450 ppm at 6 feet ($\cong$ 2 m) bgs in P4. The highest concentration of TCE in the LS/GT was at an equivalent order of magnitude: 360 ppm at 18 feet ($\cong$ 6 m) bgs in P3. The concentration of TCE in monitor well 5025 (1.7 ppm), sampled the same week in which the profile samples were acquired, is shown as a bar extending the length of the screened interval. This well is five feet ($\cong$ 1.5 m) away from P2 and is screened the entire thickness of the LS/GT unit. It has been demonstrated [9] that *in situ* groundwater sampling devices can yield accurate, precise, and representative data. Vertical profiling data indicate the presence of both high and low permeability units with varying contaminant concentrations within the LS/GT that are not detectable using conventional monitor wells.

The highest concentration of TCE in groundwater that has been found to date is 910 ppm in a monitor well installed at the former tank location and screened in the LS/GT unit. No profile was made at that location.

Hydrogeology

Field descriptions of flow rates, turbidity, and injection pressures allowed the distinction of subunits within the MCS and LS/GT. The highest concentrations of TCE were found in the lowest permeability units. Subunit thicknesses were on the order of 1 foot ($\cong$ 0.3 m) or less.

Future Plans

An additional field effort is planned for the spring of 1996. The distribution of hydraulic properties at the site will be characterized using nested drive point piezometers with 1 foot ($\cong$ 0.3 m) screens, slug tests in the piezometers, and permeability testing on cores. Additional drive point profiling is planned to characterize contaminant distribution along transects perpendicular to the hydraulic gradient as determined from the piezometer study. Results of the profiling will be used to identify locations for coring. This will be another attempt to locate the suspected DNAPL sources.

CONCLUSIONS

The conceptual site model for contaminant distribution at this site remains incomplete. The results of future field work should lead to better understanding of the hydraulic controls on contaminant distribution.

Technology Demonstration

The equipment and methodology worked to obtain groundwater samples in fine-grained, low permeability unconsolidated materials. Intervals of varying hydraulic properties and contaminant concentrations within larger hydrogeologic units were identified.

Locating DNAPL

Solute concentrations indicative of DNAPL were found, although DNAPL was not sampled. The location in which the highest concentrations of TCE in groundwater from the LS/GT unit have been obtained using a conventional monitor well was not profiled. More variable and higher concentrations of TCE than are present in the closest monitor well were identified using the profiling technique.

ACKNOWLEDGMENTS

This research was cooperatively funded by the University Consortium Solvents-in-Groundwater Research Program, The MITRE Corporation, and Air Force Base Conversion Agency Operating Location A. Field investigations, geologic interpretation, groundwater modeling, and remedial engineering conducted by geologists and engineers of Roy F. Weston, Inc., are the basis of our understanding of this site.

REFERENCES

[1]   U.S. Environmental Protection Agency, Guidance for Evaluating the Technical Impracticability of Ground-Water Restoration (Interim Final), Office of Solid Waste and Emergency Response Directive 9234.2-25, September 1993.

[2]   Pitkin, S. E., "A Point Sample Profiling Approach to the Investigation of Groundwater Contamination," Unpublished Master of Science thesis, University of Waterloo, Waterloo, Ontario, Canada, 1994, 242p.

[3]   Weston, Roy F., Inc., IRP Site 32/36 Source Area Remedial Investigation Technical Report, U.S. Air Force Installation Restoration Program (IRP) Stage 3C, Administrative Record File for Pease Air Force Base (AFB), February 1992.

[4]   Weston, Roy F., Inc., Soil Removal at IRP Site 32 (Building 113) Informal Technical Information Report, U.S. Air Force IRP Stage 3, Administrative Record File for Pease AFB, April 1991.

[5]   Weston, Roy F., Inc., Site 32/36 Feasibility Study Report, U.S. Air Force IRP, Administrative Record File for Pease AFB, October 1993.

[6]   Weston, Roy F., Inc., Pease Air Force Base Site 32 Technical Impracticability Evaluation, U.S. Air Force IRP, Administrative Record File for Pease AFB, March 1994.

[7]   Broholm, M. M., Ingleton, R. A., and Cherry, J. A., "The Waterloo Drive-Point Profiler - Detailed Profiling of VOC Plumes in Groundwater Aquifers," Transport and Reactive Processes in Aquifers, Dracos and Stauffer (eds), Balkema, Rotterdam, 1994, pp 95-100.

[8]   Pitkin, S. E., Ingleton, R. A., and Cherry, J. A., "Use of a Drive Point Sampling Device for Detailed Characterization of a PCE Plume in a Sand Aquifer at a Dry Cleaning Facility," National Ground Water Association Eighth National Outdoor Action Conference on Aquifer Restoration, Ground Water Monitoring and Geophysical Methods, Minneapolis, 1994.

[9]   Blegen, R. P., Hess, J. W., and Denne, J. E., "Field Comparison of Ground-Water Sampling Devices," National Water Well Association Second National Outdoor Action Conference on Aquifer Restoration, Ground Water Monitoring and Geophysical Methods, Las Vegas, 1988.

Jeffrey A. Farrar [1]

**RESEARCH AND STANDARDIZATION NEEDS FOR DIRECT PUSH TECHNOLOGY
APPLIED TO ENVIRONMENTAL SITE CHARACTERIZATION**

REFERENCE: Farrar, J. A., ''Research and Standardization Needs for Direct
Push Technology Applied to Environmental Site Characterization,''
Sampling Environmental Media, ASTM STP 1282, James Howard Morgan, Ed.,
American Society for Testing and Materials, 1996.

ABSTRACT: Research and standardization needs in the area of direct push
sampling and testing is reviewed. Direct push testing and sampling is a
rapidly growing area in environmental site characterization. Direct
push active soil gas sampling techniques are well accepted for
environmental site investigations in the vadose zone. Standardization
efforts are underway. Better guidance is needed on purging, sealing,
and avoidance of cross contamination. The cone penetration test is
gaining popularity due to it's excellent stratigraphic mapping
capability and water pressure information. Research is needed to refine
hydrostatic pressure measurement capability and prediction of hydraulic
conductivity. Discrete point water sampling using protected screen
samplers has gained wide acceptance. Research is needed concerning
water sample protection, effective sealing, use of vacuum in low
hydraulic conductivity soils. Comparison studies should be done in the
laboratory and in the field. The greatest area of research interest is
in development of chemical sensors for use in direct push testing.
Several promising methods using fiber optic laser technologies are being
developed. Successful technology transfer to everyday application in
private industry will require sensors be developed into fully tested,
rugged, simple modules to be housed in a penetrometer body. Full scale
testing under laboratory controlled conditions will help in validation
of advanced sensor systems.

KEYWORDS: Direct Push Technology, Site Characterization, Cone
Penetrometer, Water Sampling, Insitu Testing

INTRODUCTION

The purpose of this paper is to review research and standardization
needs with respect to subsurface investigations performed using "direct
push technology." ASTM has a newly formed task group D-18.21.01 on
direct push technology. Direct push methods of subsurface investigation

---

[1] Technical Specialist, Earth Sciences Laboratory, Bureau of Reclamation,
PO Box 25007, D-8340, Denver, Colorado, 80225, Head Task Group D-18.21.01
on Direct Push Technology, Vice Chairman subcommittee D-18.02

of soils include both sensors and sampling devices advanced by direct insertion of sampling rods without borehole excavation. Sampling or sensing devices are attached to rods that extend to the surface. The rods and test devices are generally pushed by hydraulic methods with or without the inclusion of vibration, rotation, or impact. The insertion of the testing device displaces soil and forms a seal of compacted soil around the rods.

Task group 18.21.01 is interested in developing standards related to direct push testing methods applied to subsurface environmental investigations. Devices such as push in wellpoints, continuous or discrete depth soil samplers, and cone penetrometers with sampling and testing devices shall be considered. Special considerations for sampling or testing of soil and contained pore fluids will be addressed.

Direct push testing and sampling technology is a rapidly growing area of interest as reflected by the presentations in this symposium. There are many advantages in use of direct push insertion techniques over conventional drilling methods. Some advantages include, rapid advancement, real time testing, discrete interval testing, minimal disturbance to environment, and lack of soil cutting generation. The prime driving force for the growth in this area is the lack of soil cuttings or waste that must be characterized and disposed of during environmental site characterization. Facilities with hundreds of conventional drilling wells generate hundreds of barrels of soil waste that must be characterized and disposed of. Disadvantages to direct push sampling and testing include the inability to drive through certain soils (gravels, tills, cemented soils, and partially weathered and unweathered rock) and the lack of soil sampling which may make interpretation of site stratigraphy more challenging. Additional limitations include the inability to measure temporal changes with single sampling events and small sample volumes requiring multiple sampling events.

Review of research activities will also be provided. This review is from the perspective of the end users and focused toward development of reliable proven systems. The tendency of researchers today is toward advanced chemical sensor systems while the more basic issues may be neglected. There are many technical disciplines involved in the research programs requiring better coordination and communication. Most of the research to date has focused on site studies while there is a lack of laboratory controlled studies.

SOIL GAS SAMPLING

Dircet push soil gas sampling in the vadose zone is well accepted in industry. Thousands of sites have been tested and there is an abundance of small direct push driving systems present in industry. Many research studies have shown that this technique is a valuable field screening tool. Most of the research focuses on operation of chemical detection devices. Several types of testing equipment are available. Some systems use tubing to the sampling point while others just apply vacuum to the rods themselves. There is a need for better guidance on avoidance of cross contamination, decontamination of equipment, maintenance and evaluation of annular seals, soil types applicable for testing, and purging issues during multiple sampling events.

Standardization for soil gas sampling in the vadose zone is under the jurisdiction of ASTM task group D-18.21.02. They have issued a

guide for soil gas sampling in the vadose zone [1] and currently are
working on specific standard test methods for active direct push
sampling approaches.

CONE PENETROMETERS

The cone penetration test (ASTM D-3441) [2] has been in use by the
geotechnical engineering profession for over 50 years. These engineers
have developed detailed correlations to engineering properties of soil
through detailed research. Shear strength and compressibility
correlations are developed by combinations of laboratory controlled
chamber tests, and field studies. The geotechnical engineering
community holds regular symposiums to discuss penetrometer research
progress [3,4]. Cone testing has developed into rapid, economic,
dependable, testing systems. The cone test's major limitation is it's
inability to penetrate hard, cemented, or lithified materials such as
shales and coarse soils such as gravels. The cone test is best applied
in softer alluvial or lacustrine soils.
    Figure 1 shows a schematic drawing of the typical electronic cone
penetrometer. Typical electronic penetrometers have five channels of
information of which four channels are, tip resistance, sleeve
resistance, inclination, pore water pressure. The dimensions of the tip
and sleeve have been standardized along with the pore pressure
measurement locations. The advancement rate is also standardized at 2
cm/sec. In this schematic, the fifth channel is used for a resistivity
module. Other sensors such as seismic geophones, temperature, acoustic
emissions sensors, radiation scintilators, and pressuremeters have been
explored by the geotechnical engineering community. Addition of
multiple sensors is generally impractical due to increased complexity of
the apparatus.
    The cone penetrometer test is unsurpassed in it's ability to
provide detailed information on subsurface stratigraphy. Fortunately
for the environmental industry the cone is also a superior tool for
determining hydrogeologic information. Penetration resistance depends
on the drainage conditions. Drainage of excess pore water pressures in
the soil that are created during sounding depends on hydraulic
conductivity. The cone data easily distinguish between low hydraulic
conductivity clays and high hydraulic conductivity sands. Sand layers
as thin as 10 cm can be mapped in detail that sometimes stuns the site
geologist.
    Addition of a piezometer element allows for further confirmation
of layering and for measurement of at rest hydrostatic pressure for
calibration of ground water models. The dynamic pressures created
during penetrometer advancement have been correlated to soil type and
engineering properties. Measurement of dynamic pressures can be
difficult, especially if the system becomes cavitated due to dilation or
penetration in the unsaturated zone. During pauses in testing,
dissipation tests can be performed in sand layers. In free draining
sands, the excess pressure equalizes to hydrostatic pressure rapidly (15
min). Dissipation times depend on conditions around the cone and
initial dynamic pressures. Dissipation behavior is sometimes erratic in
dirty sands, finely layered soils, overconsolidated soils, or cemented
materials. The piezometer for a piezocone must be designed to withstand
very high pressures during push and there it's accuracy at low
hydrostatic pressures is low (± 500 mm). This resolution may not be

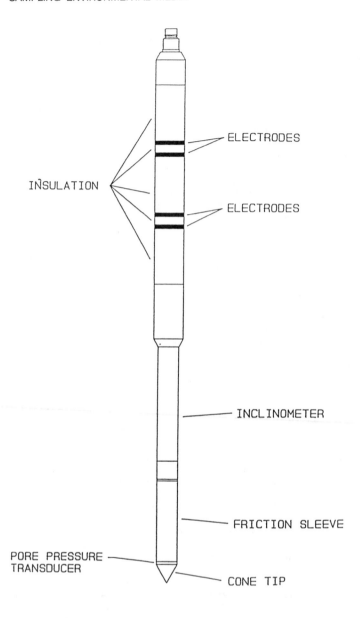

Figure 1.   Schematic of a typical piezocone penetrometer with a resistivity module.

sufficient to calibrate ground water models in level environments where
head differences of only ± 50 mm may be important to detect gradients
for a small site. Research is needed to develop a low pressure
sensitive piezocone for measurement of small water heads and to provide
better guidance in interpreting dynamic and dissipation pressures.

Research in the application of cone penetration for ground water
investigations has focused on determination of hydraulic conductivity,
k. The cone penetration test, alone, can distinguish from very high (<
10^{-4} cm/sec) to very low (< 10^{-6} cm/sec) hydraulic conductivity soils.
Hydraulic conductivity can be estimated using the relationship between
tip resistance and friction ration shown on Figure 2. This relationship
alone may be sufficient for performing reliable ground water modeling.
In critical cases, it may be necessary refine the estimates by site
specific comparison to field and laboratory tests. The piezometer
element can be used to estimate k in a limited range by performing
dissipation tests (Fig 3). In the dissipation test, the high excess pore
pressure in the soil is monitored with time and the time required to
reach 50% dissipation is recorded. The primary economic limitation is
the time to conduct the dissipation test, with some clays requiring
hours to reach 50% dissipation. Dissipation tests are not frequently
performed due to the increase time of testing. As shown on Figure 3, the
existing information on dissipation tests shows that the uncertainty in
estimate of k is about two orders of magnitude, ± 10 cm/sec [5]. To
refine the correlation, a site specific study could be conducted to
compare laboratory hydraulic conductivity (ASTM D-5084 [6]) to the field
estimates. Research is now focusing on injection tests of high
hydraulic conductivity materials, such as sands, using pumps and
pressure sensors. Field studies can be performed by conductivity
comparison to soil cores. Laboratory studies should be conducted in the
laboratory using actual penetrometer insertion under differing stress
conditions to model the disturbed zone around the penetrometer.

Standards for electronic friction cone and piezocone penetration
testing are under jurisdiction of ASTM subcommittee D-18.02. A new
standard is currently in the balloting process. Task group D-18.21.01
will develop a standard practice for performing cone tests for
environmental site characterization.

DYNAMIC PENETROMETERS

One method to overcome pushing limitations with the quasi-static
cone penetrometer is to add impact forces to the rods. Use of impact
hammers operating at fairly high frequencies can help with additional
penetration capability through hard and coarse soils. Geotechnical
engineers, particularly in Europe, use dynamic penetrometer tests to
estimate soil type and engineering properties through empirical
correlations [3]. These tests are performed with steel pointed
penetrometers and forces are measured at the surface. The penetration
resistance or number of hammer impacts per depth can be correlated to
soil types, i.e., gravels, sands, and clays. Dynamic penetration
resistance also depends partially on drainage conditions.

Many smaller direct push driving systems using high frequency
impact are finding popularity in the U.S. for wellpoint installations,
soil gas sampling, etc. These systems use high frequency impacts that
are too fast to count. However, it may be possible to standardize the
impact energy and frequency and measure static downforce to advance the

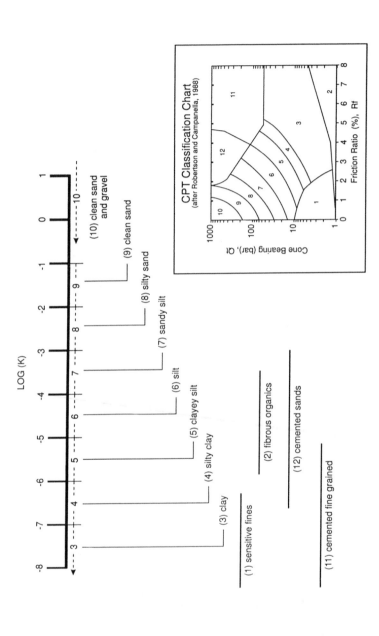

Figure 2.   Simplified soil classification chart showing soil types and hydraulic conductivity zones [5].

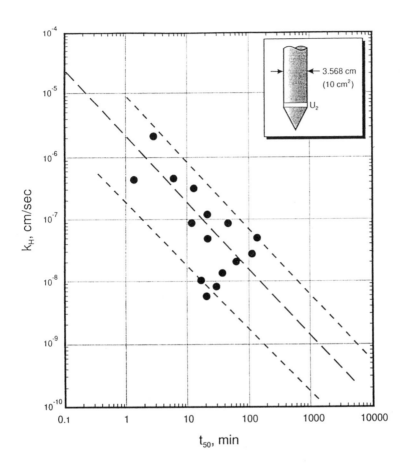

Average values of laboratory derived horizontal coefficient
of permeability (k_H) and CPTU t_{50} for U_2 pore pressure location

Figure 3.   Hydraulic conductivity versus dynamic pore pressure 50%
dissipation time [5].

rcds. In this manner reliable correlations of penetration resistance to soil types could be developed. Even the new vibrosonic drilling methods could benefit from force and rate measurements.

DIRECT PUSH WATER SAMPLING

Direct push water sampling is gaining great popularity and acceptance in environmental site characterization. Simple sampling systems consist of simply rods with an expendable tip operated in a grab sampling mode. More complex systems have samplers equipped with chambers and/or pumps. Our task group has classed these samplers as either exposed screen or protected screen samplers. The protected screen samplers may be pushed into discrete layers to be sampled and thus provide excellent point measurements. One limitation to direct push water sampling is that fine grained soils with low hydraulic conductivity take too much time to extract water. In these cases, soil sampling may be necessary. Another limitation is excessive silt content in metals sampling applications.

One of the first exposed screen water sampling devices built into a cone penetrometer contained pumps, tubing, and chambers for drawing in water and retrieving it at the surface for chemical analysis. This penetrometer never made it into production due to purging and cross contamination issues. The developers performed the appropriate chamber testing under laboratory conditions that showed unacceptable extraction and purging times for bringing water to the surface. Since then, other exposed screen systems have been developed and deployed [5,7]. For any multiple event sampling device with flow cells, such as that shown on Figure 4, [7], purging and cross-contamination issues will always require site specific evaluation. In field screening cases were a lower level of accuracy are acceptable, multiple sampling events with exposed screen sampling devices and flow cells may be useful. Another reason for lack of widespread acceptance of flow cell systems is the use of multiple sensors in the system. As instruments become more complicated they become more difficult to maintain and calibrate.

Today, most ground water sampling events are with protected screen, discrete point water sampling devices. These systems can provide very accurate point concentration measurements especially when they are accompanied with on-site chemical analysis. These systems can be just as rapid as multiple sampling devices if purging requirements for multiple sampling systems are long. There are a variety of devices from simple expendable tips and filters on drive rods, to samplers with chambers for retrieval of the water sample. The samplers with chambers have varying degrees of sealing to prevent volatilization of organic compounds. Samplers with complex sealing systems may be more difficult to operate in the field due to the increased amount of moving parts, and more difficult maintenance and decontamination requirements. One area of research need are to evaluate the need for sealed sample chambers in an attempt to reduce volatilization of organic compounds. After research is completed, it may show that the simpler systems are sufficient and desired in screening applications where high accuracy is not required.

Applied research has been done to compare the performance of several well know sampling systems [8,9]. The studies have been "comparison or demonstration" studies at actual field sites. These studies are generally inconclusive due to site variability and the

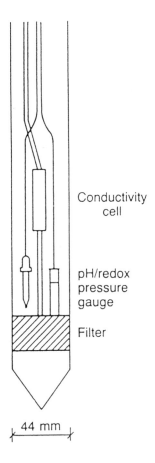

Figure 4.   Schematic of an exposed screen sampling device equipped
with a flow cell [7].

inability to know exact chemical concentrations present during testing.
Comparison studies should be done under laboratory conditions where
specimens can be prepared at known concentrations. Additional research
is required in the area of insertion depth to maintain annular sealing,
use of vacuum for low hydraulic conductivity soils, and sealing
requirements for permanent installations (wells).

Task group D-18.21.01 is currently balloting a standard guide for
direct push water sampling applied to environmental investigations. The
next step is to write standard test methods for specific equipment
configurations.

DIRECT PUSH SOIL SAMPLING

Most of the direct push soil sampling systems deployed today are
discrete point piston type samplers. The soil sampler is pushed to the
depth of interest, a piston is unlocked, and the sampler is then pushed
through the length of sample chamber. Double tube continuous samplers
are available. Double tube systems are usually operated in a similar
fashion to wireline drilling systems and are particularly advantageous
when intensive continuous soil logging is required. The outer tube of
the double tube system protects the core and allows for easy hole
completion.

Soil sampling has been researched by the geotechnical engineering
community where undisturbed sampling is required for the estimation of
engineering properties. The methods and techniques are not well known.
One important finding of geotechnical research studies is that large
diameter samples can be taken with less disturbance. Recommended
diameters range from 75 to 125 mm using conventional drilling equipment.
These sizes are well beyond capabilities of most direct push equipment.
Therefore, if undisturbed samples are required for estimation of
hydraulic conductivity where undisturbed structure is important, they
should be obtained using conventional drilling. However, for soil
logging and chemical tests, smaller sizes are acceptable. One area of
research need is the types of liners (brass, stainless, plastics, and
Teflon) suitable for storing soil samples and decontamination
requirements for samplers.

There is a great need in ASTM to develop soil sampling standards.
Currently, there are sampling standards for rock coring, disturbed
sampling using augers (D-1452), standard penetration test (SPT)(D-1586),
ring lined barrel (D-3550), and thin wall push tubes (D-1587). The
current ASTM standards are generally focussed on geologic or
geotechnical end users. Standards for continuous augers, piston
samplers, and wireline multipurpose core barrels are needed both for
geotechnical and environmental applications.

SENSOR DEVELOPMENT

As mentioned earlier, there are many sensors that have been
developed for the cone penetration test penetrometer. The resistivity
module has been in use in the geotechnical engineering community for 20
years. Engineers have used the resistivity module to estimate the
porosity of sands for strength and flow deformation behavior. Current
research has advanced to predicting stress dilatency of sands. The
typical resistivity module as shown on Figure 1 consists of a Wenner
array of electrodes operating at about 10^1 Hertz. This penetrometer is

useful in water quality studies where electrical conductivity in the saturated zone is changed by the chemicals of concern. It has been particularly useful in salt water intrusion studies. Standardization of resistivity modules is now needed. Additional research could be performed to measure conductivity at two frequencies for evaluation of dielectric constant of soil.

Temperature sensors may be useful in rare applications. Heat is generated in sands during cone penetration process and then dissipation tests must be performed. Thermal dissipation tests may not be of value if they require long testing times that slow production.

An area of intense research has been the development of advanced chemical detection sensors for use in direct push systems [10]. In the rush to get chemical data, new technology is being applied in the direct push fields. Most of this work is funded by the Department of Defense and Energy [10]. The largest area of interest is related to the use of fiber optic systems and lasers. Laser induced fluorescence (LIF) has already been successfully deployed in prototypes. Work is expanding into other forms of spectroscopy. A schematic of an LIF probe is shown on Figure 5. This probe consists of the cone penetrometer, LIF components, and a tip grouting system. Other sensors such a video imaging, radar, time domain reflectrometry, opto-electronic, and optical supramoleclues are under varying phases of assessment and prototype development [10]. There have even been proposals to miniaturize gas chromatographs and mass spectrophotometer systems within a penetrometer. Most of the research with these penetrometers has been first in lab bench tests and then in the field as demonstration programs. Full scale testing under laboratory controlled conditions will help in validation of these systems. These full scale tests should include penetrometer insertion to evaluate the effects of a disturbed - crushed zone around the penetrometer.

While it is highly desirable to couple sensors on cone penetrometers there are also the disadvantages of increased complexity and risk of expensive penetrometer loss if broken. Successful technology transfer to everyday application in private industry will require sensors be developed into proven, rugged, simple modules to be housed in any penetrometer body. Multiple sensors housed within a cone penetrometer body will probably not find spread usage due to the above limitations so it's probably best to design cone penetrometers with only a single additional sensor. More complex systems may have to abandon the cone penetrometer and stand alone.

GENERAL RESEARCH NEEDS

While there is considerable interest in the new sensor development, the more basic research into simple issues is being neglected. Many basic research needs have been addressed earlier.

One general area of research need is grouting of direct push holes. The necessity for grouting of very small diameter holes has greatly increased the cost of testing. The current state of practice for grouting of direct push test holes was summarized by Lutenneger in a recent ASTM symposium [11]. While retraction grouting through sacrificial tips or friction reducers provides the best hole sealing, it also may increase the complexity of the penetrometer design as shown in Figure 5. Regulators must use common sense to avoid specifying

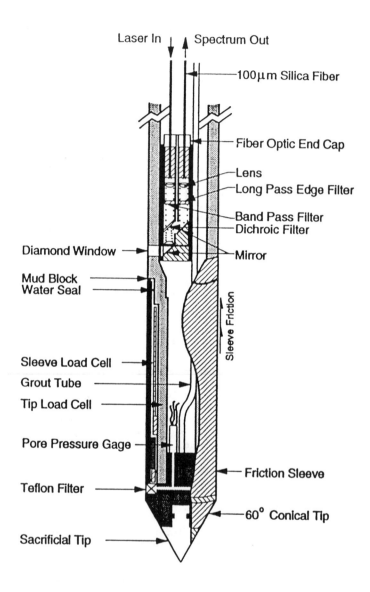

Figure 5.   Schematic of a laser induced fluorescence penetrometer
[12].

difficult grouting techniques. In a single aquifer system, there is no
hydrostatic head potential inducing cross contamination. Often
sloughing, squeezing, and caving of the small diameter direct push holes
should not allow them to effectively transfer contaminants. The only
case where grouting is critical is when direct push holes encounter
perched aquifers. Even in these cases, prompt reentry grouting may be
acceptable. There is a need for developing mixture designs for cement
or bentonite based grouts for the direct push industry because of the
extremely small diameter of the grouting tubes. Mixture designs used
for conventional monitoring well installations cannot always be used for
direct push grouting.

There is a definite need for better coordination for direct push
research and development. More meetings and symposiums directly devoted
to direct push testing and sampling would allow for better coordination
of researchers, especially with the multiple disciplines involved.
These meetings should be sponsored by major technical professional
societies whose publications are widely distributed. Those involved
with development of new sensors, must develop them into simple rugged
testing systems validated by full scale laboratory and field tests.
Federal government agencies need to complete technology transfer by
publication of findings in peer review journals, not obscure agency
reports.

CONCLUSION

Research and standardization needs in the area of direct push
sampling and testing have been reviewed. Direct push testing and
sampling is a rapidly growing area in environmental site
characterization. ASTM subcommittees' D-18.21.01, -18.21.02, and -18.02
are actively developing standards for this type of testing.

Direct push active soil gas sampling techniques are well accepted
for environmental site investigations in the vadose zone.
Standardization efforts are underway. Better guidance is needed on
purging, sealing, and avoidance of cross contamination.

The cone penetration test is gaining popularity due to it's
excellent stratigraphic mapping capability and water pressure
information. ASTM is currently developing standards. Research is
needed to refine hydrostatic pressure measurement capability and
prediction of hydraulic conductivity.

Impact driving systems used for direct push insertion could
benefit if the driving systems were standardized. In this manner it
would be possible to correlate penetration resistance to soil types.

Direct push water sampling techniques are gaining popularity.
Discrete point sampling using protected screen samplers has gained wide
acceptance. Exposed screen flow cell devices have been used less
frequently due to their inherent complexity and suffer from purging and
cross contamination issues. Research is needed concerning sample
protection, effective sealing, use of vacuum in low hydraulic
conductivity soils. Quality assurance protocols can be studied through
detailed comparison studies. Comparison studies should be performed in

the laboratory and in the field.

Direct push soil sampling can be performed either discretely or continuously. ASTM needs additional standardization of soil sampling methods both for conventional geotechnical and environmental applications. There currently are not standards for several commonly used sampling devices such as the piston sampler.

The greatest area of research interest is in development of chemical sensors for use in direct push testing. Several promising methods using fiber optic laser technology are being developed. Successful technology transfer to everyday application in private industry will require sensors be developed into rugged simple modules to be housed in any penetrometer body. Full scale testing under laboratory controlled conditions will help in validation of these systems.

One area of research and standardization is for grouting of direct push holes. Cumbersome grouting requirements often slow the rate of testing and increase the cost of investigations. There is a need for guidance on when and how to grout these small diameter holes.

Finally, there are many disciplines involved in development of direct push testing and sampling systems. There is a need for better research and standardization coordination through widely publicized meetings and symposia devoted specifically to this area of interest.

REFERENCES

[1]   Annual Book of ASTM Standards, "Standard Guide for Soil Gas Monitoring in the Vadose Zone." D5314-92, Section 4, Construction, Volume 04.08, Soil and Rock; Dimension Stone; Geosynthetics, American Society for Testing and Materials, 1916 Race Street, Philadelphia, PA 19103, USA., 1992.

[2]   Annual Book of ASTM Standards, " Standard Test Method for Deep, Quasi-Static, Cone and Friction-Cone Penetration Tests of Soil," D-3441-86, Section 4, Construction, Volume 04.08, Soil and Rock; Dimension Stone; Geosynthetics, American Society for Testing and Materials, 1916 Race Street, Philadelphia, PA 19103, USA., 1992.

[3]   Proceedings of the First International Symposium for Penetration Testing, ISOPT-1, DeRuiter(ed.) Balkema, Rotterdam, ISBN 90 6191 8014., 1988.

[4]   Proceedings of the Second European Symposium on Penetration Testing, Amsterdam, May 24-27, 1982.

[5]   Woeller, D.J., Weemees, I., Kokan M., Jolly, G., and P.K. Robertson, "Penetration Testing for Groundwater Contaminants," ASCE Geotechnical Engineering Congress, Special Technical Publication, American Society for Civil Engineers, New York, June, 1991.

[6]   Annual Book of ASTM Standards, " Standard Test Method for

Measurement of Hydraulic Conductivity of Saturated Porous
Materials Using a Flexible Wall Permeameter", D-5084-90, Section
4, Construction, Volume 04.09, Soil and Rock; Dimension Stone;
Geosynthetics, American Society for Testing and Materials, 1916
Race Street, Philadelphia, PA 19103, USA, 1994.

[7]   Olie, J.J., Van Ree, C.C.D.F., and C. Bremmer, "Insitu Measurment
by Chemoprobe of Groundwater from Insitu Sanitation of Versatic
Acid Spill," Geotechnique, 42, No. 1, 1992, pp 13-21.

[8]   Zemo, D., et al., "Field Comparison of Analytical Results from
Discrete-Depth Groundwater Samplers: BAT Enviroprobe and QED
HydroPunch". Proc. NGWA Outdoor Action Conf., 1992, pp. 341-355.

[9]   Blegen, R.P., et al., "Field Comparison of Groundwater Sampling
Devices," Second National Outdoor Action Conference and Aquifer
Restoration, Groundwater Monitoring and Geophysical Methods, Las
Vegas, Nevada, National Water Well Association, Dublin, OH, 1988,
23p.

[10]   Bowders, J.J and Daniels, D. D., "Summary Report of the Workshop
on Advancing Technology for Penetration Testing for Geotechnical
and Geoenvironmental Site Cahracterization," June 14-15, 1994,
Department of Civil Engineering, University of Texas, Austin,
Texas, July, 1994.

[11]   Lutenneger, A.J., and D.J. DeGroot, "Sealing Cone Penetrometer
Holes to Protect the Subsurface Environment," Hydraulic
Conductivity and Waste Contaminant Transport in Soils, ASTM STP
1142, David E. Daniel and Stephen J. Trautwein, Eds, American
Society for Testing and Materials, Philadelphia, 1993.

[12]   Applied Research Associates Inc., Breifing Package for Workshop on
Advancing Technology for Penetration Testing, (reference [10]
above.), 1994.

Sampling Media

Delph A. Gustitus,[1] Peter M. Blaisdell[2]

INVESTIGATING THE PRESENCE OF HAZARDOUS MATERIALS IN BUILDINGS

REFERENCE: Gustitus, D. A. and Blaisdell, P. M., ''Investigating the Presence of Hazardous Materials in Buildings,'' Sampling Environmental Media, ASTM STP 1282, James Howard Morgan, Ed., American Society for Testing and Materials, 1996.

ABSTRACT: Environmental hazards in buildings can be found in the air, on exposed surfaces, or hidden in roofs, walls, and systems. They can exist in buildings in solid, liquid, and gaseous states. A sound methodology for investigating the presence of environmental hazards in buildings should include several components. The first step in planning an investigation of environmental hazards in buildings is to ascertain why the investigation is to be performed. Research should be performed to review available documentation on the building. Next, a visual inspection of the building should be performed to identify and document existing conditions, and all suspect materials containing environmental hazards. Lastly, samples of suspect materials should be collected for testing. It is important to sample appropriate materials, based on the information obtained during the previous steps of the investigation. It is also important to collect the samples using standard procedures.

KEYWORDS: hazardous materials, contaminants, environmental sampling, buildings, investigation, methodology, bulk sampling.

INTRODUCTION

The presence of hazardous materials in buildings is a significant environmental concern for building occupants, users, owners, and construction workers. Hazardous materials can be found in all types of building construction and in existing buildings of all ages. The presence of such hazards may be well established for some structures, where shown on building drawings or identified through previous investigations. However, for many structures, environmental hazards are not known to be present.

The environmental concern is exposure of environmental hazards to building occupants and users, which could result in adverse health effects. Adverse health effects have been linked to human exposure to several materials commonly found in buildings. The presence of environmental hazards in a building does not necessarily mean that the

---

Principal, The Gustitus Group, Architecture/Consulting, 6535 North Ashland Avenue, Chicago, Illinois 60626.

President, Kendall, Taylor & Company, Inc., Architects/Environmental Consultants, 381 Boston Road, rica, MA, 01821-1803.

health of building occupants is at risk. Many hazardous materials found
in buildings, such as asbestos, do not present a concern unless they are
damaged, disturbed, or deteriorated [1]. However, other materials found
in buildings can present an environmental concern to occupants simply by
their presence in the building.

ENVIRONMENTAL HAZARDS IN BUILDINGS

The term 'hazardous' is generally used to describe a substance or
situation that has been established as being harmful to humans and their
welfare. In the construction industry, materials are considered to be
hazardous if, during construction or during normal use or occupancy of a
structure, exposure could result in adverse health effects to humans. The
science of toxicology, or the study of poisons, is an inexact science [2].
It aims to establish levels of reasonable confidence of no adverse
effects, rather than the doses at which adverse effects will occur with
certainty in humans. Exposure levels that will cause adverse health
effects in humans can vary greatly among the population. One person could
have severe health effects from the same exposure that could result in no
adverse health effects on another person.

Various U.S. government departments are continually working to
establish and revise the levels of exposure corresponding to an acceptable
risk of adverse human health effects. Acceptable exposure levels have
been established for a wide variety of materials. However, it is
important to note that there are numerous materials and organisms that are
potentially hazardous to humans, depending on the specific conditions and
hazards. Environmental regulations are constantly being created, reviewed
and revised. Regulations do not exist for all materials or organisms that
are known or suspected of causing adverse health effects in humans.
However, there are numerous materials used in buildings that are known or
suspected human carcinogens.

Environmental hazards in buildings can be found on exposed surfaces,
in the air, or hidden in roofs, walls, and systems. They can exist in
buildings in solid, liquid, and gaseous states. Solids are currently the
most common environmental hazards found in buildings. Gaseous hazards in
buildings are typically the result of off-gassing from other materials.
Threats to building occupants by liquid hazards are less common, but they
can be found primarily in building systems, such as mechanical, fire
suppression, or electrical components and/or systems.

The mere presence of environmental hazards in buildings does not
always present a health risk to building occupants. Some materials are
hazardous only when disturbed, damaged or deteriorated. However, other
materials can be hazardous simply by their presence in the building. A
thorough risk analysis should be performed to determine the health threat
to occupants by hazardous materials in a building.

An often overlooked consideration regarding environmental hazards is
potential exposure during a natural event, such as a flood or earthquake,
or a disaster, such as a building fire. Such occurrences can result in
the release of hazardous materials or biological organisms which
contaminate the building and the surrounding environment. Suddenly,
users, occupants, and even the general public may be exposed to
environmental hazards that previously presented no risk [2]. Therefore,
when considering the risks of environmental hazards, an assessment of the
risk of such occurrences should be a component of the evaluation of the
potential environmental hazards.

Solid Contaminants or Hazards

The most commonly encountered hazardous materials in buildings are
asbestos and lead. These materials are solids that were typically used as
components of other building products. Asbestos is a naturally-occurring
mineral that was used in over 4,000 known products, including:
insulation, floor tile, fireproofing, joint compound, sealants, textured

paint, vinyl sheets, plaster, concrete wall panels, mastics and roofing shingles [1, 4]. Lead is a naturally occurring element that was commonly used in lead based paint coatings, in solders, and in automobile fuel [2].

Generally, neither asbestos or lead present a direct health threat if the products that contain them are maintained in good condition, and are not likely to be disturbed or damaged [4]. However, when these materials are damaged, for example during maintenance procedures, or disturbed, such as during building renovation work, contaminants can be released, presenting a potential health risk to building occupants.

Asbestos presents a health threat by the release of fibers into the air which can be breathed in by humans. An asbestos containing building material (ACBM) that is friable (can be crumbled, pulverized, or powdered by hand pressure), is more likely to release harmful fibers into the air than nonfriable materials. However, when nonfriable ACBMs are cut, drilled, sanded, or broken, such as during repair or renovation work, harmful fibers can also be released.

Lead presents a health hazard from ingestion of a lead containing material, or inhalation of lead containing dust or particles [5]. Lead based paint can be a hazard when the condition of the paint is deteriorated and it is peeling, chipping, and flaking. Also, during repair or renovation work, lead containing dust can be projected into the air during sanding or scraping, or lead vapors can become airborne during paint removal with heat guns. Most commonly, however, lead dust generated by friction surfaces, such as along window jambs of operable sash, presents a significant risk to children.

Liquid Contaminants

PCBs--PCBs (polychlorinated biphenyls) are manufactured polymers which were widely used in electrical equipment such as transformers, capacitors, switches, and voltage regulators [6]. They were used in insulating liquids, synthetic rubber, plasticizer, flame retardants, floor tile, printer's ink, coatings for paper and fabric, brake linings, paints, automobile body sealants, asphalt, adhesives, and other similar products. In buildings, PCBs were most commonly used as a dielectric in capacitors of lighting ballasts, as a liquid coolant in electrical transformers (often called ASKARAL), and as a component of building sealants.

The manufacture and processing or distribution of PCBs was essentially banned after April, 1979 in a response from the U.S. Environmental Protection Agency (EPA) to The Toxic Substances Control Act (TSCA) of 1976 [3]. TSCA does not affect the use of equipment already containing PCBs in totally enclosed systems. Therefore, PCBs that were in use or installed did not require removal and disposal under TSCA.

A thorough understanding of the health effects from human exposure to PCB's is not widely known. The National Institute for Occupational Safety and Health reports that PCBs are potentially carcinogenic and teratogenic in humans. PCBs are subject to transplacental movement, and have been shown to cause liver damage and chloracne in humans as a result of penetration through the skin [7].

Chemical Contaminants--A wide variety of chemicals are used in buildings from cleaners used during maintenance, to refrigerants used in mechanical equipment. Adverse health effects can result from exposure to chemical contaminants. However, there are numerous different conditions that could be discussed regarding this topic, which is beyond the scope of this paper.

Gaseous Contaminants

Radon--The most common source of indoor radon is uranium in the soil or rock on which buildings are built. Radon enters the buildings through dirt floors, cracks in concrete walls and floors, and floor drains. When radon becomes trapped in buildings, concentrations build up, and exposure risk increases. Radon can pose problems for both new and existing buildings. The predominant adverse health effect of radon exposure is lung cancer [8].

Biological Aerosols--Adverse health effects can result from exposure to biological aerosols, which include airborne contaminants from viruses, bacteria, fungi (molds), protozoa, arthropods, and mammals (pets) [9]. A discussion of the variety of sampling techniques for all of the different types of biological aerosols is beyond the scope of the paper. However, a discussion of molds follows.

Molds--Water leakage into building walls, floors or ceiling can result in environmental contaminants that pose a health concern for building occupants. Water leakage can occur in any building, from a residence, to the tallest high-rise, for a variety of reasons. Water leakage has been seen by the authors to be due to lack of proper maintenance of exterior walls and roofing systems, condensation, and construction defects such as window problems. Additionally, water can be released inside of a building during a fire when sprinklers are activated, or if plumbing pipes leak, which can result in several floors of a building being wetted.

If water leakage is severe, it can accumulate in roof, wall, and floor construction, resulting in moist conditions that are difficult to dry. Molds frequently grow in such moist locations. Spores of molds can be inhaled. When they gain access to the human respiratory tract, they can produce disease. Typical health effects from molds include upper and lower respiratory problems, hypersensitivity, or allergic reactions [9]. Certain molds produce mycotoxins which can cause disease symptoms in humans, called mycoses. Molds also produce and emit volatile organic compounds (VOCs) which can result in adverse health effects [9].

Airborne Chemical Contaminants--Airborne chemical contaminants can include volatile organic compounds (VOCs) such as toluene or vinyl chloride, inorganic compounds such as ammonia or carbon monoxide, pesticides, aldehydes and ketones such as formaldehyde, and even environmental tobacco smoke [10]. Indoor air contamination can originate from many sources, including various indoor activities, use of appliances, and outgassing of construction and furnishing products and materials [10].

Indoor air contamination can be enhanced by improper construction practices that do not allow for proper ventilation of the vapors. Contamination can result from off-gassing of in-situ materials and products, such as off-gassing of formaldehyde from carpeting, or from tasks performed within an indoor space, such as vapors from cleaning products or photocopy machines.

Possible adverse health effects range from headaches, dizziness, sleepiness, and watery eyes, to breathing difficulties or even death [11]. However, it is difficult to assess the potential environmental hazards posed by indoor air contaminants. As discussed above, exposure levels for which adverse health effects will occur in humans are difficult to establish. Humans appear to have widely varying tolerances to chemical exposure. Therefore, contaminant levels that may not cause adverse health effects in one person may result in significant health effects in another person.

Investigations of hazards presented by indoor air contaminants in buildings are being requested more often. Government regulations regarding indoor air quality are currently being developed, and the importance of remediating these hazards is becoming continually more important and publicized.

INVESTIGATION METHODOLOGY

A systematic methodology for identifying environmental hazards is prudent for a successful risk minimization strategy. Sampling of environmental hazards in buildings is an important component of a successful methodology.

The same basic procedures can be followed to investigate the presence of various environmental hazards and contaminants in buildings. Several important procedures should be performed prior to sampling so that the sampling program will be most informative.

Planning

The first step in planning an investigation of environmental hazards in buildings is to ascertain why the investigation is to be performed, and to establish the parameters for what environmental hazards are to be investigated. An investigation can be performed to verify that the health and safety of building occupants is not being threatened. This type of investigation may be part of an ongoing operations and maintenance program at the building, or it could be motivated by an event like a building fire or flood. An investigation could also be performed to determine whether construction materials intended to be disturbed during repair or renovation work contain environmental hazards. Compliance with governmental regulations such as the Americans with Disabilities Act (ADA), or high-rise sprinkler building code requirements, could be another reason for performing an investigation. Further reasons for performing investigations of environmental hazards include purchase and sale, or the demolition of a structure. The reason for an investigation can affect the approach, sample locations, sample quantities, or sampling techniques.

Cooperation is required for performing an investigation of environmental hazards. Environmental hazards are a highly emotional issue which can easily frighten building occupants. It is important that the building owner understands the seriousness of potential exposure to the environmental hazard being investigated. Cooperation must also be obtained from building engineering, maintenance, and operations personnel. Educating building personnel on the planned investigation will go a long way toward establishing the level of cooperation needed. All parties should be prepared to discuss the investigation in an informative manner that does not cause undo anxiety to building occupants.

Research

An important component of any investigation of environmental hazards in buildings is a basic understanding of where such hazards are likely to be found. This can be accomplished by performing research on the history of construction for the building. A review of drawings and specifications is typically a good start. Other commonly reviewed documents include previous consultant reports, remodelling records, shop drawings, water leakage reports, material submittals, or surveys of building occupants. This research should reveal the age of the building, the type of construction, typical details, construction materials used, the original intended use of the building, and any adverse health concerns reported by building occupants.

Research should also be performed to become familiar with typical materials used in the construction industry at the time the building was built. This is important to understand the context in which the building was constructed. For example, it may be possible to determine whether a specific material used in the building contained an environmental hazard by reviewing product literature from the time the building was constructed.

If no survey of building personnel or occupants exists, it is good practice to perform such a survey. Often times, building occupants are aware of problem conditions, or have experienced adverse health effects at some time, which may provide insight into the investigation. Interviews should be performed with building occupants, engineers, maintenance staff, and property managers. Questions should include whether they have seen evidence of suspect hazardous materials or experienced adverse health effects while at the building. Specific answers should be obtained as much as possible. An occupant survey should be performed as part of the research phase of the investigation so that information revealed from the survey can be used in subsequent steps.

After the research is completed, the results should be reviewed to identify potential environmental hazards and generally where they might be located in the building. At this point, it is also important to begin assessing how such materials could be detected during sampling.

Visual Inspection and Documentation

Following research and review, a visual inspection of the building should be performed. Visual inspection is important for an environmental hazard investigation because, more often than not, existing conditions in buildings differ from the original construction shown on drawings. Visual inspection and documentation of the building or structure should be performed to verify existing building conditions, to determine the general condition of building materials, and to locate potential environmental hazards.

Results of the visual inspection should include the following:
- building name
- project number
- date of inspection
- name of inspector
- inspection location for each specific sheet
- differences between original drawings and existing conditions
- condition of existing building materials
- visible signs of water damage or other deterioration
- visible signs of friable materials
- location of suspect materials containing environmental hazards
- description of suspect materials

For this work, it is important to develop good survey sheets to be used in the field for quick documentation of existing conditions. The results of the visual inspection must be accurate and reliable.

While most visual observation will be performed in readily accessible areas, it may be necessary to observe concealed conditions. If the research results reveal suspect materials containing environmental hazards in hidden areas, alternate techniques must be used to inspect the hidden areas. Concealed construction can be observed through the use of borescope probes, inspection openings, mirrors, video technology, or other suitable techniques. However, there are various protocols for random sampling for environmental hazards, including ASTM standards. If sampling is being performed for asbestos in a school building, the AHERA standards are very specific in terms of number and locations of samples. For lead based paint testing, the U.S. Department of Housing and Urban Development (HUD) has specific protocols for random sampling. Various states also have protocols for lead testing, including specific surfaces to be sampled in-situ, or the taking of paint chips for laboratory atomic absorption (AAS) testing.

Selecting sample locations

Often, the most difficult part of an investigation is the selection of sample locations and quantities. Sampling is a necessary procedure to test for the presence of environmental hazards in building materials identified as "suspect" during the visual inspection. The goal of sample selection is to have reasonable confidence that, after the samples are taken and tests are performed, sufficient information will have been gathered to determine whether an environmental hazard exists in the areas inspected. Information obtained during the visual inspection is used in conjunction with results of the initial project research to plot the most advantageous and necessary sample locations.

For many projects where budget concerns are significant, sample locations can be planned for two different phases. The initial sampling phase might give a general overview of the types of conditions that will be encountered at the building. The second phase of sampling could then be used to determine the extent of the hazardous materials in greater detail. Building owners frequently request this type of a phased approach for investigative work.

When selecting sample locations for environmental hazards in buildings, it is important that all materials that could contain an environmental hazard are included, even if the investigation is phased. Sample locations should also be representative of conditions observed in the building. Consideration should also be given to different construction details and techniques that are observed.

It is also important to consider whether field or laboratory testing is to be performed on the sample. The type of testing can be an important factor in selecting sample locations. For example, if all samples are to be laboratory tested, then physical samples will be required to be removed from the building. This may present a problem for building owners where a hole in a wall or ceiling may result from the sampling process.

The number of samples to be selected will vary greatly among buildings and various investigation parameters. It is recommended that enough samples be collected to obtain an accurate representation of all existing building conditions and materials. If two areas are of similar construction and materials and are located adjacent to one another, but one area has been altered or repaired, both areas should be sampled since the repaired area may differ from the unrepaired area. The U.S. Environmental Protection Agency (EPA) recommends a minimum of three bulk samples when investigating for the presence of asbestos.

It is generally most desirable to building owners to take bulk samples in hidden, concealed, or service areas of the building to minimize disturbance to occupied areas. In many cases, it is possible to sample in these locations. However, in other cases, samples must be taken from occupied spaces. Sample locations should consider locations that are most relevant to determining the presence of environmental hazards, rather than the convenience of obtaining the samples.

When planning for encountering environmental hazards during repair or renovation work, it is important to fully understand what work is scheduled to be performed. All materials to be disturbed should be considered for sampling. Knowledge of building systems and components is critical in evaluating suspect materials for testing. Repair or renovation details should be reviewed. The proposed construction processes, tools, and materials should also be reviewed. Materials disturbed during repair or renovation procedures present a significant risk for exposure to environmental hazards. If the structure is to be partially demolished or razed, all materials need to be evaluated for the presence of environmental hazards.

Sampling

An important question to answer when sampling is "What materials can physically be sampled in the building?" Construction materials are most commonly sampled to verify the presence of or to identify environmental hazards in buildings. Sampling of construction materials is referred to as bulk sampling. Bulk samples are sent to a laboratory where desired tests are performed. Other materials used in buildings, such as refrigerants in mechanical systems, liquid fuels from storage tanks, and lighting ballasts can also be sampled for laboratory identification testing. Air sampling is another useful sampling method, typically used for detecting airborne contaminants and particulates.

Lastly, building occupants can be sampled to determine whether environmental hazards have affected occupants. For example, sampling people has been an effective method for determine the presence of lead based paint hazards in homes and schools. Children are tested for blood lead levels in many parts of the country as part of standard health testing procedures, based upon the 1991 "Lead-Based Paint Poisoning Prevention Act (LBPPPA). High blood lead levels indicate an exposure source to be located.

For this section of the paper, bulk sampling methodology will be

presented.   Due to the variety of alternate sampling procedures and
methods, air sampling, liquid sampling and others are not further
discussed in this paper.

Preparation--Preparing for a field visit to obtain bulk samples
requires a significant amount of planning.   Each specific sample location
should be reviewed for possible access problems, unusual conditions or
other requirements.   Before going to the site, sample data sheets should
be prepared to document information obtained during sampling.   The data
sheets should also include building floor plans or elevations so that
sample locations can be accurately documented.

Protective Equipment--Appropriate personal protective equipment must
be worn at all times to ensure that the technician performing sampling
will not be unnecessarily exposed to the potential hazard above
permissible levels.   The need for protective clothing should be assessed
on a case-by-case basis, since it may not be required for all work.   In
most cases, a minimum of a face mask or respirator would be recommended
when performing bulk sampling, since in taking the sample, the material is
being disturbed, presenting possible airborne contamination.

Documentation--During sampling, it is important to fully document
the sample information.   It is likely that numerous samples will be taken
for a single project, and it will be critical that information for each
sample is available and orderly.   Typical sampling information could
include:

- location of sample
- date
- project name
- project number
- name of person performing sampling
- put written information tag on sample, and include the tag in photos for reference
- number of photos taken
- location of photos
- observations of the sample area prior to sampling
- observations of the physical material sample:
    - Size, color and material of the sample
    - What is the condition of the sample?
    - Is it damaged?
    - Is it water stained?
    - Are there other materials on the sample which could interfere with laboratory test results?
    - sketch of construction detail where sample was taken indicating precisely which materials were sampled and where in the detail.

Bulk Sampling--For most bulk samples, the physical sampling
procedure is fairly simple in that a small piece of the suspect material
is removed from the construction.   However, it is important to have
qualified, trained personnel performing the sampling. Asbestos inspectors
must be trained and certified in accordance with EPA and State standards.
Lead testing also requires training and certification in some states, and
federal requirements regarding lead testing and inspection are
forthcoming.   The EPA is currently working on accreditation of lead
consultants and contractors.
Generally, hand tools are sufficient to obtain bulk samples.   For
example, when sampling for laboratory tests to determine the presence of
asbestos in pipe insulation, utility knives, core borers, or chisels would
be sufficient for removing the specimen.   Similarly, when sampling for the
presence of mold, a drywall saw can be used to obtain a gypsum wallboard
sample.
Before the sample is removed, photographs should be taken of the
existing conditions.   Once the specimen is removed, it should be

photographed again, then placed in a container as recommended by the laboratory who will be performing the tests. Certain container materials can contaminate a specimen so that test results would be invalid. For most materials sampled as bulk, plastic bags are sufficient, provided they can be fully sealed to prevent release of contaminants to the air. Glass jars are also commonly used for containers.

Each specimen should be marked with the project identification, date and sample identification. Other information can be documented on the sample data sheet. For certain conditions, it may be advantageous to take additional photographs of the sample location, or other specific conditions uncovered after the sample is removed.

For investigation of the presence of certain materials such as asbestos, quality control samples are required in addition to those selected for the investigation. The quality control samples are taken adjacent to a project sample, but they are submitted to an alternate laboratory for testing to confirm the results with the project sample. Quality control sampling and testing is a good practice where laboratory tests are performed on the sample.

When a bulk sample is taken, a void is left in the source material or construction. In occupied buildings, it is important that the void be filled to prevent contamination to other areas of the building. Most buildings have significant air movement within construction assemblies and open spaces through natural air movements that occur. It is possible that contaminants could be spread to other portions of the building unless the voids are covered. Therefore, a necessary part of bulk sampling is repairs to patch the holes after the sample is removed. In some instances, it is beneficial to leave the void open, such as for others to inspect the existing conditions. Alternate techniques, such as plastic sealed around the opening, can be used to prevent contamination, while still allowing subsequent access to the sample location by others.

Testing of samples is dependent on the environmental hazard to be identified. For asbestos, polarized light microscopy is an identification test commonly performed in the laboratory. For lead, laboratory atomic absorption tests are commonly performed. In addition, in-situ X-ray Florescence (XRF) tests, or chemical reaction tests, such as sodium sulfide, can be used to determine the presence of lead. For organic compounds such as PCBs, infrared spectroscopy (IR) is commonly used for identification. For biological aerosols, such as fungi or mold, simple microscopic inspection of the bulk samples can reveal the presence of molds. Further laboratory culturing is necessary to identify the molds.

Air sampling--Air sampling is typically used for clearance sampling after asbestos abatement projects. For asbestos, sampling consists of drawing air through a filter at a known rate, capturing the asbestos fibers in the filter. The rate of air flow is important for comparison with current regulatory requirements which are usually described in fibers per volume of air. The filter is then sent to a laboratory for microscopic analysis of the fibers.

Air sampling can be performed using personal samplers or by using ambient monitoring equipment. When choosing a sampler for collection of a viable biological aerosol, the size of the particles, the relative fragility of the organisms, and the expected concentrations must all be considered. Some organisms must remain viable during sampling and analysis so that they can be identified. This requirement may affect the sampling equipment of collection media. Viable microorganisms have been collected using culture plate impactors, liquid impingers, membrane filter cassettes, and high-volume electrostatic devices [9].

CASE STUDIES

The authors have investigated numerous buildings. In some cases, environmental hazards were a primary component of the project from the beginning. In other cases, environmental hazards were identified during the course of the investigation, and subsequently became a significant component of the project. The following case studies illustrate typical situations where environmental hazards were investigated and sampled. Each author has presented two of the following four case studies (hereinafter generically referred to as writer).

Asbestos

In 1993, one of the writers was contacted by a local architectural firm, advising that they had been engaged by a non-profit agency to design a gut-rehab renovation project to a 19th century nine story manufacturing building in Boston to house the non-profit agency in a new home. The writer was asked by the architect to perform an asbestos survey for the non-profit agency who was in the process of buying the building. The situation was a typical purchase and sale investigation, where the buyer wanted to assure themselves that they did not have hazardous materials on site.

The writer made arrangements with the seller of the building to do a walk-through of the structure to identify any suspect ACBM's that would need to be bulk sampled and analyzed in a laboratory for asbestos fibers. The Owner of the building advised the writer that this was a "fools errand" in that he had removed all of the asbestos from the building in 1988, five years earlier. The writer's office identified 83 suspect ACBM for sampling and analysis, and received permission from the client, the non-profit agency (buyer), to obtain the samples and have them analyzed. It is important to note that all 83 ACBM were miscellaneous materials like: ceiling tile, floor tile, mastics, caulking compounds, tile grout and glazing putties, and; surfacing materials such as: wall and ceiling plaster. The writer could find no evidence of Thermal System Insulation (TSI) such as pipe or boiler insulation.

The writer prepared a bulk sampling plan, and hired a local industrial hygene firm to take the bulk samples and analyze them using Polarized Light Microscopy (PLM) in accordance with U.S. Environmental Protection Agency requirements. Of the 83 samples that were taken, 13 tested positive for asbestos. Most importantly, 8 of 15 wall plaster samples contained asbestos fiber. Other materials testing positive for asbestos were caulks, putties and floor tile.

Upon notifying the client of the initial findings, the writer was authorized to take an additional 30 bulk samples of plaster. Of the 30 samples, 16 of them tested more than 1 percent asbestos. Of 10 ceiling plaster samples, 2 tested positive, however, it was apparent that all of the wall plaster was positive. Of a total of 45 plaster samples, 24 were positive.

At this point, the non-profit agency halted the purchase and sale process, and demanded that the seller either adjust the sales price to remove the ACBM (estimated at $750,000 @ $10/s.f.) or else remove the ACBM before concluding the purchase and sale. It should be noted that the purchase price of the building was in the $2.25 million range.

The Seller was livid, and called the industrial hygene consultant used in 1988 and discussed how the asbestos plaster could have been missed. The writer reviewed the 1988 asbestos inspection report and learned that only 8 bulk samples had been taken (7 of TSI and 1 of ceiling tile), and that no samples of the plaster had been taken for analysis. The Seller's 1988 industrial hygene consultant requested permission to take parallel bulk samples and have them analyzed to see if the writer's findings were correct. The Seller's 1988 consultant advised that they would use "point counting" as opposed to the PLM analysis used on the writer's samples. The Seller's 1988 consultant claimed that point counting was state of the art, and a much superior analytical technique than the one used by the writer's laboratory. The writer accompanied the Seller's 1988 consultant to each of the 45 locations where plaster samples were originally taken, and observed that dual samples were obtained by a licensed inspector. The Seller's 1988 consultant advised the Seller that they would have an independent laboratory analyze their samples and report the results from the point counting analysis method.

Subsequently, the Sellers attorney faxed the writer lab results which indicated that only 5 of the 45 samples had been analyzed, and amazingly, all 5 indicated no asbestos present. A meeting with the non-profit agency and Seller was called, and the parties asked the writer how the results could be so different. It was explained that of the 45 samples taken by the writer, 21 did not have asbestos fiber. The parties hypothesized that perhaps the Seller's 1988 consultant had "cherry picked" 5 of the 45 selected samples for analysis to protect himself.

The non-profit agency and Seller decided to have the writer prepare a testing protocol for an independent industrial hygene laboratory to take bulk samples and perform an analysis of samples to confirm previous test results. The results were predictable. The independent laboratory confirmed the writer's initial results and debunked the 1988 consultant's attempt to vindicate themselves. In the end, the Seller had all of the asbestos plaster removed from the building during the summer of 1994 as a condition of the purchase and sale with the non-profit agency.

Lead Based Paint

In 1994, one of the writers was hired by a large housing authority to prepare contract documents for the removal of lead paint from the interior of 61 units in a 291 unit, federally funded, family housing development. A lead inspection had been performed by a local industrial hygene consultant using XRF tests and AAS technology on a random testing protocol set forth by HUD.

Field investigations revealed that most of the abatement work could be accomplished by replacement of components as opposed to deleading the components in place. The advantage of this approach was minimization of lead dust, and the hope of obtaining waivers from the State's department of health to allow the tenants to only vacate the units for one day while the work was occurring, saving the cost of having to relocate and house tenants in other quarters for an extended period of time.

Part of the scope of work included abatement of some 80 pressed metal door frames. The writer was given control of a typical two bedroom unit to perform field investigations for determining appropriate methodologies for abatement. The existing construction was wood frame and drywall. It was anticipated that wall and floor construction would be cut out to view the jamb and floor clips. During the field investigation it was also necessary to determine architectural details for the replacement of the frame, door, and hardware with a pre-hung assembly. The walls and floors would then be patched and matched after the deleading work was completed.

The field investigation started by opening the walls and floor with a hammer and chisel. It was quickly determined that the door frames were nailed with heavy gauge, welded metal anchor clips to board floor sheathing. The floor sheathing was covered with either sheet flooring or carpeting, on 3/4 inch thick tongue and groove plywood, on 1/8 inch thick vinyl asbestos tile (VAT), on 1/4 inch thick luan plywood, on the floor sheathing. The VAT was sampled and tested, and it was determined to contain 17% chrysotile asbestos fiber. The writer reported these findings to the housing authority, and advised them that disturbing the asbestos during the door replacement project would be cost prohibitive. It was decided to delead the door frames in place, and to incur additional costs to relocate tenants during the work.

As a pilot program, the writer tried to remove the lead based paint using a dustless needle gun. This process so distorted the frame it was determined to be inappropriate. Another pilot program with caustic paste proved to be a highly viable option in terms of time and cost, and had the added benefit of not creating lead dust.

The State would not grant waivers for tenant occupancy when chemical paint strippers are used. However, a cost analysis proved that by temporarily housing tenants in empty units in the housing complex, the revised procedure of deleading the frames in place with caustic paste, in lieu of the cut/patch and match approach first contemplated, was an attractive option. Waivers were still sought and granted for units where caustic strippers were not being used.

This lead project continues to this date with additional testing of 235 of the units, with the objective of issuing State letters of compliance. Unfortunately, the State and HUD requirements don't match, forcing compliance with the more stringent requirements of both. Even the testing protocols differ. In Massachusetts, the protocol for lead testing is XRF, with confirmatory testing by sodium sulfide, or sodium sulfide testing alone. For HUD, sodium sulfide testing is not allowed, and confirmatory testing of XRF results is done by AAS on paint chips. In Massachusetts, permission from the State is necessary to use AAS of paint chips for confirmatory testing.

Further complicating the issue is the type of deleading that must be done. The pressed metal frames, for instance, are a good example of the difference in approach between federal and state standards. Under Massachusetts standards, as long as the paint is intact, the frames would not have to be deleaded. Under HUD standards, the frames have to be deleaded regardless of the existing condition of the paint.

Mold

In 1992, one of the writers was requested to investigate mold growth that was observed on interior surfaces of gypsum wall board, in mechanical ductwork, on suspended ceilings, inside wall cavities, on wall fabrics and paint surfaces, and in carpet padding. Mold growth developed after a fire occurred in the building and the water sprinkler system activated, put out the fire, but wetted several floors of this high-rise building. An investigation was performed to determine locations where molds were growing on visible surfaces as well as hidden surfaces, such as in wall cavities.

A significant amount of testing and evaluation is necessary to determine the extent of a mold problem, and how to eliminate mold contamination from the building. Samples of the interior construction materials were obtained from the building and examined microscopically for mold contaminants. Samples were located where molds existed on visible surfaces as well as at locations where sprinkler water was believed to have wetted the construction. At several areas, molds were observed through inspection openings to be growing within wall construction and between two layers of gypsum wallboard. It has been reported that under moist conditions, mold between layers can propagate to the outer wallboard surface.

After several different molds were determined to be present, representative samples of the molds were sent to a biological laboratory for identification. Several different types of molds, including aspergillus niger, aspergillus tamarii, and penicillium citrinum were identified during this investigation. This testing was performed to determine the potential health effects of identified molds on building occupants. Molds can be remediated by cleaning them with biocides, bleach, or other effective products. However, all molds must be removed for a fully effective abatement project. It may not be possible to remove molds from porous surfaces or hidden surfaces in some cases, which can necessitate removal of the material contaminated with mold in order to abate the environmental hazard.

This case study illustrates one condition that allowed mold to grow. In other circumstances, it may be necessary to investigate the cause of conditions that are allowing molds to grow. For example, water leakage may be occurring through distressed brick masonry, and before an effective mold investigation and remediation project can be performed, the distressed masonry conditions must be identified and repaired.

Coal Tar Pitch

In 1993, one of the writers performed an investigation involving indoor air contamination. An older retail store was being rehabilitated for a new tenant, who, for aesthetic reasons, desired an exposed roof structure on the interior. After the rehabilitation was completed, occupants complained about indoor air contamination. To investigate this condition, the writer visited the building and sampled several suspect construction materials in the space. The materials were taken to the laboratory and were placed in an oven overnight, in an effort to accelerate off-gassing. None of these initial samples revealed an answer. Additional samples were obtained from concealed areas, and from roofing materials above the space. This sampling revealed that the source of the contamination was coal tar pitch used as a roofing material. Off-gassing vapors from the coal tar pitch were penetrating the roof decking at joints, and entering the interior space. At certain exposure levels, coal tar pitch is a carcinogen, causing cancer in humans, but it is commonly used as a roofing material on low-slope roofs. To resolve this condition, options were presented to the owner, who selected the least costly option of installing a new exhaust system to increase the interior air changes in the space. This type of environmental hazard was not even remotely anticipated during the design and construction documents phases that occurred during the rehabilitation of this building. This case study illustrates that environmental hazards can be hidden so that building occupants are not affected, and then, at some later time, they can present a potential problem.

CONCLUSIONS

Environmental hazards exist in many buildings. These hazards can be present on exposed surfaces, in the air, or hidden in roofs, walls, and in building systems. Sampling of suspect materials containing environmental hazards can be performed for subsequent testing or identification of the materials. A systematic methodology should be used when performing an investigation of environmental hazards in buildings. An important component of such an investigation is planning. Proper planning can mean the difference between a successful investigation, or an unsuccessful one. The actual physical act of sampling can be greatly simplified if good planning is performed. For buildings, bulk sampling is commonly used to detect the presence of environmental hazards in construction materials. Air sampling can also be used to determine the presence of airborne contaminants and particles. Sampling methods will depend on the goal of the investigation, and access to sample locations.

REFERENCES

[1]   Exposure Evaluation Division, Office of Toxic Substances, Office of Pesticides and Toxic Substances, U.S. Environmental Protection Agency (EPA), Guidance for Controlling Asbestos-Containing Materials in Buildings, EPA 560-5-85-024, Washington, D.C., 1985.

[2]   Gustitus, D. A., and Scheffler, M. J., "Proposed Methodology for the Identification and Remediation of Environmental Hazards During the Preservation or Rehabilitation of Older or Historic Structures," Standards for Preservation and Rehabilitation, ASTM STP 1258, Stephen J. Kelley, Ed., American Society for Testing and Materials, Philadelphia, PA 1995.

[3]   Gustitus, D.A., and Scheffler, M.J., "Addressing Environmental Hazards: Investigation and Design Issues in the Rehabilitation and Repair of Existing Buildings," Designing Healthy Buildings: Indoor Air Quality, The American Institute of Architects, Washington, D.C., 1992.

[4]   Pesticides and Toxic Substances, U.S. EPA, Managing Asbestos in Place: A Building Owner's Guide to Operations and Maintenance Programs for Asbestos-Containing Materials, EPA 20T-2003, 1990.

[5]   National Institute for Occupational Safety and Health, Centers for Disease Control, Public Health Service, U.S. Department of Health and Human Services, NIOSH Alert, Preventing Lead Poisoning in Construction Workers, DHHS publication no. 91-116a, April, 1992.

[6]   "PCBs in Fluorescent Light Fixtures," pamphlet by U.S. EPA Region 10, Air & Toxics Division, May, 1993.

[7]   Sittig and Marshall, Handbook of Toxic and Hazardous Chemicals, Noyes Publications, Park Ridge, NJ, 1981, pp 560-562.

[8]   U.S. EPA, and U.S. Consumer Product Safety Commission, The Inside Story, A Guide to Indoor Air Quality, EPA 402-K-93-007, September, 1993.

[9]   Office of Environmental Assessment, U.S. EPA, Indoor Air-Assessment, Indoor Biological Pollutants, EPA 600/8-91/202, January, 1992, pp. 1-1 to 1-10.

[10]   "Project Summary, Compendium of Methods for the Determination of Air Pollutants in Indoor Air," EPA/600/S4-90/010, May, 1990, pp. 1-5.

[11]   "What You Should Know About Combustion Appliances and Indoor Air Pollution," pamphlet by CPSC, EPA, and American Lung Association.

Kevin Ashley,[1] Paul C. Schlecht,[2] Ruiguang Song,[3] Amy Feng,[3] Gary Dewalt,[4]
and Mary E. McKnight[5]

ASTM SAMPLING METHODS AND ANALYTICAL VALIDATION FOR LEAD IN PAINT, DUST,
SOIL, AND AIR

REFERENCE: Ashley, K., Schlecht, P. C., Song, R., Feng, A., Dewalt, G.,
and McKnight, M. E., ''**ASTM Sampling Methods and Analytical Validation
for Lead in Paint, Dust, Soil and Air,**'' Sampling Environmental Media,
ASTM STP 1282, James Howard Morgan, Ed., American Society for Testing
and Materials, 1996.

ABSTRACT: ASTM Subcommittee E06.23 on Abatement/Mitigation of Lead
Hazards has developed a number of standards that are concerned with the
sampling of lead in environmental media, namely paint, dust, soil and
airborne particulate. An ASTM practice for the collection of airborne
particulate lead in the workplace has been published. New ASTM
standards for the collection of dry paint film samples, surface soil
samples, and surface dust wipe samples for subsequent lead analysis have
also been promulgated. Other draft standards pertinent to lead sampling
are under development. The ASTM standards concerned with lead sample
collection are accompanied by separate sample preparation standard
practices and a standard analysis method. Sample preparation and
analytical methods have been evaluated by interlaboratory testing; such
analyses may be used to assess the efficacy of sampling protocols.

KEYWORDS: airborne particulate, analysis, lead, paint, sampling, soil,
surface dust.

[1]Research Chemist and [2]Supervisory Chemist, U.S. Department of
Health and Human Services, Centers for Disease Control and Prevention,
National Institute for Occupational Safety and Health, Cincinnati, OH
45226.

[3]Statistician, Computer Sciences Corporation, 5555 Ridge Avenue,
Cincinnati, OH 45213.

[4]Senior Scientist, QuanTech, 1911 N. Fort Myer Drive, Suite 1000,
Rosslyn, VA 22209.

[5]Research Chemist, U.S. Department of Commerce, National Institute
of Standards and Technology, Gaithersburg, MD 20899.

The subject of leaded paint abatement and related activities has
received much attention in recent years. Interest in the mitigation of
lead hazards has been fueled largely by the potential hazard that lead
poses to the developmental capabilities and health of young children,
and by federal initiatives to reduce lead hazards. According to the
Centers for Disease Control (CDC), millions of children are believed to
be at risk of lead poisoning [1]. Also, according to the U.S.
Department of Housing and Urban Development (HUD), over 50 million homes
in the United States contain lead-based paint [2]. Occupational

exposure to lead is another concern, apart from childhood exposures to residential sources. Such occupational exposures may be of even more importance when considering the increased rate of lead abatement activities in residences.

In 1990, HUD published interim guidelines describing testing and abatement procedures for public and native American housing [3]. These guidelines have recently been revised [4], and are also applicable to private U.S. housing. During the development of the interim guidelines, HUD identified a need for consensus standards for environmental lead sampling methods, to ensure consistency and quality in its nationwide efforts to reduce lead hazards in housing. So in 1991, HUD requested that the American Society for Testing and Materials (ASTM) form a subcommittee to develop consensus standards concerning the identification and mitigation of lead hazards in housing. Thus ASTM Subcommittee E06.23 on Abatement/Mitigation of Lead Hazards was created to develop needed standards [5].

Key standards in the assessment of lead hazards include those for sample collection, sample preparation, and analysis of lead in various environmental matrices. Within E06.23, Task Group E06.23.16 on Laboratory Test Methods has developed ASTM standards for the sampling of lead in paint, surface dust, soil, and airborne particulate. These sample collection standards are accompanied by separate standards describing sample preparation and analysis for lead in the various media. It is expected that these and other applicable ASTM standards will be referenced in federal publications and regulations that are promulgated in response to legislative acts such as the Residential Lead-Based Paint Hazard Reduction Act of 1992 (also known as 'Title X') [6].

In this paper, the content of the environmental lead sample collection standards is outlined in detail. Also, performance data for associated sample preparation and analysis standards from interlaboratory testing of environmental lead samples are presented. The ASTM sample collection protocols for lead in paint, dust, soil, and airborne particulate may be evaluated indirectly by using the performance results obtained from interlaboratory testing. The statistical design criteria that would be used in describing a sampling protocol for inspection, risk assessment, or other purpose are outside the scope of this article.

ASTM SAMPLE COLLECTION PROTOCOLS

As mentioned above, ASTM standards describing the collection of paint, surface dust, soil, and airborne particulate lead samples have been published. These standards describe sampling protocols for collection of single samples, and do not address the sampling design criteria (that is, number and location of samples) that are used for inspection, risk assessment, or other purposes. However, it is generally recommended that sufficient numbers of representative samples be obtained so as to provide for valid conclusions. Sampling and statistical design issues are to be covered in a separate ASTM standard which is under development within subcommittee E06.23.

Sampling of Paint

A new standard that describes the collection of dried paint samples for subsequent lead determination, ASTM practice E 1729, has been published [7]. The standard is intended for use in the identification of potential lead hazards in dried paint films on surfaces, and covers the collection of paint samples from buildings and related structures. In order to determine the lead concentration in paint, collected samples are first prepared for analysis according to ASTM practice E 1645 [8], and then analyzed by atomic spectrometry according to ASTM test method E 1613 [9].

ASTM E 1729 describes the collection of dried paint samples from areas of known dimensions by using a heat gun, cold scraping, or coring methods. The standard is intended to be used to collect paint samples for subsequent determination of lead on either an area basis (milligrams of lead per area sampled) or concentration basis (milligrams of lead per gram of collected paint, or weight percent of lead), or both.

The collection procedure described in ASTM E 1729 involves the delineation of a known sample area, with subsequent removal of all paint within that area down to the substrate. Removal of paint layers is accomplished with a number of methods that utilize cutting and scraping tools. A variety of methods are needed to achieve paint sample removal under wide ranging field conditions which are commonly encountered. Collection methods include heat gun-assisted removal, coring, and cold scraping. Paint is collected into a collection tray or funnel fabricated from paper and tape. Use of a collection tray or funnel is necessary since it is generally not possible to prevent the sample from breaking apart during sample collection by scraping and cutting. After collection into the funnel or tray, the sample is transferred to a rigid walled, resealable sample container for transport to the laboratory.

Since there are safety hazards associated with its use, employing a heat gun is only recommended for instances in which the paint is very difficult to remove via cold scraping or coring. A heat gun is especially useful on metal surfaces, for it is extremely difficult to remove paint layers from metal surfaces by cold scraping or coring.

Use of a coring device can reduce the potential for inadvertent collection of paint from adjacent areas that can be disturbed during the collection process. If a coring tool is used to collect paint film samples, there is no need to delineate the sample area, since the sample area is defined by the dimensions of the coring device. Also, the coring method precludes the use of a sample collection funnel or tray, since during collection the sample is adhered to a piece of tape, which is placed upon the paint surface prior to use of the coring device. After using the coring tool, the paint sample is transferred directly to a rigid-walled resealable sample container for transport to the laboratory.

An ideal paint sample contains no substrate. Collection practices are written to give a priority to collecting all of the paint while omitting inclusion of substrate. However, in practice, it has been found that the requirement to include no substrate in the paint sample is nearly impossible for many real-world materials. In such cases, it is necessary to minimize the amount of substrate included within the sample as much as possible to minimize dilution or contamination of the paint sample. Inclusion of substrate should not affect results for paint samples to be analyzed on an area basis.

Sampling of Soil

A new standard governing the collection of soil samples for subsequent lead analysis has been published [10]. This standard, ASTM practice E 1727, covers the collection of soil samples using coring and scooping methods. The practice is intended for use in the collection of soil samples in an around buildings and related structures for the subsequent determination of lead concentration (micrograms of lead per gram of soil sample), as described in the HUD guidelines [3,4].

Both coring and scooping methods are included in the standard since coring devices cannot be used on all soils. Coring methods are useful for collection of soil samples from dense, hard, or sticky soils. Scooping techniques are effective for collection of soil samples from soft, powdery, loose, and sandy soils. Coring methods are not recommended for loose, sandy soils, while scooping methods are not recommended for very hard or frozen soils. A variety of coring or scooping tools may be used.

In using either the coring tool or scooping method, only the top portion (ca. 1.5 cm depth) of the soil is collected for subsequent laboratory analysis. The reason for this is that it is mainly the top portion of soil which is thought to contribute to potential human exposures to lead. In either technique, be it scooping or coring, the soil is sampled to a known depth. Scoop sampling may be conducted by using a spoon, or by scooping directly into an appropriate sample collection container. Soil samples are to be transported to the laboratory in sealed containers.

Soil samples are not to be sieved nor ground in the field; these sample preparation steps are to be conducted in the laboratory prior to sample digestion and analysis. The preparation of soil samples is described in ASTM practice E 1726 [11], and analysis of digested soil samples is covered in ASTM test method E 1613 [9].

Sampling of Surface Dust

A new standard which covers the collection of surface dust using wipe sampling for subsequent lead analysis, ASTM practice E 1728, has been published [12]. This standard is based largely on a National Institute for Occupational Safety and Health (NIOSH) method [13] for surface wipe sampling. The practice is intended for use in the collection of surface dust samples for subsequent lead determination on a loading basis, that is, micrograms of lead per area sampled. The standard describes the collection of wipe samples of settled dust from hard surfaces from areas of known dimension, using moistened disposable towellettes and a specified wiping pattern.

Problems may be encountered in wipe sampling of too large or too small a sampling area, so minimum and maximum recommended sampling areas (100 cm^2 and 1000 cm^2) are prescribed in the standard. If the sampling area is too large, the wipe materials may not be able to withstand the sample collection procedure; also, collection efficiencies may be undesirably low. If the sampling area is too small, there may be insufficient lead loading for analytical purposes. Also, the collected sample may not be large enough to be representative of the environment being evaluated for surface dust lead contamination. Other potential problems may arise from the presence of background lead in the wipe material; hence, it is stressed that the wipe material chosen for use must be evaluated for its suitability in the laboratory [14]. Also, wipe sampling is generally not recommended for soft surfaces; vacuum sampling techniques, such as that described in ASTM D 5438 [15], are preferred for sampling dust from soft surfaces such as carpets and upholstery.

Within the wipe sampling practice (ASTM E 1728), procedures are described for template-assisted sampling and confined space sampling of surface dust. Generally, a wetted wipe material is passed three times over the delineated area with manual pressure. The wipe is folded between each pass so that an unexposed portion of the wipe is always used for sample collection. After the entire delineated area is sampled, the wipe is placed into a rigid-walled, sealable container for transport to the laboratory. Compositing of wipe samples is not recommended, since this may lead to serious problems in subsequent digestion and analysis of the wipes (for example, low lead analytical recoveries). Wipe samples are digested in the laboratory according to ASTM practice E 1644 [16], and then analyzed for lead content by ASTM test method E 1613 [9].

Sampling of Workplace Air

A standard describing the collection of airborne particulate lead during work activities, ASTM practice E 1553 [17], has been published. The practice is intended for use in protecting workers from high concentrations of airborne lead. The standard is based on NIOSH methods for monitoring airborne lead concentrations [18], and may be used to ensure that lead levels in workplace air do not exceed the Occupational Safety and Health Administration (OSHA) permissible exposure limit (PEL)

for airborne lead [19]. Air sample concentrations are to be expressed
in units of micrograms of lead per cubic meter of sampled air (usually
for an 8-hour workday).
The workplace air sampling practice involves the use of battery-
powered personal sampling pumps to collect airborne particulate onto
cellulose ester membrane filters. The air filters are housed within
plastic filter holders, which are attached to the sampling pump by means
of flexible tubing; filter cassettes may be two- or three-piece. The
sampling pump and filter holder are attached to the body of a worker by
means of clips, with the filter holder arranged inlet end down within
the worker's personal breathing zone.
According to the ASTM standard [17] and NIOSH methods [18],
workplace air samples are collected at a known flow rate of between 1
and 4 liters per minute for a full 8-hour workday. Thereafter, the
filter holders (containing the exposed air filters) are placed into
sealable plastic bags for transport to the laboratory. In the
laboratory, the air filter samples may be digested according to ASTM
practice E 1741 [20], and can then be analyzed for lead content by ASTM
test method E 1613 [9].

DETERMINATION OF LEAD IN PAINT, DUST, SOIL, AND AIR SAMPLES

Aside from sample collection protocols for lead in paint, dust,
soil, and air, laboratory techniques for the preparation and analysis of
lead in environmental samples (developed by ASTM subcommittee E06.23)
have been published. Separate sample preparation procedures were
written for each of the four environmental matrices (paint, dust, soil,
and air), while a single analytical test method was promulgated which
covers all four media. Since the emphasis of this work is on the ASTM
sample collection protocols, only a cursory discussion of the ASTM
sample preparation practices and the ASTM test method for lead in paint,
dust, soil and air is presented in this paper.

Sample Preparation

ASTM sample preparation protocols for the extraction of lead from
environmental samples are generally similar in terms of the acids and
high temperatures used to solubilize lead. Specific differences within
the sample preparation practices are attributed to the unique nature of
each environmental medium of concern (i.e., paint, dust wipes, soil, and
air filters). Lead in all four media may be extracted essentially
quantitatively by means of a mixture of concentrated nitric acid and
hydrogen peroxide, and the use of a hotplate to effect the high
temperatures required (ASTM practices E 1726 [11], E 1741 [20], E 1644
[16], and E 1645 [8]). Some media, e.g., paint and air filters (ASTM E
1645 [8] and E 1741 [20], respectively) are amenable to microwave
digestion in a mixture of concentrated nitric and hydrochloric (or
other) acids. The ASTM sample digestion procedures were adapted from
existing NIOSH [18] and U.S. Environmental Protection Agency (EPA) [21]
protocols.

Analysis

Once lead becomes solubilized by extraction, sample aliquots from
all four environmental matrices may be analyzed by the same analytical
method, ASTM test method E 1613 [9]. This test method describes
procedures for the determination of lead in sample aliquots by any one
of three spectrometric methods: flame atomic absorption
spectrophotometry (FAAS), graphite furnace atomic absorption
spectrophotometry (GFAAS), and inductively coupled plasma atomic
emission spectrometry (ICP-AES). This standard was developed from
existing NIOSH [18] and EPA [21] methods.

INTERLABORATORY EVALUATION OF PROTOCOLS FOR LEAD ANALYSIS

Methods that are equivalent to the aforementioned ASTM standards for the preparation and analysis of environmental samples for lead determination were evaluated by interlaboratory testing. A subset of data from the Environmental Lead Proficiency Analytical Testing (ELPAT) program [22], which is overseen by the EPA and NIOSH, was used to evaluate the performance of ASTM sample digestion and lead analysis protocols. Recovery and precision results from laboratories participating in ELPAT which utilized the ASTM procedures were examined for this purpose. Knowledge gained from this analysis may be used to evaluate sampling protocols, since sample preparation and analysis must be characterized initially.

The ELPAT program evaluates the performance of laboratories conducting environmental lead analysis. Proficiency test samples are prepared by an American Industrial Hygiene Association (AIHA) contractor using real-world paints, dusts, and soils. These samples are sent quarterly to participating laboratories, and the performance of those laboratories is evaluated by NIOSH. Data from Round 6 of the ELPAT program [22] were used in this study.

Sources and Processing of ELPAT Samples

Quality control samples that are used for ELPAT are collected from real-world sources, that is, paints, surface dusts, and soils. These matrices are thoroughly homogenized before samples are sent to participating laboratories.

Paint samples for ELPAT Round 6 were prepared from paint chips collected from a number of sites. Paints were ground to a maximum particle size of 120 μm.

Soil samples used in ELPAT Round 6 came from lead-contaminated soils in a variety of locales. The soils were first dried at room temperature (25 °C), then sterilized by heating at 163 °C for a minimum of 2 hours, and finally sieved to a maximum particle size of 250 μm.

Dust samples on filter media used in ELPAT Round 6 were prepared by gravimetrically loading Whatman 40 filter paper with sterilized (gamma-irradiated) household and post-abatement dust, which had been sieved to a maximum particle size of 250 μm. The loaded filters were moistened with 0.5 mL of 3% hydrogen peroxide solution to prevent mildew growth.

Evaluation of Results from ELPAT

Within the ELPAT program, lead analysis procedures used by participating laboratories can be categorized by the two factors of sample preparation technique and analytical method. A subset of ELPAT data from laboratories that used sample preparation and analysis procedures which were equivalent to the ASTM protocols was utilized for interlaboratory evaluation. The purpose of this evaluation is to: 1. evaluate the effects of the two factors (i.e., sample preparation and analysis); 2. estimate the bias between different sample preparation techniques and analytical methods; and 3. estimate the precision associated with the sample preparation techniques and analytical methods. As mentioned above, data from Round 6 of the ELPAT program (conducted during the Spring of 1994 [22]) were used in this study.

The sample preparation techniques which were examined by interlaboratory evaluation included hotplate and microwave strong acid digestion procedures. The analytical methods evaluated included FAAS and ICP-AES; GFAAS was not included since too few laboratories used this procedure in the ELPAT program to allow for statistical evaluation. The number of laboratory analyses corresponding to each combination of sample preparation technique and analytical method in ELPAT Round 6 are listed in Table 1 for each sample matrix (paints, soils, or dust-loaded filters).

TABLE 1--<u>Numbers of laboratory analyses corresponding to each combination of sample preparation technique and analytical method.</u>

Sample Prep. Technique	Sample Type	Analytical Method FAAS[a]	ICP-AES[b]
Hotplate Digestion	Paint	111	68
	Soil	102	66
	Dust	107	61
Microwave Digestion	Paint	9	13
	Soil	10	10
	Dust	7	10

[a]Flame atomic absorption spectrophotometry.
[b]Inductively coupled plasma atomic emission spectrometry.

The model used to test the effects of the two factors (i.e., sample preparation and analysis) and their interaction is given by the following two-way, crossed classification, with interaction, fixed effects expression [23]:

$$y_{ijk} = \mu + T_i + M_j + (TM)_{ij} + e_{ijk} .$$

Here y_{ijk} is the measured value; $i = j = 1, 2$; $k = 1, \ldots, n_{ij}$ (where n_{ij} = total number of laboratory analyses corresponding to the ith sample preparation technique and the jth analytical method); μ is the mean value, T_i is the ith sample preparation technique; M_j is the jth analytical method; and e_{ijk} is an error term (which accounts for the overall experimental error associated with sample preparation and analysis). In the above expression, T_1 and T_2 represent hotplate and microwave sample preparation techniques, respectively; M_1 and M_2 represent FAAS and ICP-AES analytical methods, respectively; and $(TM)_{ij}$ is the interaction between sample preparation technique and analytical method.

This model was used for each of 12 combinations of the 3 sample types and 4 lead concentration levels of each sample matrix (i.e, paint, dust, and soil). Outliers are excluded from this analysis. (A result is identified as an outlier if the value in question is 3 or more quantile ranges away from the 75% quantile or the 25% quantile [22].) Analysis of variance (ANOVA) results showing p-values obtained from tests of the above model (for samples from ELPAT Round 6) are given in Table 2.

The ANOVA results (Table 2) show that, in general, the effects of sample preparation technique and analytical method are insignificant. The implications are that the two sample preparation techniques and the two analytical methods are essentially equivalent statistically. The results shown in Table 2 also illustrate that there is no significant interaction between the sample preparation technique and the analytical method. This implies that any combination of sample preparation technique and analytical method is essentially statistically equivalent to any other combination.

However, the two analytical methods FAAS and ICP-AES may give rise to significant, albeit small, differences in mean values for certain samples [22]. Differences which have been observed in analytical results from certain sample matrices (e.g., paints and dust-loaded filters) may be attributed to sample-specific matrix effects. See for example the last two entries in the fourth column of Table 2, where significant statistical differences are observed for dust samples.

TABLE 2--List of p-values obtained from testing the effects of sample preparation technique, analytical method, and their interaction.

Sample Type	Sample Batch[a]	T[b]	M[c]	(T*M)[d]
Paint	1	0.523	0.228	0.776
	2	0.385	0.262	0.430
	3	0.214	0.445	0.512
	4	0.107	0.519	0.379
Soil	1	0.844	0.073	0.375
	2	0.395	0.777	0.480
	3	0.217	0.549	0.858
	4	0.815	0.562	0.926
Dust	1	0.637	0.305	0.928
	2	0.438	0.173	0.943
	3	0.534	0.046[e]	0.650
	4	0.869	0.037[e]	0.388

[a]Each batch for each sample type corresponds to a different lead concentration.
[b]Factor corresponding to sample preparation technique.
[c]Factor corresponding to analytical method.
[d]Term corresponding to interaction between T and M.
[e]Significant at 5% level.

To investigate further the differences in results obtained from the two analytical methods, bias estimates were computed by using the following expression (using data from both sample preparation techniques):

Bias = $(\mu_1 - \mu_2)/[(\mu_1 + \mu_2)/2]$,

where μ_1 and μ_2 are the mean values determined from FAAS and ICP-AES analyses, respectively. Computed bias estimates, as well as relative standard deviations (RSDs) for each analytical method and each sample type, are presented in Table 3.
    The observed bias (Table 3) is significantly different from zero for the dust-loaded filter samples: FAAS yields higher mean values than does ICP-AES by ca. 4.5%. For soils, although the difference between the two method means is not significant, FAAS means are consistently higher (ca. 1 to 3%) than means from ICP-AES. For paint samples, the observed bias appears to depend on the lead concentration level in the respect that the FAAS mean is higher than the ICP-AES mean at low lead concentrations, while the reverse is observed at high lead loadings. In all cases, estimated bias is quite low (< 5%), which is indicative of the essential equivalence of the two analytical methods over the lead concentration ranges within the various media.
    Estimates of RSDs for each analytical method and each sample matrix (Table 3) range from ca. 7% to ca. 17%. The observed RSD values are quite low for interlaboratory results from real-world samples, and are in general compliance with laboratory accreditation guidelines targeted by the Federal Interagency Lead-Based Paint Task Force, Subcommittee on Methods and Standards [24].
    Usually the highest RSDs are found for the lowest lead concentrations (Table 3); this is not an unexpected result. For certain samples (e.g., paint), higher RSDs observed at higher lead concentrations may be due to matrix influences. These results

TABLE 3--Estimates of bias and precision from ELPAT results: FAAS
and ICP-AES analyses.

Sample	Bias (%)	FAAS[a]			ICP-AES[b]		
		μ^c	RSD[d]	n[e]	μ	RSD	n
paint 1	3.2	0.56	8.3	113	0.54	11.7	77
paint 2	-1.3	0.88	13.2	115	0.90	11.5	77
paint 3	-3.8	3.63	11.9	116	3.76	10.5	76
paint 4	3.4	0.10	10.6	112	0.09	14.9	75
	0.4 ←(pooled estimate of bias)						
soil 1	2.2	1475	7.0	108	1443	9.7	75
soil 2	0.9	104	14.7	106	103	16.9	76
soil 3	1.5	775	8.2	110	764	10.2	74
soil 4	1.4	639	12.9	112	630	10.6	75
	1.5 ←(pooled estimate of bias)						
dust 1	4.3	333	13.5	109	319	16.2	71
dust 2	4.8	101	13.2	111	96	13.8	70
dust 3	4.5	1010	9.6	110	966	12.2	70
dust 4	4.3	707	11.2	112	677	12.2	69
	4.5 ←(pooled estimate of bias)						

[a]Flame atomic absorption spectrophotometry.
[b]Inductively coupled plasma atomic emission spectrometry.
[c]Mean lead concentration; units for paint: % Pb by weight; units for
   soil: μg Pb/g of sample; units for dust: μg Pb.
[d]Relative standard deviation (%).
[e]Number of analyses.

illustrate the importance of using real-world matrices for proficiency
test samples. Interlaboratory RSDs for synthetically prepared lead-
containing samples which are prepared from lead nitrate solutions, such
as those used in the NIOSH Proficiency Analytical Testing (PAT) program
[25], are much lower. Real-world sample matrices such as those used in
the ELPAT program allow for a more realistic test of laboratory
performance than do synthetic samples.

CONCLUSION

New ASTM standards for the collection of paint, surface dust,
soil, and workplace air samples for subsequent lead determination,
developed by ASTM Subcommittee E06.23 on Abatement/Mitigation of Lead
Hazards, have been published recently. An overview of the content of
these standards, and the rationale behind their development, has been
presented in this paper. It is intended that these new standards will
be used in the detection and identification of lead hazards in
residences, commercial structures, and other locales. Additional draft
standards are under development by ASTM Subcommittee E06.23 to address
sampling for and on-site testing of lead in environmental media of
concern.

The ASTM sample collection protocols are accompanied by ASTM
laboratory sample preparation and analysis procedures. Methods that are
equivalent to the ASTM sample preparation and analysis procedures have

been evaluated by interlaboratory proficiency testing. The
interlaboratory evaluation has shown that the sample preparation and
analytical protocols perform extremely well, as evidenced by excellent
interlaboratory agreement. These standards may be useful for the
evaluation of other sample collection and sampling design protocols that
are under consideration within ASTM Subcommittee E06.23.

ACKNOWLEDGMENTS

We are indebted to the members of ASTM Subcommittee E06.23 and
others who assisted in the development of standards pertaining to
sampling and analysis of lead in environmental media. Special gratitude
is extended to Richard Baker, Bruce Buxton, Sharon Harper, Meredith
Hunter, Walt Rossiter, and Kenn White for their contributions. David
Bartley and Aaron Sussell of NIOSH provided helpful comments on the
draft manuscript.

DISCLAIMER

Mention of company names or products does not constitute
endorsement by the Centers for Disease Control and Prevention (CDC) or
the National Institute of Standards and Technology (NIST).

REFERENCES

[1] CDC, Strategic Plan for the Elimination of Childhood Lead
    Poisoning; U.S. Department of Health and Human Services, Public
    Health Service, Centers for Disease Control, Atlanta, GA, 1991.

[2] HUD, Comprehensive and Workable Plan for the Abatement of Lead-
    Based Paint in Privately Owned Housing; U.S. Department of Housing
    and Urban Development, Washington, DC, 1990.

[3] HUD, Lead-Based Paint: Interim Guidelines for Hazard
    Identification and Abatement in Public and Indian Housing; U.S.
    Department of Housing and Urban Development, Washington, DC, 1990.

[4] HUD, Guidelines for the Evaluation and Control of Lead-Based Paint
    Hazards in Housing: Final Report; U.S. Department of Housing and
    Urban Development, Washington, DC, 1995.

[5] Ashley, K., and McKnight, M. E., "Lead Abatement in Buildings and
    Related Structures: ASTM Standards for Identification and
    Mitigation of Lead Hazards," ASTM Standardization News, Vol. 21,
    No. 12, pp. 32-39, 1993.

[6] Title X, Residential Lead-Based Paint Hazard Reduction Act of the
    Housing and Community Development Act of 1992 (Public Law 102-
    550). Federal Register; U.S. Government Printing Office:
    Washington, DC, 1992.

[7] ASTM E 1729, "Standard Practice for Field Collection of Dried
    Paint Samples for Lead Determination by Atomic Spectrometry
    Techniques," in Annual Book of ASTM Standards; American Society
    for Testing and Materials, Philadelphia, PA, 1995.

[8] ASTM E 1645, "Standard Practice for Preparation of Dried Paint
    Samples for Subsequent Lead Analysis by Atomic Spectrometry," in
    Annual Book of ASTM Standards; American Society for Testing and
    Materials, Philadelphia, PA, 1995.

[9] ASTM E 1613, "Standard Test Method for the Analysis of Digested Samples for Lead by Inductively Coupled Plasma Emission Spectrometry (ICP-AES), Flame Atomic Absorption (FAAS), or Graphite Furnace Atomic Absorption (GFAAS) Techniques," in ASTM Standards on Lead-Based Paint Abatement in Buildings; American Society for Testing and Materials, Philadelphia, PA, 1994.

[10] ASTM E 1727, "Standard Practice for Field Collection of Soil Samples for Lead Determination by Atomic Spectrometry Techniques," in Annual Book of ASTM Standards; American Society for Testing and Materials, Philadelphia, PA, 1995.

[11] ASTM E 1726, "Standard Practice for the Sample Digestion of Soils for the Determination of Lead by Atomic Spectrometry," in Annual Book of ASTM Standards; American Society for Testing and Materials, Philadelphia, PA, 1995.

[12] ASTM E 1728, "Standard Practice for Field Collection of Settled Dust Samples Using Wipe Sampling Methods for Lead Determination by Atomic Spectrometry," in Annual Book of ASTM Standards; American Society for Testing and Materials, Philadelphia, PA, 1995.

[13] Eller, P. M., and Cassinelli, M. E., Eds., NIOSH Manual of Analytical Methods (4th ed.), Method No. 9100; U.S. Department of Health and Human Services, Public Health Service, Centers for Disease Control and Prevention, National Institute for Occupational Safety and Health, Cincinnati, OH, 1994. DHHS/PHS/CDC/NIOSH Publication No. 94-113.

[14] Millson, M., Eller, P. M., and Ashley, K., "Evaluation of Wipe Sampling Materials for Lead in Surface Dust," American Industrial Hygiene Association Journal, Vol. 55, pp. 339-342, 1994.

[15] ASTM D 5438, "Standard Practice for Collection of Floor Dust for Chemical Analysis," in Annual Book of ASTM Standards; American Society for Testing and Materials, Philadelphia, PA, 1994.

[16] ASTM E 1644, "Standard Practice for Hot Plate Digestion of Dust Wipe Samples for Determination of Lead by Atomic Spectrometry," in Annual Book of ASTM Standards; American Society for Testing and Materials, Philadelphia, PA, 1995.

[17] ASTM E 1553, "Standard Practice for Collection of Airborne Particulate Lead During Abatement and Construction Activities," in ASTM Standards on Lead-Based Paint Abatement in Buildings; American Society for Testing and Materials, Philadelphia, PA, 1994.

[18] Eller, P. M., and Cassinelli, M. E., Eds., NIOSH Manual of Analytical Methods (4th ed.), Method Nos. 7082 & 7105; U.S. Department of Health and Human Services, Public Health Service, Centers for Disease Control and Prevention, National Institute for Occupational Safety and Health, Cincinnati, OH, 1994. DHHS/PHS/CDC/NIOSH Publication No. 94-113.

[19] Code of Federal Regulations, "Lead Exposure in Construction: Interim Final Rule" (29 CFR Part 1926.62). Federal Register; U.S. Government Printing Office, Washington, DC, 1992.

[20] ASTM E 1741, "Standard Practice for Preparation of Airborne Particulate Lead Samples Collected During Abatement and Construction Activities for Subsequent Analysis by Atomic Spectrometry," in Annual Book of ASTM Standards; American Society for Testing and Materials, Philadelphia, PA, 1996.

[21] EPA, "Standard Operating Procedures for Lead in Paint by Hotplate-
or Microwave-Based Acid Digestions and Atomic Absorption or
Inductively Coupled Plasma Emission Spectrometry" (EPA 600/8-
91/213); U.S. Environmental Protection Agency, Office of Research
and Development, Research Triangle Park, NC, 1991.

[22] Schlecht, P. C., and Groff, J. H., "ELPAT Program Report:
Background and Current Status" (Quarterly Report), Applied
Occupational and Environmental Hygiene, Vol. 9, pp. 529-536, 1994.

[23] Miller, J. C., and Miller, J. N., Statistics for Analytical
Chemistry; Ellis Horwood Ltd.: Chichester, U.K., 1984; Ch. 6.

[24] EPA, "Laboratory Accreditation Program Guidelines: Measurement of
Lead in Paint, Dust, and Soil" (EPA 747-R-92-001); U.S.
Environmental Protection Agency, Office of Pollution Prevention
and Toxic Substances, Washington, DC, 1992.

[25] Schlecht, P. C., and Groff, J. H., "Proficiency Analytical Testing
Program" (Quarterly Report), Applied Occupational and
Environmental Hygiene, Vol. 9, pp. 393-394, 1994.

William S. Brokaw,[1] G. Stephen Hornberger[2]

MUNICIPAL SOLID WASTE SAMPLING AND CHARACTERIZATION

REFERENCE: Brokaw, W. S. and Hornberger, G. S., ''Municipal Solid Waste Sampling and Characterization,'' Sampling Environmental Media, ASTM STP 1282, James Howard Morgan, Ed., American Society for Testing and Materials, 1996.

ABSTRACT: The character and composition of solid wastes need to be determined for a variety of objectives. These objectives include developing design criteria for resource recovery facilities, determining performance efficiency, and evaluating source reduction initiatives (e.g. recycling).

The techniques described in this paper are for the collection and characterization of municipal solid waste and can be employed at resource recovery facilities, transfer stations and sanitary landfills. The results have been used in conjunction with emissions testing and ash analyses to calculate a mass balance for an incinerator at a resource recovery facility.

The technique utilized is an adaptation of several related ASTM methodologies developed for sampling and analyzing refuse derived fuel. Representative waste samples are manually collected from trash collection vehicles (rolloff containers, compactor trucks, etc.) as they unload. The discrete samples are collected and composited during a predetermined period. Each composite sample is mixed and characterized on-site and a portion recovered and prepared for laboratory analyses.

A significant amount of data can be generated from the sampling such as the mass and percent of recyclable materials, heating value, and elemental composition (proximate and ultimate parameters). The sampling technique can be readily modified to obtain the desired data.

KEYWORDS: municipal solid waste, waste stream, recycling, characterization, resource recovery

---

[1]Scientific Associate, Environmental Management Division, Recon Environmental Corp, Raritan NJ 08869
[2]Senior Chemist, AquaAir Analytical Division, Recon Environmental Corp, Raritan, NJ 08869

Municipal solid waste is a highly heterogenous media composed of discarded consumer goods, print media, construction debris and organic wastes. The character of the waste stream and elemental composition need to be determined in order to understand and utilize the material or to reduce the quantity through recycling efforts. At a resources recover facility the collection of waste samples can be conducted in coordination with air emission testing and ash analyses for mass balance calculations and measuring incinerator efficiency.

Existing ASTM Standards D5115 (test method for collecting waste samples from a conveyor belt), E889 (test method for measuring throughput of resource recovery unit operations), and E1107 (test method for composition of purity of solid waste materials stream) provide the basis for the sample collection method described in this paper.

SAMPLE COLLECTION

Representative sampling is conducted on a real time basis (i.e. collecting samples upon disposal as opposed to sampling by means of borings or test pits). The real time method of collection provides data on the waste as it is generated during the sampling period (e.g. summer).

Samples are collected at regular intervals, coinciding with the arrival of waste loads at the disposal facility (a landfill, transfer station or incinerator). One sample is collected per unit (such as each truck or group of trucks). A sample is typically collected by selecting and retrieving a full garbage bag from the truck load. If the load contains loose materials a pitch fork or shovel is also used. The number of samples to be collected is contingent on the number of loads per day that arrive at the facility Most facilities are limited to logging in one load at time creating a manageable waste stream.

A discrete sample typically consists of 10-15 kg of waste which are composited into a sample size of about 120 kg. Larger samples sizes become increasingly difficult to manage. As the selected truck unloads, its contents are visually examined and categorized as residential, commercial, industrial or construction wastes. The discrete sample that is collected should be proportioned to the approximate percentage of the various wastes being disposed. For example if the load consists of half household wastes and half office waste then, the sample should also be half and half.

The sampling methodology described herein is not suitable for bulk wastes (such as building debris or sewage sludge) but these materials can be accounted for though observation and weigh tickets.

CHARACTERIZATION

The waste samples are accumulated and composited for a predetermined period of time. At large or busy facilities the period may be as short as an hour. At smaller facilities the time period may be extended or determined by the volume of the waste stream (i.e. a set number of trucks arriving at the facility). During the selected collection period the collected wastes are composited on a tarp or in a truck bed. They are then manually mixed with a rake and shovel. The waste pile is quartered and two quarters are discarded. The remaining pile is quartered again and one of the quarters is bagged and labeled for transport to the laboratory. The other three quarters are used for onsite characterization.

The wastes are characterized according to composition of the material (wood, plastic, glass, etc.) with certain classifications being further divided (e.g. ferrous/ non-ferrous metal). Table 1 provides a breakdown of typical classifications.

TABLE 1 -- Typical Waste Classifications

Combustible	Non Combustible	Potentially Hazardous
Newspaper	Ferrous Metals	Batteries
Corrugated	Non-Ferrous Metals	Oil Filters
Cardboard	Glass	Aerosol Cans
Chip Board	Rock, Brick, Concrete	Paint
Other Paper	Soil, Sand,	Solvents
Wood	Fines (*)	Pesticides
Plastics (**)		Other Chemicals
Rubber		Compressed Gas
Food Wastes		Cylinders
Yard Wastes		Pressure Treated
Other Organic Wastes		Wood

(*) Fines consists of small particles of material (usually less than 2 cm diameter) which are relatively indistinguishable and may be combustible or non combustible

(**) Various grades or categories of plastics found in the wastes stream are difficult to distinguish in the field but may be retained for later characterization

The wastes are segregated according to the selected classifications and weighed accordingly. The weights are recorded and then the materials discarded.

The remaining quarter is transported to the laboratory for analyses.

The results of waste classification analyses have demonstrated the success of recycling efforts in Long Island, New York. The results have also illustrated the seasonal impact of yard wastes at disposal facilities.

PREPARATION FOR LABORATORY ANALYSES

At the laboratory the sample container is opened and the wastes visually examined. Depending on the size of the original pile this final quarter may need to be divided again to obtain a reasonably sized sample for analyses. A final sample size of about 10 kg can be reasonably handled by the laboratory. Potentially dangerous materials such as compressed gas cylinders, are segregated before further processing. Weights and composition are recorded and the material is retained as part of the sample.

ASTM Method E829 (Preparing Refuse Derived Fuels [RDF] Laboratory Samples For Analysis) is followed for preparation of the samples.

The laboratory samples are initially equivalent to RDF-1 fuels, in the "as discarded" form. The samples are weighed, and then placed in a low temperature, mechanical draft drying chamber at approximately 24° C for 24 hours (mixing at 12 hours intervals facilities the drying). After 24 hours the wastes are again weighed to determine "air dried" moisture content. The drying period may need to be extended based on the nature of the waste. To be assured that the wastes are thoroughly dried, the samples should remain in the oven until two consecutive weights are consistent (similar to the procedure described in ASTM Standard E 790).

The dried wastes are then reduced in size by shredding using a wood chipper. The dried, shredded wastes are equivalent to RDF-2 and RDF-3 fuels. RDF-2 fuels are wastes processed to a course particle size and RDF-3 fuels are the combustible waste fraction processed to particle sizes, 95% weight passing 2-in^2 screening.

Prior to shredding, bulk materials such as steel cans and bottles are removed to avoid damage to shredder and to prevent a safety hazard. The bulk materials are weighed and retained as part of the sample.

The shredding greatly increases the homogeneity of the sample. The wastes are more uniform in size (about 2 square cm) and can be more readily mixed and composited. The wastes may be passed through the shredder several times to create a uniform sample.

Following shedding and mixing, the sample is quartered and milled using a chopping mill such as a WILEY # 4 mill (A hammermill also works well for this application.   Several manufacturers produce suitable mills including FITZPATRICK COMPANY and MICROPUL CORP.) The WILEY # 4 mill consists of enclosed, rotating blades which effectively chop the sample into smaller pieces. A series of screens are used until the sample consists of particles 2 mm or less. A vacuum is applied to the mill to pull the milled sample through the mill screen resulting in a sample equivalent to an RDF-4 fuel (combustible waste fraction processed into powder form). ASTM Method E 829 prescribes a final sample size capable of passing through a 0.5 mm or smaller screen. However, experience has demonstrated that using a screen less than 2 mm results in clogging and prevents final processing of the sample.

The final sample is a moisture-stable, powder-like material and is suitable for most laboratory chemical analyses such as residual moisture, ash content, fixed carbon, heating value, and elemental elements (metals, carbon, hydrogen, nitrogen, sulfur, chlorine, and oxygen) or TCLP waste classification. Due to the heating and aeration, the sample is not suitable for determining volatile content or similar analyses.

RESULTS

The results of sampling and analyses provide the raw data necessary to classify the wastes and make the desired correlations. As with all activities involving the collection of data, a statistical determination needs to be made as to the quality of the data. The statistical analysis of the data is beyond the scope of this paper.

CONCLUSIONS

All sampling events are subject to limitations imposed by time and economics. Waste generation is subject to both seasonal and social variabilities. Yard and garden wastes increase during the growing season and community efforts affect the quantity of recyclable materials or hazardous wastes in the waste stream.

The cost of the on-site sample collection described above is relatively minimal, limited to labor costs and inexpensive equipment (scales). No drill rigs, backhoes or instrumentation are required and no special skills are involved. Sample preparation is more involved and requires specialized equipment such as a drying chamber and a laboratory mill. Laboratory analyses can be as extensive as desired or as allowed by a budget.

The simplicity of the sampling allows for collection by non-technical personnel with minimal training and the low costs allow numerous sampling events and long term collection of data. The results are useful for a wide range of evaluations including measuring the efficiency of recycling efforts, designing or monitoring mass-burn incinerators and evaluating the amount of waste material generated by a community or region. The sampling techniques and analyses can also be readily modified as needed based on the final application of the data.

REFERENCES

[1]   Horberger, G. S. and Walters R. A., Report of Solid Waste and Ash Testing at A Resource Recovery Facility, Recon Environmental Corp, Raritan NJ, 1994

John M. Mason[1] and William J. Gabriel[2]

USE OF PETROGRAPHIC ANALYSIS IN ENVIRONMENTAL SITE
INVESTIGATIONS: A CASE STUDY

REFERENCE: Mason, J. M. and Garbriel, W. J., **"Use of Petrographic Analysis in Environmental Site Investigations: A Case Study,"** "Sampling Environmental Media, ASTM STP 1282," James Howard Morgan, Ed., American Society for Testing and Materials, 1996.

ABSTRACT: Geologists conducting environmental site investigations in compliance with Federal and State regulations are often forced to rely on limited site and/or region specific geologic and hydrogeologic data. These sites are often located in industrial areas where surficial geologic features are obscured. The purpose of this paper is to describe the use of a simple, economical petrographic analysis method performed in conjunction with an environmental site investigation to better characterize subtle changes in subsurface geology. The method differentiates changes in rock type, classifies the rock samples, and provides detailed geologic descriptions for stratigraphic correlation. The results of the analyses were instrumental to understanding the site geology and ground water transport mechanisms.

KEYWORDS: petrographic analysis, classification, staining, geologic model, site investigation

---

[1]Project Hydrogeologist, Earth Sciences, O'Brien & Gere Engineers, Inc., 5000 Brittonfield Parkway, P.O. Box 4873, Syracuse, New York 13221.

[2]Managing Hydrogeologist, Earth Sciences, O'Brien & Gere Engineers, Inc., 5000 Brittonfield Parkway, P.O. Box 4873, Syracuse, New York 13221.

143

One of the objectives of site investigations is to characterize the subsurface geological conditions, evaluate ground water flow, and assess contaminant transport. The investigations generally require a detailed assessment of both the overburden and bedrock underlying the site and surrounding area. However, geologic characterization of bedrock can be difficult in areas where subtle changes in rock types occur with depth and where local marker beds are absent. In these areas, detailed descriptive techniques, such as petrographic analyses, may be useful tools to provide a better understanding of subsurface geology. This paper describes a basic and cost-effective method to perform petrographic analyses utilizing simple laboratory procedures to aid in the identification of subtle changes in carbonate and siliciclastic rock types. A case study where these methods proved useful is also presented.

BACKGROUND

A site investigation was conducted in the Salem physiographic province of central Missouri to evaluate regional and local geologic conditions, assess the ground water flow conditions, and characterize the ground water quality. Bedrock in the area consists of mixed siliciclastic and carbonate rocks of Ordovician age. In the vicinity of the site, caves, springs, faults, sinkholes, and other geologic features have been documented. Regional ground water flow across the study area is to the northeast.

The site is located in a small city with a population less than 10,000. Municipal drinking water wells are located within the city limits and one is situated north of the site. The municipal drinking water well north of the site is cased from the ground surface to approximately 106 m (350 ft) and is open borehole to approximately 259 m (850 ft). The city did not operated this municipal well during the course of the site investigation. Dual tube air rotary drilling techniques were used to advance bedrock boreholes to depths of up to approximately 165 m (550 ft) to characterize the subsurface geology. The use of the dual tube drilling method allowed the collection of representative rock cutting samples from the subsurface without extraneous rock cutting from the rest of the borehole. Rock samples from the borehole were collected at 1.5 m (5 ft) intervals and described in the field. During drilling, subtle changes in rock type were noted and were difficult to discern in the field. Upon the completion of the boreholes, ground water monitoring wells were installed at depths of approximately 46 m (150 ft) (shallow), which is the first encountered ground water zone, 99 m (325 ft) (intermediate), and 165 m (550 ft) (deep). The well locations and depths were selected to evaluate the horizontal and vertical ground water flow and ground water quality.

Ground water elevations were measured in the monitoring wells and used to generate ground water flow maps. Figures 1 and 2 illustrate the shallow and intermediate ground water flow conditions, respectively. The shallow ground water elevation map (Fig. 1) reveals a complicated ground water flow pattern in the area of the site. The northern flow component illustrated on Figure 1 is contrary to the regional flow direction. The intermediate ground water flow direction (Fig. 2) is, however, consistent with the regional flow direction. Analysis of the intermediate

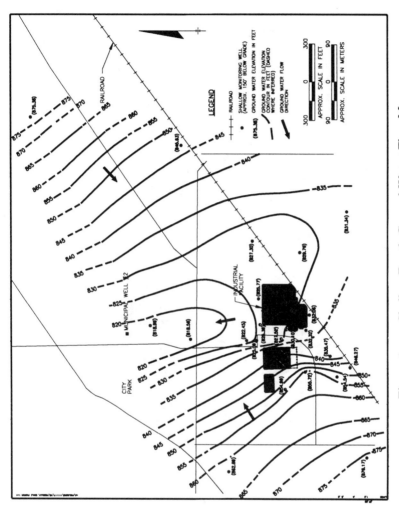

Figure 1   Shallow Depth Ground Water Flow Map

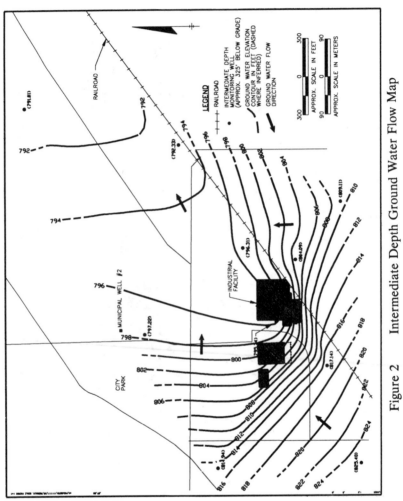

Figure 2   Intermediate Depth Ground Water Flow Map

ground water flow conditions indicates that the hydraulic gradient increases in the vicinity of the site.

To understand the complexity of the hydrogeologic conditions, an attempt was made to develop a comprehensive site geologic model. However, due to the subtle changes in rock types and lack of distinctive marker beds, the descriptive geologic logs prepared in the field were insufficient to provide the detail necessary to develop the model. Therefore, to provide the necessary detail and to avoid the costs associated with coring rock, a petrographic analytical method was devised to re-examine and describe the rock samples collected during drilling using simple analytical techniques. The method differentiates changes in rock types based on the texture and composition, classifies the rock samples, and provides detailed geologic descriptions for stratigraphic correlation. These analyses proved useful in identifying the subtle changes in rock types which, when correlated across the site, provided a basis for the development of the geologic model.

METHODS OF STUDY

During drilling, rock chip samples were collected at 1.5-m (5-ft) intervals from the bedrock boreholes. The collected samples were placed in labelled plastic bags and sent to the petrographic analysis laboratory. Once received, a portion of the sample was placed on a Petri dish, or like container, and air dried for approximately 1 hr. The sample was then examined with a petrographic microscope to identify the percentage of siliciclastic and carbonate material using visual estimation percentage charts [1]. The siliciclastic and carbonate components were described and classified separately. The methods described below provide the systematic approach used to classify the siliciclastic and carbonate components.

Siliciclastic Classification Methods

To describe the siliciclastic components of the sample, a binocular microscope was used with objective lenses of 1X and 4X and an ocular lens of 10X. The percentages of quartz, feldspar, and rock fragments were estimated using the visual estimation charts. The Wentworth scale [2] was used to describe the grain size of the components. Once these components were estimated, the siliciclastic portion of the rock sample was classified in accordance with the Folk system [3].

Carbonate Classification Methods

Carbonates were identified using a hardness test and reaction to a solution of 10% hydrochloric acid (HCl). Mohs scale of hardness, which ranks minerals from 1 to 10 with increasing hardness, was used to initially distinguish carbonates from siliciclastics. Carbonates typically have a hardness of less than 4 and siliciclastic minerals, such as quartz and feldspar, typically have a hardness greater than 5 [4]. To identify the carbonates further, staining techniques were used [5]. The stain used was mixed by first preparing a solution of 0.2% HCl by adding 2 mL of concentrated laboratory grade HCl to 998 mL of distilled water. One g of alizarin red-S and 5 g of potassium ferrocyanide were dissolved in 1 L of the 0.2% HCl solution. This stain will last approximately 1 d. The carbonate sample was immersed in a 2% HCl solution for approximately 20 s and then rinsed with distilled water. The carbonate sample was then immersed in the staining solution for approximately 4 min, rinsed with distilled water and allowed to air dry. Using this method, calcite stains pink with the color changing to purple or blue with increasing iron (Fe^{2+}) concentrations up to about 5%. Iron-rich dolomite stains turquoise and iron-poor dolomite does not stain. Subsequent to staining, the binocular microscope was used to assess the fabric and texture of the carbonate component, if any. However, due to the small size of most of the rock cuttings, the descriptions of the fabric and texture were limited. The carbonate sample was then classified in accordance with the Dunham system [6].

Systematic Description of Rock Types

The sample was systematically described, in descending order of percentage, indicating color, major grain size or rock type, classification of sample with noted reactions with stains and HCl solutions, and any other information concerning the sample. For example, a rock sample may be described as follows:

> 70% Brown, very fine to fine SANDSTONE (95% Quartz, 5% Feldspar {Quartzarenite}), subrounded grains
> 25% Light grey, CRYSTALLINE LIMESTONE (reacts with HCl, stains pink {Calcite})
> 5% Dark grey, CHERT, laminations noted

The rock sample descriptions were used to prepare a petrographic analysis log for selected bedrock boreholes. Once described, the samples were grouped into rock types based on composition. For example, the groups may be assembled as follows:

> Quartzarenite Sandstone
> Crystalline Limestone
> Laminated Chert

The rock types identified using these methods can be correlated to facilitate the preparation of site-wide stratigraphic cross-sections.

CASE STUDY

Six primary rock types were identified and described from the rock samples collected during installation of the bedrock boreholes and ground water monitoring wells at the site. These included siliciclastic sandstone, chalky dolomite, crystalline dolomite, calcareous sand, chert, and clay. The identified rock types were described as follows:

- Siliciclastic sandstone

The siliciclastic sandstone is medium brown to red brown in color and ranges from very fine to medium grain size. Individual grains are subrounded to subangular in appearance. The siliciclastic sandstone is composed of varying amounts of quartz, feldspar, and rock fragments and are classified from an arkose to a quartzarenite. Grains are cemented by either dolomite or quartz. Minor amounts of oolitic sand occur in some rock samples. The ooids have a porcelain surface texture and are generally spherical.

- Chalky dolomite

The chalky dolomite is white to tan in color with light brown staining and some micropitting. The dolomite appears to consist of silt size grains and is classified as a mudstone or wackestone. In a number of samples, staining revealed the presence of iron dolomite and minor amounts of calcite.

- Crystalline dolomite

Three distinct crystalline dolomitic rock types have been identified and separated based on crystal size and associated minerals. The crystalline dolomitic rock types include: a light brown to medium grey, fine to medium crystalline dolomite; a light grey, medium to coarsely crystalline dolomite with associated chert; and a brown to medium grey, fine to medium crystalline dolomite with associated quartz druse. Dolomite crystals are interlocking and typically do not stain, indicating a low iron content.

- Calcareous sand

The calcareous sand is medium to coarse grained and is light grey to light brown in color. The sand is composed of varying amounts of chert or quartz grains and calcite or dolomite grains and is classified as a litharenite. In rock samples, chert or quartz range from 25% to 95% of the sample whereas calcite or dolomite grains range from 5% to 75%. The grains are subangular to angular in appearance and are moderately spherical.

- Chert

The chert ranges in color from white to medium grey and is aphanitic. In a number of samples, the chert appears to have replaced the original rock type and preserved the fabric. Laminated and ripple cross-stratified fabrics have been noted in the chert.

- Clay

The clay is red brown in color and exhibits plastic behavior. Sand to cobble size fragments of slight to severely weathered dolomite also occur in the clay. Specific clay minerals were not identified.

To assess the lateral distribution of rock types across the site, a geologic cross section was prepared along the line shown in Figure 3. The major rock type for each 1.5-m (5-ft) interval was indicated on the graphical log.

Correlations of the rock types were performed to construct a site geologic model. During correlations, the contact between the fine to medium crystalline dolomite and the chalky dolomite indicates a vertical offset of approximately 23 m (75 ft) over a distance of approximately 426 m (1400 ft) between 8D and 14DD. This offset could be due to stratigraphic changes or structural influences, although a structural model was favored due to the magnitude of the offset. A structural (fault) geologic model (Fig. 4) was selected due to the presence of faults in the vicinity of the site and absence of folding in the area.

Relative changes in vertical and horizontal permeability, caused by the fault displacement, can be used to explain the complex shallow ground water flow directions and abrupt change in hydraulic gradient in the intermediate depth ground water zone. The presence of faults to the north and south of the site also support a structural (fault) geologic model.

VERIFICATION OF THE GEOLOGIC MODEL

In addition to analyzing geologic and ground water flow data, an aquifer performance test was conducted using a municipal well located approximately 275 m (900 ft) north of the site. The municipal water supply well was pumped at approximately 1136 L/min (300 gal/min) for 43 hours and drawdown measurements were collected in the intermediate and shallow ground water monitoring wells.

Based on the drawdown information collected from the intermediate ground water monitoring wells, a cone of influence map was prepared (Fig. 5). A review of Fig. 5 indicates a slight elliptical pattern around the pumping well which suggests an inhomogeneous aquifer and anisotropic flow. To assess the presence of anisotropic flow, a tensor diagram was constructed [7]. Transmissivity and storativity values calculated from the drawdown of the intermediate monitoring wells were used to calculate a directional transmissivity. The directional transmissivity was then plotted with the angular direction of the monitoring well from the pumping well. Isotropic flow will be represented on the resultant diagram as a circle. The tensor diagram prepared for the site (Fig. 6), however, was elliptical which indicates anisotropic flow. The long axis of the ellipse, indicating the preferential flow direction, is oriented northwest-southeast.

Although the pumped well was cased from the ground surface to approximately 106 m (350 ft), drawdown also occurred in the shallow (46 m or 150 ft) monitoring wells which indicates a connection between the shallow and intermediate zones. As

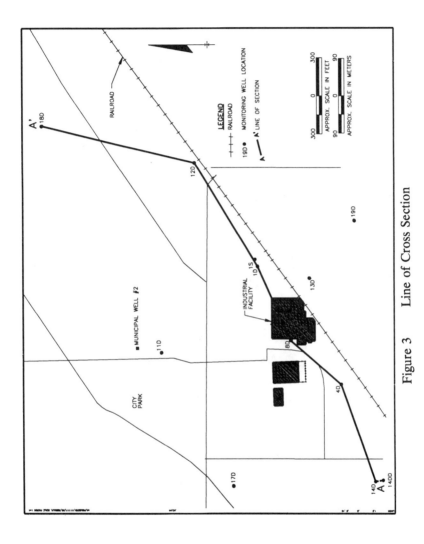

Figure 3   Line of Cross Section

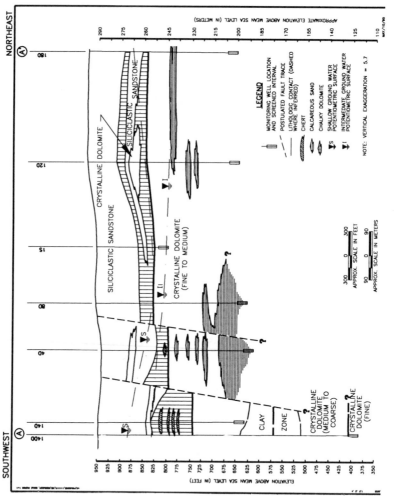

Figure 4   Cross Section A-A'

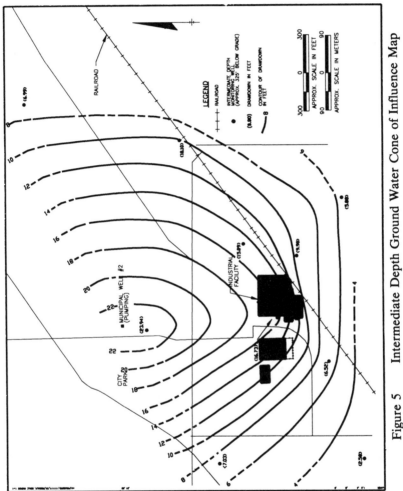

Figure 5   Intermediate Depth Ground Water Cone of Influence Map

illustrated on Figure 7, the drawdown in the shallow monitoring wells occurred in the vicinity of the interpreted faults. The northwest-southeast orientation of the drawdown in the shallow monitoring wells is consistent with the preferential flow direction in the intermediate zone.

These results support the interpretation of a structural (fault) model and suggest the orientation of the fault in the northwest-southeast direction which is consistent with faults mapped to the north and south of the site. The increased vertical permeability along the fault trace allows shallow ground water to recharge the intermediate zone and explains the north-south trending depression shown in Figure 1. Similarly, the change in horizontal permeability resulting from the vertical displacement of the fault can be used to explain the increase in hydraulic gradient noted in the intermediate zone in the vicinity of the fault.

CONCLUSION

Simple, economical petrographic analyses can be used to enhance geologic site investigations in areas where subtle changes in rock type are encountered and not easily identified using standard field methods. In the case study presented, petrographic analysis was instrumental in the development of a comprehensive site geological model. Additional information, such as aquifer performance test data, were used as supporting evidence for the developed geologic model. The geologic model was used to describe geologic and ground water flow conditions and can be used to assess potential contaminant flow pathways.

REFERENCES

[1]   Terry, R.D. and G.V. Chilinager 1989. "Comparison chart for estimating percentage composition", American Geological Institute, Data Sheet No. 23.1.
[2]   Wentworth, C.K. 1922. "A scale of grade and class terms for clastic sediments", Journal of Geology, Vol. 30, pp. 377-392.
[3]   Folk, R.L. 1974. Petrology of Sedimentary Rocks. Austin, Texas: Hemphill Publishing Company.
[4]   Dietrich, R.V. 1989. "Mineral hardness and specific gravity", American Geological Institute, Data Sheet No. 13.1.
[5]   Leeder, M.R. 1982. Sedimentology: Process and Product. London: George Allen & Unwin Ltd.

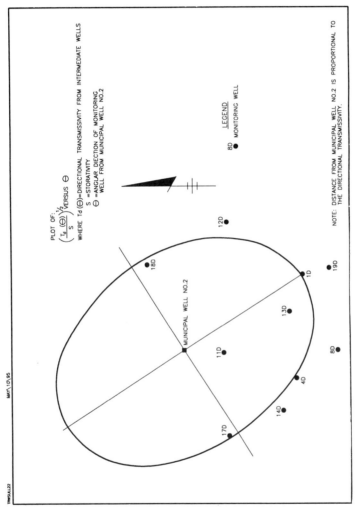

Figure 6   Tensor Diagram (Not to Scale)

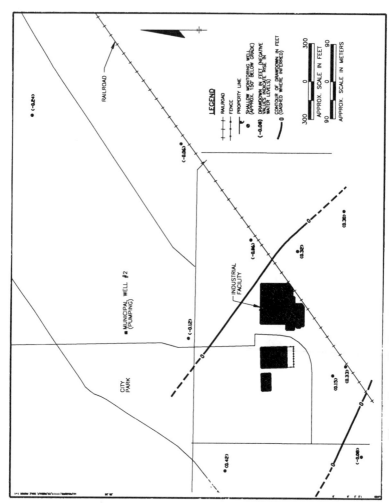

Figure 7   Shallow Depth Ground Water Cone of Influence Map

[6]   Dunham, R.J. 1962. "Classification of carbonate rocks according to depositional texture", American Association of Petroleum Geologists, Memoir No. 1, pp. 108-121.
[7]   Maslia, M.L. and Randolph, R.B. 1986. "Methods and computer program documentation for determining anisotropic transmissivity tensor components of two-dimensional ground-water flow", United States Geologic Survey Water-Supply Paper 2308, pp. 1-16.

Stuart A. Smith

MONITORING BIOFOULING IN SOURCE AND TREATED WATERS: STATUS
OF AVAILABLE METHODS AND RECOMMENDATION FOR STANDARD
GUIDE

REFERENCE: Smith, S. A., "**Monitoring Biofouling in Source and Treated Waters:
Status of Available Methods and Recommendation for Standard Guide,**" Sampling
Environmental Media, ASTM STP 1282, James Howard Morgan, Ed., American Society
for Testing and Materials, 1996.

ABSTRACT: Biofouling control in ground-water systems depends upon early detection
as part of a maintenance program. However, standard methods of analysis for biofouling
components have not provided reliable detection in operational use. Improved methods
have been useful in evaluating and predicting biofouling in ground water systems. Filtration and flowcell methods improve sample quality collection, enhancing the usefulness of
microscopy. Time-series sampling improves the usefulness of culture analysis of biofouling
waters. Prepackaged cultural methods provide practical on-site culture analysis. Used together, such methods provide early warning of biofouling, permitting timely judgments on
relative severity, type, and location of the problem. These improvements in methods are
being incorporated into standard practices. Although fully quantitative methods of assessing biofouling are not currently available, present methods can be useful to monitor
ground water systems for operational purposes. Method development is at the point where
a standard guide for biofouling monitoring in operational practice is possible.

KEYWORDS: biofouling, iron bacteria, microscopy, sample collection, ground-water
systems, cultural analysis

INTRODUCTION

Among applications of microbiological analysis, biofouling monitoring of groundwater systems is a case study of methods in transition attempting to meet user needs. In
contrast to available detection methods for microflora of concern to human health, convenient methods yielding information useful to formulate a response to biofouling have not
been readily available for operators of ground water systems. This   paper will review the
current status of methods for biofoul   monitoring, improvements that are underway, and

Microbiologist and hydrogeologist, S.A.Smith Group, Ada, OH 45810-0088 USA

prospects for the development of standard practices for routine maintenance monitoring of wells, water treatment systems, and associated pipelines.

The Operational Problem

It is now widely recognized that biofouling is the first- or second-most costly deteriorating factor for ground-water-source systems [1,2,3,4,5]. Symptoms of biofouling are clogging, corrosion, and alteration of water quality in pumped samples. Biofouling shares top trouble-making honors with abrasion and clogging by silt and sand in ground-water-source systems for water supply, long-term dewatering, and on-site treatment to remove contamination [1,4]. In fact, the two problems are frequently interactive [2,4].

In addition to clogging, and corrosion, the presence of biofouling is a significant factor in altering the quality of raw or treated ground water. For this reason, biofouling is significant to the representativeness of ground water samples from wells. Iron biofilms are known to efficiently remove Fe and Mn, as well as other metals, from solution, resulting in possible false negatives [5,6,7,8]. Our own experience is that (1) low total Fe and (2) total Fe:Mn ratios of samples near 1:1 or upset (Mn > Fe) correspond to microbially active alluvial aquifers with significant Fe biofouling in wells.

The Role of Biofouling Monitoring in Maintenance

Successful control of biofouling and associated changes in water quality, such as increased turbidity or alteration of metals content, depends very much on preventive or early warning monitoring. To be treated effectively, biofouling has to be detected as early as possible. This is especially the case in maintaining monitoring and pumping wells and associated systems for ground-water clean up, for which rehabilitation options are highly restricted [4].

Biofouling analyses ideally are one important component of a maintenance program for wells and associated systems [2,3,4,9]. The purpose of the maintenance monitoring program is the same as in ground-water monitoring: to detect changes in quality in time to perform some mitigating action.

At a minimum, a preventive-maintenance monitoring program should perform regular analyses to answer the following questions:

(1) Is a deteriorating condition (such as biofouling) occurring?

(2) When do we need to take some action, if any?

(3) What action(s) are most likely to be effective in controlling or mitigating the problem?

(4) Additionally, if control measures are employed, are they working?

For biofouling, this means information for making reasonable judgments on, for example, the potential for biofoulingthe type(s) of biofoul   occurring, its effects on the system, and how it can be controlled. Biofoul   monitoring is ideally combined with other types of analyses (hydraulic performance, physico-chemical water quality, physical examinations of equipment and deposits) [2,4,9].

BIOFOULING ANALYSIS: STANDARD METHODS AVAILABLE AND RELEVANCY IN MAINTENANCE MONITORING

Existing standard methods for analyzing aspects of biofouling are described in ASTM Test Method for Iron Bacteria in Water and Water-Formed Deposits (D 932) and Section 9240 - Iron and Sulfur Bacteria, *Standard Methods for the Examination of Water and Wastewater* (18th Ed.). Both D 932 and Section 9240 also provide sampling recommendations, but ASTM standards D 887 (Sampling Water Formed Deposits) and D 3370 (Practices for Sampling Waters) also provide guidance in biofouling sampling.

D 887 is more relevant for sampling to analyze deposits such as material clogging a pump. This discussion will focus on methods for preventive and diagnostic monitoring of the water phase (pumped and wellbore).

Review of Standard Methods

Microscopic examination and analysis--The presence of filamentous (e.g., *Leptothrix* or *Crenothrix*) or stalked (*Gallionella*) iron bacterial forms is accepted as a positive indicator that biofouling is present. Examination by light microscopy has traditionally been the method of choice for confirming and identifying these "iron bacteria" (e.g., ASTM D 932 and Section 9240).

However, the absence of such visible structures does not necessarily mean the absence of biofouling:

(1) Samples may not include enough recognizable materials to provide the basis for a diagnosis of biofouling.

(2) Samples examined may not include the filamentous or stalked bacteria normally searched for in such analyses. It is generally understood that the morphologically distinct types are only part of the biofouling present.

(3) It has always been difficult to correlate the results of testing for iron bacteria, for example, and the degree of deterioration of components of the systems they inhabit.

In addition, the existing standard microscopic tests (ASTM D 932 and Section 9240) are specified as being qualitative. Attempts have been made to quantify degrees of biofouling by microscopy. Mathematical models to calculate growth rates on surfaces are limited largely to types that do not form stalks or filaments [10]. Using microscopy to determine biofilm thickness has been proposed [11], but considered to be unreliable for Fe-Mn biofilms due to the presence of metal oxide inclusions and the variable length of attached stalks or filaments [12].

An available semiquantitative method using microscopy is described by Barbic et al. [13]. In this approach, numerous immersed microscopic slides are observed for signs of biofouling. An index is expressed based on (1) the number of slides that exhibit attached bacteria or particles and (2) the density of attachment. While providing a numerical index (that can be analyzed mathematically), the methodology for assigning an index value is by nature dependent on the judgment of the individual microscopist.

Culturing methods for detection--Media for Fe-precipitating bacteria have been used with mixed success. Standard methods cultivating media described (Section 9240) that could supplement microscopy may not support many microorganisms that cause fouling [9,12,14].

An additional problem is that no reported effort has been made to standardize these media with reference cultures from well water. Thus the efficiency of recovery of iron-precipitating bacteria from ground water samples remains unknown at present [12].

Cultural enrichment media for heterotrophic Fe- and Mn-precipitating bacteria presented in Section 9240 are seldom used among researchers working with Fe-biofouling problems. The environmental conditions for growth in these media do not seem to match up well with groundwater environmental conditions. The modified Wolfe's medium for *Gallionella* enrichment is more commonly used with varied results [12,15].

Useful Mn-precipitation media are virtually nonexistent in common practice [9] although they are being refined according to some researchers [16]. The available sulfur oxidizer media are still non-isolating, enrichment media. These weaknesses are all recognized by the *Standard Methods* Section 9240 joint technical group, and the JTG intends to improve and define these media formulations in future editions (beyond 19th Edition). ASTM does not have standard practices for enrichment of these microorganisms.

Employment of all of these culture methods require preparation from raw materials (no packaged media), sterilization, and maintenance in a microbiological laboratory by a skilled person.

Reasons for Limited Use of Biofouling Monitoring

In routine monitoring of the environment of any engineered system, sampling and analyses are conducted by busy technicians rather than research scientists. Methods used should

(1) Reliably provide results to the level of detection needed to make operational decisions.

(2) Be relatively easy to use.

(3) Precisly characterize any chemical deposits that may be a secondary maintenance consideration.

In the case of monitoring for biofouling indicators, experience shows that the available standard sampling and analytical tools for biofouling detection have not met these criteria for usefulness.

It is for these reasons that, while the value of early detection in controlling biofouling is widely recognized, and methods exist for detection of many forms of biofouling, maintenance monitoring methods have been slow in development and adoption in operational use. Consequently, biofoul   monitoring has only occasionally been incorporated into the routine practice of water operations and environmental laboratories.

RECENT DEVELOPMENTS IN BIOFOULING MONITORING METHODS

The period since the 1980s has seen accelerated progress in improving biofoul monitoring methods. Two issues have been addressed in recent developments: (1) sample collection to provide better samples and more information and (2) improved ease of use in operational setting.

Sampling Methods

Random pumped sampling--Pumped (grab) sampling (e.g., D 3370) is the easiest way to obtain samples for analysis from wells or sample taps in pipelines. However, if pumping fails to detach and suspend biofilm particles, they will not be available for collection and analysis.

The absence of iron bacterial structures in samples taken this way and examined using ASTM D 932 or Section 9240 procedures may simply mean that the bacteria remained undisturbed and attached, not that they are actually absent in the sampled system. A presumption of "probably absent" (per D 932) may be a false negative.

The reason for this is that after a period of sustained pumpling, biofilms will yield very litt of the turbid material usually necessary for microscopic examination. For this reason, analyses of samples taken after prolonged pumping may fail to detect the presence of biofilm components. If the sampling technician arrives at 10:00 a.m. and the pumps started at 6:30 a.m., biofilm particles shed in the early pumping would be downstream and not available for collection and analysis. In addition, bacteria captured in any single pumped water sample (and recognizable by microscopy) are likely to represent only a tiny fraction of the population and diversity of organisms that comprise the biofilm.

Filtration or centrifugation as recommended in standard methods increases the odds of recovery of material useful for microscopic identification. Improvements in the sampling further enhance the odds.

Time-series sampling--Cullimore [17] describes a time-series pumped-sampling procedure (similar to familiar procedures for ground-water quality analysis) that attempts to overcome the randomness of grab sampling. Cullimore's procedures involve taking advantage of the phenomenon that biofilm detachment occurs preferentially on start-up after a period of rest. The pump is shut down for 2 hours to several days to permit the biofilm to reestablish. Samples are taken:

(1) just prior to shut-down,
(2) immediately after restart,
(3) a few hours later, and optionally,
(4) some days later.

This approach, which includes taking replicates of samples at each sampling step, helps to overcome the statistical limitations of random pumped grab sampling for culture analysis. This information can be used in making an assessment of the microbial ecology in the well and the aquifer adjacent to the well [17]. In addition to the time steps, analysis of samples taken at various points in a ground-water-source system permits the development of a profile of the microbial ecology of the system.

Collection on slides or coupons--Grab samples, including those taken in time-series, remain unreliable for microscopic analysis. Large numbers of samples may need to be collected and processed. This is especially the case if monitoring is commenced prior to heavy biofouling build up.

Where indicators of biofouling are not abundant, some method is needed to provide enough sample to view or otherwise analyze mineralogically or chemically. Collection of biofilm on immersed surfaces can provide essentially intact biofilms for analysis. These methods are also adaptable for collection of samples of inorganic encrustations. Several experimental designs for collecting biofilm organisms have been presented in the literature since at least the 1920s. Recent methods and equipment for biofilm coupon collection are summarized in [9,18,19].

Among coupons available, glass slides have the advantages of being readily available, inexpensive, and noncorrodible. They can also be directly examined microscopically, or sampled for chemical or mineralogical analysis and for cultural recovery of microorganisms. Other coupon materials can be substituted for glass to provide more realistic surfaces to be examined by electron microscopy or evaluated for encrustation and corrosion potential.

Collection surfaces can be placed at various locations in the water system. For collection of biofilm samples from wells or pipelines, collectors may be placed directly in the well bore or in the water stream pumped from the well.

In-well or in-stream biofilm collectors are practical only if there is access to the well bore or pipeline itself, which is usually difficult. For example, the inserter devices refined by Smith and Tuovinen [19] was developed to permit better assurance of slide recovery from tight wells with submersible pumps or similar tight accesses.

Side-arm collection devices provide a means of collection even on systems which are only accessible by sampling ports. Among sidearm collectors, the flowcell slide collector system of [9] has been quite satisfactory in practical use. Water is permitted to trickle slowly past slides housed in a filled chamber, so that biofilms can develop on the slides. These devices have proven to be readily field-serviceable, as well as being adaptable for a variety of collection applications and use as laboratory models [12].

In-line filter collection--Filter cartridges, such as those used for collecting *Cryptosporidium* or *Giardia* or other particulates (microscopic particulate analysis, MPA) in a flowing water stream, or membrane filters may be substituted for the flowcell slide approach, as specified in D 932 for iron bacteria or in the U.S. Environmental Protection Agency (EPA) "consensus method" for MPA [20].

Another approach is that of Howsam and Tyrrel [21], in which a small trickle-through sand filter (moncell) is employed. Water is trickled through the moncell filter and biofouling material trapped starting at the in-flow end. Moncells are simpler than the flowcell, and readily renewable. These devices have also proven practical in operational use in England, and provide useful models of biofilm development in filters. Their value in realistically modeling the biofouling potential of individual wells has to be evaluated on a case by case basis according to the developers (P. Howsam, pers. comm.).

Sampling time periods--The question of how long to expose the slides, filter, moncell, etc. in the sampling environment is difficult to answer. D 932 provides for exposure of filters for 24 hours. Our experience is that this is likely to be too short for early detection of biofouling indicators, although long enough in systems with advanced biofouling. Based on the literature and experimentation [4,9], slide exposures to trickling flows (~1 L/min) of one to four weeks may be necessary to establish biofilms on slides. The exposure time should be determined based on experience with the specific system to be sampled.

Another approach is to link exposure to flow volume rather than time. The EPA MPA sampling process consists of passing a metered 1900-5700 L (500 to 1500 gal) through a 1 μm nominal pore-size filter [20,22]. Decisions on the volume to filter are based on experimentation (J. Clancy, pers. comm.). This sampling approach would permit a quantification of biofouling indicators such as filaments in the pumped water stream per a fixed volume such as 1 m^3.

Analytical Methods Improvements

Microscopy--Little has changed in microscopy to improve on ASTM D 932 or Section 9240 for the purpose of biofouling analysis. Normal light and simple wet mounts have proven to be sufficient for the identification of filamentous or stalked iron-, manganese-, and sulfur-related bacteria. One weakness of the existing microscopic standard methods is that the illustrations of microorganisms provided for comparison to those in samples can be misleading. Recent improvements in practice have been in sample collection as described previously, which provide more complete samples for analysis.

Hybrid methods--A subcategory of cultural methods are field-usable enrichment procedures to increase the potential for successful detection by microscopy. For example, Alcalde and coworkers [23,24] describe a simple enrichment and staining technique to enhance the numbers and visibility of filamentous bacteria on glass slides. Cullimore [17] provides a review of several such methods, including the "GAQC" method in which corn whiskey is the carbon source and a washer the source of iron. While easy to use, all these methods are

(1) Presence-absence only
(2) Not likely to represent the environment of growth in the system
(3) Not likely to provide information on the full scope of the existing biofouling.

Pre-packaged cultural methods (1): BART method--Currently, the most commonly used cultural approaches for routine biofouling monitoring purposes is the Biological Activity Reaction Test (BART™) Method (Droycon Bioconcepts Inc., Regina, Saskatchewan). BART tubes contain dehydrated media formulations and a floating barrier device, which is a ball that floats on the hydrated medium of the sample. These devices and their proposed use are described in detail in [17]. They can be used as an enrichment method to provide a presence-absence (P-A) or semiquantitative (MPN) detection of bio-fouling factors [4,9,17,25].

BART tubes are available in a variety of media mixtures. The IRB-BART™ test, for example, is designed to recover microaerophilic heterotrophic Fe- and Mn-precipitating microorganisms. The exact formulations are proprietary at present, but derived from described formulations. The IRB-BART medium is derived from the W-R iron bacteria medium [9], for example.

This method, which is gaining wide operational acceptance as a means of detecting and characterizing biofouling symptoms, was found to provide useful qualitative information in well biofouling events in field trials [9].

To use the BART, the person conducting testing simply opens the sterile outer tube, fills it with a water sample, and fills the inner tube (containing dried media) to the marked line (about 15 mL). The cap of the tube is replaced, the tube is labeled, and the sample is permitted to incubate at ambient temperature until a reaction occurs. Interpretation is based on the:

(1) Changes in the appearance of the inoculated tubes (both at initial reaction and as reactions change over time), and

(2) days of delay (d.d.) or time (usually days) until a noticeable reaction (or change in previous appearance) occurs.

Interpretation depends on the appearance of the tubes when reactions occur. Various types of appearance are assigned numbers, which are read from a chart. For example, the IRB-BART exhibits approximately 10 reaction types. While analyst judgment is necessary for interpretation, reactions types for samples taken from known systems over time vary only slightly [9]. When the reaction type is combined with the d.d., a two-number code is produced. For example, a code of 4.5 indicates a reaction type 4 that occurred 5 days after inoculation.

These data (d.d. and reaction type) are compared to charts that aid in interpretation of the results [17] in terms of

(1) "Aggressivity" of the biological activity (based on d.d. and pegged to empirical experience of the developers [17]). Rapid reaction times indicate more aggressive microbial activity.

(2) Microbial physiological types present (based on reaction type). For example, for the purpose of interpreting microbial ecology, IRB-BART reaction type numbers may be grouped (D.R. Cullimore, pers. comm.) into four groups: A = reactions 1, 2, 3, 4; B = reaction 5; C = reactions 6, 7; and D = reactions 8, 9, 10.

• The Group A reactions are produced by miscellaneous nonfilamentous heterotrophic Fe-precipitating bacteria. These are bacteria are recoverable with the modified W-R medium [9] and precipitate Fe.

• Group B is associated with anaerobic gas-producing bacteria.

• Group C reactions are associated with *Klebsiella* and *Enterobacter*.

• Group D reactions are associated with facultatively anaerobic pseudomonads.

· In field tests [9], filamentous IRB were absent from reacted IRB-BART tubes, except for rare fragments that may have been captured during sampling.

Changes occurring subsequent to the first reaction provide information on microflora likely to be present in the sample. This information can be used by an analyst familiar with the subtleties of the method interpretation to assess the clogging, corrosive, and biotransformation reactions that can be expected in the system. A software package (BART-SOFT) is now available from the manufacturer to aid in the interpretation

Pre-packaged cultural methods (2): MAG method--A similar system has been developed independently by M.A. Gariboglio, La Plata, Argentina [24,26]. These "MAG" tests for heterotrophic iron-related bacteria (IRB) and sulfate-reducing bacteria (SRB) consist of a prepared liquid medium contained in small septum bottles such as those used for packaging injection serum. The MAG medium for iron-related heterotrophic bacteria (BPOM-MAG) utilizes ferric ammonium citrate (like W-R and R2A+FAC [9]) while the SRB medium (BRS-MAG) utilizes Postgate C medium under a reducing atmosphere, supplemented with iron filings [26].

A 10-mL sample is taken using a syringe and injected through the septum into the medium and permitted to incubate, as with the BART methods. Inoculation of a single bottle provides a presence-absence (P-A) result. Dilution to extinction provides a semiquantitative (MPN) result [26]. No further interpretation of MAG reactions parallel to the BART microbial ecology interpretations have been put forward by the developer.

In small-scale tests now under way, BRS-MAG methods provide P-A results that seem comparable to SRB-BART results, but further testing is necessary to make definite conclusions about their comparability. Comparability between BPOM-MAG and IRB-BART methods is unknown at present.

RECOMMENDATIONS FOR USE OF METHODS IMPROVEMENTS IN OPERATIONAL PRACTICE

Field and laboratory studies [9,12] support both previous work suggesting the practical use of BART or MAG methods [17,24,25,26] and the value of improvements in sampling methods for microscopic analysis in maintenance monitoring and diagnosis. These conclusions are being further supported in practical use that is not yet well documented in public literature.

Recommendations for Choices

Analytical choice considerations--All the available analytical methods have their limitations. Fortunately, the strengths and weakness of microscopy and cultural enrichment appear to be complementary [9], so are best used in tandem:

(1) Cultural methods reveal the nonfilamentous biofouling component not easily identified under the microscope, however,

(2) Microscopy remains an essential part of a complete biofouling monitoring system because it is the only way to identify filamentous and stalked bacteria (that are difficult to culture).

The best use of BART or MAG methods is as quick and easy indications of:

(1) Early detection or confirmation of probable biofouling and microbially induced corrosion (MIC) conditions,
(2) The presence of viable IRB, SRB, and other types of bacteria (such as pseudomonads), and
(3) Relative "aggressivity" or activity of the biofouling and corrosion-inducing communities.

Neither BART nor MAG methods seem to provide enrichment for microscopically identifiable *Gallionella* or filamentous forms [9]. Enrichment for *Gallionella* continues to require modified standard methods [12,15]. Alcalde and Castronovo de Knott [23] do described successful enrichment for *Sphaerotilus-Leptothrix* forms on slides immersed in an undefined alfalfa-infusion medium.

As with *Standard Methods* Section 9240 heterotrophic formulations, BART and MAG media are enrichment (rather than isolating) media, and are not precise research tools. However, keeping in mind their limitations, the value of the BART or MAG tubes over previously published procedures is that they can be unpacked and used outside the laboratory by any diligent person such as a water treatment technician. This ease-of-use encourages rather than discourages use. Because they are used, more information can be collected. This is their primary benefit as monitoring tools.

Both MAG and BART methods apparently provide a more realistic environment of growth when compared to agar plates [9,17,26]. Both tube designs make provision for sample isolation from oxidation in the incubating environment and provide a redox-potential gradient in a liquid column.

Sampling refinements--Sampling methods are closely linked to analytical methods: Pumped sampling is used for cultural enrichment (MAG or BART), while collection on surfaces is preferred for microscopy samples [9,20,21,22]. Both are practical in the operational setting [4,9]. As with analytical methods, pumped and surface-collection methods are best used in tandem [9].

(1) Time-series sampling provides the capacity to determine the location of types of biofouling (e.g., on the pump, in the filter pack, or deep in the aquifer) and the relative biological activity intensity (aggressivity) in these zones based on the appearance and rates of reaction in samples taken at specific times since pumping began [17].
(2) Flowcell or in-line filtration sampling provides
   · Capacity to collect identifiable biofilm structures at low levels
   · More reliable recovery of biofilm structures
   · More or less intact biofilm samples for analysis
   · Potential for quantification of biofilm particles per volume.

Practical Issues in Maintenance Monitoring Implementation and Interpretation

Implementation strategy--Maintenance monitoring of biofouling must be implemented as part of a systematic maintenance program involving (1) institutional commitment, (2) a goal of deterioration prevention, (3) systematic monitoring as part of site maintenance procedures, and (4) method evaluation. Selection of methods normally requires some expert judgment (how much depends on the experience of the system operator). The value of monitoring, and responses to the information provided, will be site specific, and likely require adjustment based on experience.

Interpretation--There is a sharply sloping learning curve with all biofouling monitoring methods, which results in evolving recommendations [4,9,17,25]. For example, early attempts were made to link BART days-of-day values to colony-forming units (plate counts) [17,25]. This relationship has been difficult to support [9] and has evolved to become an index of aggressivity in [17] and current BART product literature.

(1) It is important to note that while BART d.d. values are no longer tied to an artificial comparison such as CFU/mL, neither is there, at present, any independent standard to define "aggressive." However, there is sufficient "field correlation" between BART or MAG reactions (especially SRB-BART or BRS-MAG) and well deterioration [4] to take these results seriously in maintenance planning. If reactions are rapid, deterioration (especially corrosion) tends to be rapid.

(2) Likewise, while exposed flowcell slides provide very good samples of iron- and manganese-related biofilms [9,12,18], it is very difficult to make judgments on biofouling rates based on biofilm development on slides [12]. A semi-quantitative index (analogous to BART aggressivity) that relates slide appearance (or other characteristics) to degrees of biofouling would be necessary to relate slide changes to changes in biofouling conditions. Slide indexing has been in fact employed for this purpose for over 20 years by Barbic et al. in monitoring biofouling in the Belgrade, Serbia, wellfield [13,27]. Such an index is likely to be inherently site-specific, however.

(3) All collection methods may have environments that differ from (1) the point of raw water inflow to the system (e.g. well screen) or (2) the sampled pipeline. The degree and importance of any such environmental alternation should be a part of the planning process for a maintenance monitoring program.

CONSIDERATION OF METHODS CAPABILITIES IN STANDARDS DEVELOPMENTS

It has been demonstrated that indicators of biofouling can be detected using available methods that are suitable for use by trained water system operators under operational conditions (not only in the academic microbiological laboratory). Wide application of these recently refined methods will require that operators and managers of water production and treatment systems accept that improved methods will improve their operations.

Standardization of practice has historically provided the seal of approval that speeds implementation of new methods.

Current Status of Biofouling Monitoring Standardization

Standard Methods--The 18th Edition (1992) of Section 9240 is unchanged from previous editions. The 19th Edition will include the flowcell sample collection procedure described in [9] and improved illustrations of the common iron bacteria. A standard procedure for the now widely accepted BART methods was not available in time for inclusion in the 19th Edition, but will be considered for the 19th Edition Supplement and beyond.

ASTM--Relevant ASTM standard practices are essentially unchanged since the mid 1980s, but are due for review. Inclusion of improvements balloted for inclusion in Section 9240 (19th Edition), as well as simple field and pre-packaged cultural approaches such as BART and MAG, would be possible and recommended.

Issues in the Acceptability of New Methods

Widespread use--The use of prepackaged enrichment methods (BART and MAG) is a good example of a widespread, accepted practice in search of a general standard method. It appears that ground-water system operators find the tests attractive and are using them, regardless of any official standard approval or disapproval. Consequently, in North America, the recommended sampling and analytical methods of the BART vendors (which are not entirely consistent) have therefore become de facto standard methods in the operational community.

Recommendations for use of BART methods are provided by the developer [17] and separately in [9], based on work up to late 1991. The large-scale acceptance of their usefulness, and considerable refinement in practice, have occurred since. The use of the MAG tests [26] is at present comparatively limited, more for market and communication reasons than for technical reasons, but important in South America.

Collection of biofouling organisms on slides and filters has a long but limited history. The practice dovetails more easily with existing microscopic identification practice, but the limited commercial availability of collection apparatus has been a limiting factor in acceptance in maintenance practice.

Independent verification--All the newer methods have limited independent testing histories. The only existing systematic study of their use in practice [9] is over four years old. Issues such as health risk are not so immediate as for improved coliform or MPA methods, for example, so the pressure for additional research has not been as great.

BART and MAG tests, with their complex, serial reactions and days-to-reaction interpretations, are at best semi-quantitative [9,26]. Fundamental questions about their interpretation await definitive comparative studies:

(1) It may be very difficult or take a long time to establish a definitive, reproducible relationship between particular reaction times and types and particular biofouling

conditions or rates (aggressivity), because of the variables involved in deterioration of any engineered system. Ground-water-source systems have so many variables that requirements for some standard of reproducibility may be practically impossible to meet.

(2) It is not well-established whether or not certain reactions are actually a result of microorganisms activity. However, this objection seems academic: As with evolution plate tectonics, the nearly universal influence of microorganisms on clogging, corrosion, and water quality alteration in ground-water-source systems is now accepted based on the preponderance of evidence [1-9, 13, 15, 17, 26-28].

Consistency in usage--Several factors may limit the establishment of universally verifiable standard sampling and analytical practices:

(1) There is the above-mentioned degree of variables in systems.

(2) Available maintenance monitoring methods require knowledge and organization to provide useful information over time. Personnel must be trained in the use of these methods, and implementation may require some expert guidance for awhile.

(3) The current methods all require subjective interpretations of factors such as the appearance of a test. As with portable "color-wheel" chemical test kits, for example, interpretation depends on the interpretation of analysts, whose opinions may vary.

For these reasons, results would be expected to vary, even for identical, standardized methods used in theoretical identical operational environments.

DEVELOPING A STANDARD GUIDE FOR OPERATIONAL BIOFOULING MONITORING

Proposal for Scope

Frameworks for a scope of a standard guide are available in [4,9]. Standardization of practice would improve the current haphazard vendor/expert-defined situation in the implementation of biofouling monitoring in operational settings. However, a rapidly improving state-of-art (further elaborated in [2,4,9,17,26-28]) and a need for local flexibility argue against a rigid standard. In the ASTM culture, the history of the development of standards for ground-water monitoring and environmental investigations provides useful experience for the development of the scope of a standard guide to operational biofouling monitoring.

As in these environmental applications, the envisioned standard guide should focus more on what information should be gathered and what to do with the information, rather than defining exactly how to do the analyses. The following are recommended components of a proposed standard guide to biofoul   monitoring in ground-water systems.

Minimum elements--The guide should provide a list of minimum elements of a biofoul.   monitoring program, including:

(1) A combination of microscopy and cultural methods in analysis (together providing information not available from one approach used alone),

(2) Sampling methods that maximize recovery and information, and

(3) A good quality records program., derived from existing ASTM standards.

Method flexibility and inclusiveness--The standard guide should have maximum flexibility and inclusiveness. By not singling out one specific technology or procedure, the standard would have more worldwide appeal, does not boost one vendor over another, and permits innovation.

(1) It should permit the use of several valid cultural analytical methods, for example, MAG, BART, or locally formulated heterotrophic methods that best fit local needs. Ideally, carbon sources in cultural formulations should bear some resemblance to C sources available in the local ground water.

(2) Alternatively, monitoring based on microscopy alone should be an option for those systems that do not have access to MAG or BART methods, or that otherwise find BART interpretation unsuitable.

(3) The guide should also be flexible in biofilm collection methods for sampling to permit local choice in practice. For example, both flowcells [9] and sand-filter moncells [22] have advantages. At the same time, existing random sampling procedures such as ASTM D 3977 remain valid procedures and need to be included.

(4) The guide should provide alternatives for semiquantitative analyses such as the Belgrade indices for microscopic evaluation of biofouling severity [13], MPN using BART or MAG bottles or BART "aggressivity" (at some point defined in terms of systems impacts, for example within a study such as [28]).

A table illustrating the benefits and limitations of various methods would be helpful. In this way, the standard would provide helpful guidance to people selecting methods.

Documentation--Flexibility and evolution in methodology require that valid analysis over time will depend on a high standard of data recording and method documentation.

(1) In any report of analyses, sampling and analytical methods must be carefully documented to permit comparison to the results of future analyses using (presumably) improved methods.

(2) The key records system component of the standard practice should also be flexible, but provide for collection of a variety of data (including physico-chemical), and permit ready data analysis. Detailed records are expected to be crucial to historical analysis over time as sampling and analytical methods improve.

Refining elements of the scope--Subsequent standards (in addition to the existing ASTM D 887, D 932, D 3370 and other relevant standards) can define specific components of the monitoring program as improvements in methods, technology, and experience suggest.

Standards Training

With the development of a standard practice for biofoul   monitoring, training would be expected to be an important component of its acceptance. This would be analogous to the promotion of ASTM standards acceptance in ground water monitoring and environmental assessment through training.

Also analogous to other environmental methods, personal judgment is an important factor in the choice of biofouling monitoring methods, their implementation, and the interpretation of results. While advanced academic education is specifically not required to use modern biofouling monitoring methods, a high degree of specific knowledge and experience greatly improves their usefulness in maintenance.

Research to Improve the Proposed Standard

The usefulness of the proposed standard would improve with further information and experience from studies (expanding on [9,27,28]) such as the following:

(1) Comparability of methods in defined and known (model) systems for early warning detection and defining biofouling processes.

(2) Correlation between factors monitored by the described methods and actual rates of deterioration such as filter clogging, incrustation build up, corrosion, or sample alteration. At present, expert judgment is important in making decisions on these factors. Analogous to the requirements of hydrogeologic or engineering analysis, this need for expert opinion may not change, but assessments could be more rapid and precise.

(3) Relationships between the presence of various types of biofouling and alteration of water quality.

(4) Once adopted, evaluation of the use of the proposed standard practice in various operational settings to identify areas of needed improvement.

Conclusions and Prospects

The development of an ASTM standard guide to biofouling monitoring for ground-water systems is now possible. It would be beneficial in increasing the acceptance of these monitoring methods, demonstrated to improve the maintenance of ground-water-source systems.

The current Committee D 18 process leading to a development of an ASTM maintenance standard guide for monitoring wells (proposed as an aid to maintaining reliable sample quality) illustrates the need for early warning maintenance monitoring for biofouling. Rehabilitative maintenance options for these well systems are very limited, but lack of maintenance can result in sample variability. In developing the maintenance standard, it is recognized that use of sensitive monitoring methods permits the choice and use of minimal-impact treatments.

Many planners, managers, and operators of ground-water-source systems recognize this already. The adoption and use of biofouling detection methods in maintenance monitoring of ground-water systems is going to proceed with or without an ASTM standard, relying on existing literature and expert advice. The situation resembles ground-water quality monitoring or the personal computer market place in the mid-1980s. The alternative to an ASTM standard practice is continuation of the current nonsystematic vendor- and expert-driven situation, or establishment of a de facto vendor-driven standard without ASTM influence.

REFERENCES

[1]   Smith, S.A. "Well Maintenance and Rehabilitation in North America: An Over-view," in Water Wells Monitoring, Maintenance, and Rehabilitation, P. Howsam, ed., E.&F.N. Spon, London, 1990, pp. 8-16.

[2]   Borch, M.A., Smith, S.A., and Noble, L.N., Evaluation, Maintenance, and Resto-ration of Water Supply Wells, American Water Works Association Research Foun-dation, Denver, CO, 1993.

[3]   Clancy, J. and Smith, S.A., "Iron Bacteria," Problem Organisms in Water: Identifi-cation and Treatment, Chapter 2, American Water Works Association, Denver, CO, 1995, pp. 7-17.

[4]   Smith, S.A., Monitoring and Remediation Wells: Problems Prevention and Cures, CRC Lewis Publishers, Boca Raton, FL, 1995.

[5]   Chapelle, F.H. Ground-Water Microbiology and Geochemistry, John WIley & Sons, New York, 1993.

[6]   Vuorinen, A., and Carlson, L., "Special of Heavy Metals in Finnish Lake Ores; Se-lective Extraction Analysis," International Journal of Environmental Analytical Chemistry vol. 20, 1985, pp. 179-186.

[7]   Mouchet, P., "From Conventional to Biological Iron Removal in France," Jour. AWWA vol. 84, 1992, pp. 158-167 .

[8]   Van Riemsdijk, W.H., and Hiemstra, T., "Adsorption to Heterogeneous Surfaces," Metals in Groundwater, H.E. Allen, E.M. Perdue, D.S. Brown, eds., CRC Lewis Publishers, Boca Raton, FL, 1993, pp. 1-36.

[9]   Smith, S.A., Methods for Monitoring Iron and Manganese Biofouling in Water Supply Wells, AWWA Research Foundation, Denver, CO, 1992.

[10]   Caldwell, D.E., Brannan, D.K., Morris, M.E., and Betlach, M.R.,   "Quantifica-tion of Microbial Growth on Surfaces," Microbial Ecology vol. 7, 1981, pp. 1-11.

[11]   Bakke, R., and Olsson, P.O., "Biofilm Thickness Measurement by Light Micros-

copy," Journal of Microbiological Methods vol. 5, 1986, pp. 93-98.

[12]   Tuhela, L., Smith, S.A., and Tuovinen, O.H., "Flow-Cell Apparatus for Monitoring Iron Biofouling in Water Wells," Ground Water, vol. 31, 1993, pp. 982-988.

[13]   Barbic, F., Bracilovic, D., Djorelijevski, S., Zivkovic, J., and Krajincanic, B., "Iron and Manganese Bacteria in Ranney Wells," Water Research vol. 8, 1974, pp. 895-898.

[14]   Smith, S.A. "Field Reports - Culture Methods for the Enumeration of Iron Bacteria from Water Well Samples - A Critical Literature Review," Ground Water vol. 20, 1982, pp. 482-485.

[15]   Hallbeck, L., "On the Biology of Iron-Oxidizing and Stalk-Forming Bacterium *Gallionella ferruginea*," Department of General and Marine Microbiology, University of Göteborg, Göteborg, Sweden, 1993.

[16]   Emerson, D., and Ghiorse, W.C., "Isolation, Cultural Maintenance, and Taxonomy of a Sheath-Forming Strain of *Leptothrix discophora* and Characterization of Manganese-Oxidizing Activity Associated with the Sheath," Applied and Environmental Microbiology, vol. 58, 1992, pp. 4001-4010.

[17]   Cullimore, D.R., Practical Ground Water Microbiology, CRC Lewis Publishers, Boca Raton, FL, 1993.

[18]   Hallbeck, E-L., and Pedersen, K., "The Biology of *Gallionella*," Proceedings of the International Symposium on Biofouled Aquifers: Prevention and Restoration, D.R. Cullimore, ed., American Water Resources Association, Bethesda, MD, 1986, pp. 87-95.

[19]   Smith, S.A., and Tuovinen, O.H., "Biofouling Monitoring Methods for Preventive Maintenance of Water Wells," Water Wells Monitoring, Maintenance, and Rehabilitation, P. Howsam, ed., E.&F.N.Spon, London, 1990, pp. 75-81.

[20]   U.S. EPA, 1992. Consensus Method for Determining Groundwaters Under the Direct Influence of Surface Water Using Microscopic Particulate Analysis. Office of Drinking Water, Washington, DC.

[21]   Howsam, P., and Tyrrel, S., "Diagnosis and Monitoring of Biofouling in Enclosed Flow Systems - Experience in Groundwater systems," Biofouling, vol. 1, 1989, pp. 343-351.

[22]   Gollnitz, W.D., Clancy, J.L., Garner, S.C., "Natural Reduction of Microscopic Particulates in an Alluvial Aquifer," Proc. 2nd International Conference on Ground

Water Ecology, American Water Resources Assn., Herndon, VA, 1994, pp. 127-136.

[23]   Alcalde, R.E., and Castronovo de Knott, E., "Occurrence of Iron Bacteria in Wells in Rio Negro (Argentina)," Proceedings of the International Symposium on Bio-fouled Aquifers: Prevention and Restoration, D.R. Cullimore, ed., American Water Resources Association, Bethesda, MD, 1986, pp. 127-136.

[24]   Alcalde, R.E., and Gariboglio, M.A., "Biofouling in Sierra Colorado Water Supply: A Case Study," Microbiology in Civil Engineering, FEMS Symposium No. 59, P. Howsam, ed., E.&F.N. Spon, London, 1990, pp. 183-191.

[25]   Mansuy, N., Nuzman, C., and Cullimore, D.R., "Well Problem Identification and Its Importance in Well Rehabilitation," Water Wells Monitoring, Maintenance, and Rehabilitation, P. Howsam, ed. E.&F.N. Spon, London, 1990, pp. 87-99.

[26]   Gariboglio, M.A., and Smith, S.A., Corrosión e encrustación microbiológica en sistemas de captación y conducción de agua - Aspectos teóricos y aplicados, Consejo Federal de Inversiones, San Martin, C.F., Argentina, 1993.

[27]   Barbic, F., Krajcic, O., and Savic, I., "Complexity of Causes of Well Decrease," Microbiology in Civil Engineering, FEMS Symposium No. 59, P. Howsam, ed., E.&F.N. Spon, London, 1990, pp. 198-208.

[28]   McLaughlan, R.G., Knight, M.J., Stuetz, R.M., Fouling and Corrosion of Ground-water Wells: A Research Study, R.P. 1/93, National Centre for Groundwater Management, University of Technology, Sydney, Australia, 1993.

Sampling Subsurface Media

Curtis A. Kraemer[1], William E. Penn[2], and Mark D. Busa[2]

USE OF THE HYDROPUNCH[3] FOR GROUNDWATER PLUME DELINEATION: A CASE STUDY

REFERENCE: Kraemer, C. A., Penn, W. E., and Busa, M. D., ''Use of the HydroPunch for Plume Delineation: A Case Study,'' Sampling Environmental Media, ASTM STP 1282, James Howard Morgan, Ed., American Society for Testing and Materials, 1996.

ABSTRACT: Historic groundwater investigations at a 300-acre manufacturing facility in Connecticut indicated that solvent and metal contaminated groundwater was migrating off site. To preliminarily investigate off-site groundwater contamination, the collection of discrete-interval groundwater samples using a HydroPunch was attempted in twenty soil borings. The borings were located along three transects perpendicular to the general direction of groundwater flow: 1) three soil borings were located near groundwater monitoring well clusters along the downgradient facility boundary, 2) ten soil borings were located along a road at the approximate midpoint between the facility boundary and a nearby river, and 3) seven soil borings were located along the banks of the river. At each soil boring, collection of the first HydroPunch sample was attempted at ten feet below the water table; sample attempts continued at 3-meter (10-ft) intervals until refusal. Samples were analyzed for volatile organic compounds and selected metals.

Geologic materials significantly influenced the sampling program success. In the fine-grained, floodplain deposits, samples could not be collected from five of the seven soil borings, with limited sampling success in the other two borings. At the two other transects, the collected sample volumes varied significantly, dependent upon the physical characteristics of the sampled materials. Analytical results from the HydroPunch samples collected along the middle transect clearly defined the horizontal extent of solvent contamination. The vertical extent of groundwater contamination was not as well characterized, because of the anisotropic nature of the aquifer materials.

KEYWORDS: HydroPunch, groundwater contamination, volatile organic compound contamination, chromium contamination, plume delineation.

[1]Principal Hydrogeologist, ICF Kaiser Engineers, Inc., 1781 Highland Avenue, Cheshire, CT 06410.

[2]Hydrogeologist and Field Geologist, respectively, United Technologies Corporation, Hamilton Standard Division, 1 Hamilton Road, M/S 2A-GG43, Windsor Locks, CT 06096.

[3]HydroPunch is a registered trademark of QED Environmental Systems, Inc., Ann Arbor, MI.

INTRODUCTION

A 300-acre aerospace manufacturing facility in Connecticut is an interim status RCRA hazardous waste storage facility. The facility entered into a Consent Agreement with the United States Environmental Protection Agency to perform a RCRA Facility Investigation (RFI) and screen remedial alternatives applicable to potential corrective action at the site. The investigation phase was designed to identify the nature, extent, and rate of releases, if any, from Solid Waste Management Units or Areas of Concern.

Investigations of groundwater hydraulics and chemical quality to date have included monitoring and sampling of the groundwater using wells screened in specific horizontal and vertical locations across the site. Results of site-wide groundwater elevation data have shown that groundwater generally flows from north to south. The water table depth across the site varies from approximately 17 to 2 m (56 to 6 ft) below the ground surface, generally from north to the south. Groundwater contaminants are primarily volatile organic compounds (VOCs) and chromium.

Groundwater sampling from a skeletal monitoring well network in 1982 defined a groundwater contaminant plume extending south of the facility, beneath residential property, toward a major river. As part of the RFI groundwater investigation in 1990, a more substantial monitoring well network was installed at the facility. South of the facility (in and around the residential property), installation of monitoring wells would have been premature based on the limited understanding of groundwater contaminant migration in this area. Therefore, to collect data to guide future investigations in the off-site area, discrete interval groundwater sampling was undertaken using the HydroPunch. The HydroPunch data assisted in the characterization of groundwater hydraulics and quality in that 1) groundwater samples were collected for laboratory analysis at specified depths, 2) logged drill cuttings were correlated with identified stratigraphic units beneath the facility, 3) HydroPunch sampler refusal was used to identify the upper portion of the glacial till, and 4) the HydroPunch groundwater sample collection success was used to qualitatively assess the hydraulic conductivity and anisotropy of the aquifer materials.

GEOLOGIC SETTING

The study area is located between the facility's southern property boundary and a major river which is dammed for hydroelectric power (Figure 1). Downstream of the dam are two river channels; one is a raceway used by the power company, the other is the natural river channel. The river has eroded through the overburden geological materials at these locations and bedrock is exposed along the river banks. The bedrock is a reddish brown arkose which locally dips to the east at about 15°. Flat-lying glacial and post-glacial sediments overlie the bedrock. Cross section A-A' (Fig 1) was constructed, using data from several sources: 1) surficial geological mapping [1], [2], 2) two seismic surveys, 3) a continuously logged borehole at monitoring well MW52, and 4) ten HydroPunch boring logs (HP304 through HP313). The stratigraphic setting is key to understanding the groundwater flow system and sampling constraints encountered during this study. A detailed discussion of site stratigraphy and interpreted depositional history is presented below.

Site Stratigraphy and Interpreted Depositional History

Holocene deposits of floodplain sediments along the riverside were observed in HydroPunch borings HP314 through HP320 (Fig 1). These

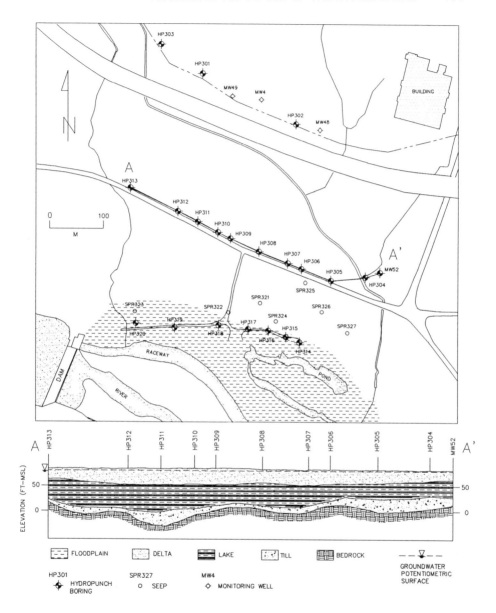

FIGURE 1 — Study area map and cross—section.
Cross—section horizontal scale is 1.5X map scale.

floodplain deposits are predominantly dark brown, fine-grained sands, silts and clays and extend to depths of 12 m (40 ft) at the above locations. The aerial extent of the floodplain deposits is approximated on Figure 1, based on soil borings, topography, and surficial mapping [1], [2]. Other Holocene river deposits lie at higher elevations, along the banks of the river and its tributaries. These deposits are yellowish brown, medium to coarse terrace sands. The northern extent of these terrace sands appears to be just north of the facility property boundary.

Glacial deposits consisting of deltaic and lake sediments underlie the terrace sands. The deltaic sediments are tan, fine sands with varying percentages of silt, which are commonly laminated and/or cross-bedded, and may contain thin clay layers (less than 5 mm thick). These sands were deposited to form a delta and are termed delta sands on Figure 1. The delta sands and the terrace sands described above are grouped as one unit (now referred to as Delta Sands) on the geologic cross-section for two reasons: 1) they exhibit similar physical characteristics (terrace sands are generally more coarse-grained), and 2) the geologic contact between these units was not resolved during this study.

The delta sands were deposited in a glacial lake on top of lake bottom sediments. These lake sediments consist of gray and reddish brown, fine sand, silt and clay arranged in rhythmic couplets (layers of fine sand and silt alternating with layers of silt and clay). These couplets have been termed "varves", representing seasonal deposition rates and sediment capacity [3]. Typically, individual varve thicknesses are 1 to 10 mm; the individual varves at the facility were generally 0.15 to 0.75 m (0.5 to 2.5 ft) thick. Although the lake sediments share similar characteristics with the floodplain deposits (silt and clay-rich, rapid dilatancy), they are distinguished by their stratigraphic position and the presence of varves.

Prior to the existence of the glacial lake, a mantle of glacial ice covered the land surface. A dense till was deposited between the ice and the land surface, and now underlies the lake sediments and mantles the bedrock surface. The till consists of a heterogeneous mixture of reddish brown clay, silt, sand, cobbles and boulders. It was encountered in every boring in the study area. In some locations the upper portion of the till was less dense and moist, whereas in other locations, the lower portion of the till was very dense and dry.

Hydrological Considerations

The depth of the water table in the study area ranged from 2.5 to 3 m (8 to 10 ft) below land surface (dashed line, Figure 1 cross-section). Shallow and deep groundwater flow south-southeast, toward the river, but away from the positive hydraulic influence of the dam. A cluster of seeps south of the roadway are identified as SPR321-SPR327 on Figure 1. The seep locations are believed to approximate the contact between the delta sands and the lake sediments (except for SPR324), and the discharge is believed to be from the more permeable delta sands.

Horizontal hydraulic conductivities were measured by aquifer hydraulic tests in each of the stratigraphic units at the facility (north of the study area). The horizontal hydraulic conductivities of the delta and terrace sands were an order of magnitude greater than the lake sediments, suggesting that the vertical groundwater flow may be retarded at the contact with the lake sediments. Although the horizontal hydraulic conductivity of the floodplain deposits was not measured, the fine-grain matrix, and the inability to recover groundwater samples, suggests it may be very low.

INVESTIGATION

Previous off-site investigations utilized screening methods, primarily a soil gas survey, to qualitatively assess the lateral extent of the off-site groundwater plume. One task of the Phase I RFI was the quantification and further definition of the extent of the off-site groundwater plume without installing monitoring wells. The installation of off-site monitoring wells would have been premature based on the limited understanding of the groundwater contaminant migration at the facility. An objective of this investigation was to decide where to install the future off-site monitoring wells during one field mobilization.

A discrete interval groundwater sampling demonstration was conducted prior to the initiation of the Phase I RFI to compare three different sampling devices; a laser-slotted lead screen auger, the HydroPunch, and a stainless steel drive well point. The outcome of this demonstration resulted in the recommendation for the HydroPunch to be used over the other alternatives. The three deciding factors were 1) time required to collect necessary sample volumes, 2) consistency of contaminant concentrations between each device, and 3) evaluation of problems encountered while using the various sampling devices. During the demonstration process, heaving sands were found to be a consistent problem for all of the sampling devices. However, installation of the HydroPunch proved to be relatively easy once heaving sands were removed from inside the hollow stem augers.

The HydroPunch sampling tool utilized during this investigation was constructed of stainless steel and Teflon[4]. The device had a stainless steel drive point, a perforated section of stainless steel pipe for sample intake, a 1-liter stainless steel and Teflon sample chamber, and adapter to attach the sampler to either cone penetrometer push rods or standard soil sampling drill rods. As the HydroPunch tool was pushed through the soil, the sample intake pipe was shielded in a watertight housing that prevented contaminated soil or groundwater from entering the sampler. The shape of the sampler and a smooth exterior surface helped minimize the downward transport of the surrounding soil and liquid as it was advanced. At the desired sampling depth, the sampler was retracted approximately 0.5 m (1.5 ft); the perforated intake pipe was exposed to the water-bearing zone, permitting groundwater to flow through the screen and into the sample chamber. During sampling, very little if any foreign material (i.e., drilling fluids, cuttings) were introduced into the zone being sampled. As the sample was collected, the drive point and the sample chamber were flush against the borehole walls, serving as "packers" to help isolate the intake screen from groundwater above and below the zone being sampled. The sample was collected under in-situ hydrostatic pressure with minimal agitation. Once the sample chamber was filled, the HydroPunch was retracted and brought to the surface. The two check valves closed due to gravity and retained the sample within the sample chamber. Upon retrieval a stainless steel and Teflon sample discharge device was inserted for transferring the groundwater sample to a sample container.

The investigation work plan called for discrete interval groundwater samples to be collected with the HydroPunch at twenty specific locations along the three transects approximately perpendicular to the anticipated general direction of horizontal groundwater flow. These transect locations (shown on Figure 1) were selected based on the following information: 1) contamination detected in nearby residential wells, 2) soil gas survey results of previous investigations, and 3)

---

[4]Teflon is a trademark of E. I. duPont de Nemours and Co., Wilmington, DE.

off-site seismic survey results depicting bedrock surface. Three
sampling locations along the northern transect were selected to
correlate data from the HydroPunch with data from existing monitoring
wells. The ten sampling locations along the middle transect were
selected along the edge of the road (within the town right-of-way),
which provided easier site access for the sampling program. The last
seven locations were selected along a non-paved access road associated
with an existing, privately owned, hydroelectric power station.

Hollow stem augers were initially used along the northern transect
to advance the boring for HydroPunch sampling, but were abandoned during
sampling at the first drill hole because of heaving sands. Mud-rotary
drilling, an alternate drilling method, was used at all three sampling
locations along the northern transect (Figure 1). Hollow stem augers
were used at the remaining seventeen sampling locations because the
saturated thickness (and the corresponding hydrostatic pressure) of the
sands were less. The drill cuttings from the hollow stem augers or in
the drilling fluid (i.e., mud rotary) were logged during the sampling
program. These drill cuttings, significantly disturbed by the drilling
process, were correlated to the previously characterized stratigraphic
units at the facility.

The first discrete HydroPunch sample interval was planned for
collection at approximately 3 m (10 ft) below the water table. There
appeared to be insufficient hydrostatic pressure to drive the
groundwater into the sample chamber within an acceptable amount of time
(30 minutes) if the HydroPunch was less than 3 m (10 ft) below the water
table. The manufacturer recommends the HydroPunch be submerged a
minimum of 1.5 m (5 ft). The HydroPunch was advanced in the same
fashion as a split spoon, as described in ASTM Standard Test Method for
Penetration Test and Split-Barrel Sampling of Soils (D 1586). In all
cases, the borehole was advanced to approximately 3 m (10 ft) below the
water table. In the three sampling locations along the northern
transect, the water table was estimated based on the water table
elevations in nearby monitoring wells. The HydroPunch was lowered to
the bottom of the borehole and then driven with a 63.5 kg (140 lb)
hammer dropping 0.76 m (30 in.). Once the HydroPunch was driven
approximately 0.6 m (2 ft), the rods were pulled up approximately 0.5 m
(1.5 ft) to expose the sample intake section. After allowing 30 minutes
for the sample chamber to fill, the HydroPunch was retrieved and the
sample transferred into the appropriate sample containers. The
boreholes were then advanced and sampled in 3 m (10 ft) intervals until
drill bit/hollow stem auger refusal or HydroPunch refusal occurred.
HydroPunch refusal was determined to be thirty blow counts, or more, for
six inches of penetration. One HydroPunch sampler was damaged early on
when trying to drive the sampler into dense glacial till (the blow
counts were slightly above fifty for six inches of penetration).

Varying lengths of time were allowed for the HydroPunch sample
chamber to fill during the first several sample collections. It was
determined that thirty minutes was sufficient time to fill the sample
chamber under most conditions. Some of the deeper stratigraphic units
(lake sediments, till) had lower hydraulic conductivities, and during
many sampling attempts the sample chamber did not fill completely.
However, it appeared that longer sample intake exposure times had
minimal affect on the volume of sample retrieved.

PRESENTATION OF DATA

A total of 95 sample attempts were made with the HydroPunch tool
at the twenty HydroPunch boring locations (HP301 through HP320, Figure
1). Adequate sample volume to satisfy the minimum sample volume
criteria (i.e., 80 ml for VOC analysis) was met at 60 of the 95 sample

locations. In more than one in three sample attempts, there was
insufficient sample volume for any analysis. Of the 60 successful
sample attempts, 48 attempts provided the target sample volume (i.e.,
190 ml for VOC and target analyte list (TAL) metals analysis) and the
remaining 12 attempts provided only the minimum sample volume required
for VOC analysis. A summary of sampling information is presented in
Table 1.

The 95 sample attempts and the success of collection of the
minimum volume requirement is categorized by geologic formation as
follows:

HydroPunch sample success vs. geology.

Geologic Unit	Sample Attempts	No. (%) of VOC & TAL samples	No. (%) of VOC only samples	No. (%) of no samples
Delta	20	15 (75%)	5 (25%)	0 (0%)
Lake	45	26 (58%)	6 (13%)	13 (29%)
Floodplain	24	4 (17%)	0 (0%)	20 (83%)
Till	6	4 (67%)	0 (0%)	2 (33%)

A number of chlorinated VOCs were detected in the HydroPunch
samples. The most frequently detected VOCs and their maximum detected
concentration were: trichloroethene (TCE) - 32 detections, 5 200 µg/l;
tetrachloroethene (PCE) - 21 detections, 680 µg/l; 1,1,1-trichloroethane
(TCA) - 41 detections, 620 µg/l; carbon tetrachloride - 13 detections,
640 µg/l; and 1,2-dichloroethene (DCE) - 20 detections, 180 µg/l. In
addition, chromium (17 detections; 155 µg/l) was the only inorganic
detected that was not attributed to naturally occurring groundwater
conditions. Table 2 summarizes the HydroPunch sample analytical results
for these five VOCs and chromium.

The HydroPunch sampling analytical results were compared to
monitoring well and seep analytical results to examine the lateral
extent of groundwater contamination as it migrates off-site from the
facility's southern boundary. TCE was selected as the representative
contaminant to assess the HydroPunch sampling results because it had the
greatest range of concentrations and it had a high frequency of
detection. An isoconcentration plot of the off-site TCE plume is
presented on Figure 2. The isoconcentration plot is based on the
highest TCE concentration detected at any discrete interval (i.e., not
depth specific) at each boring location.

The vertical extent of the TCE contamination was evaluated in the
area of the middle transect. This area was selected because the primary
purpose of the HydroPunch sampling program was to assess the extent of
off-site contamination, both laterally and vertically. The northern
transect was not used because of the limited number of HydroPunch
samples collected from this area, and the southern transect was not used
because an insufficient number of samples were collected from this area.
Figure 2 presents the vertical extent of the TCE plume along the middle
transect.

DISCUSSION OF RESULTS

The effectiveness of the HydroPunch sampling tool to collect
sufficient groundwater sample volumes for laboratory analysis varied
among the different geologic materials present, but not as widely as may
have been expected. The delta, lake and till deposits all recorded
similar percentages of successful target sample volume recoveries (i.e.,

TABLE 1--Hydropunch Sampling Summary

Boring Number	Sample ID#	Depth (m.)	Geologic Material	Sample Analysis	Boring Number	Sample ID#	Depth (m.)	Geologic Material	Sample Analysis
HP301	1	7.3	Delta	VOC/TAL	HP310	1	7.3	Delta	VOC/TAL
HP301	3	10.4	Delta	VOC/TAL	HP310	NA	10.4	Lake	NO
HP301	6	13.4	Delta	VOC/TAL	HP310	3	13.4	Lake	VOC/TAL
HP301	8	16.5	Lake	VOC/TAL	HP310	4	16.5	Lake	VOC/TAL
HP301	NA	19.5	Lake	NO	HP310	NA	19.5	Lake	NO
HP301	9	22.6	Lake	VOC/TAL	HP310	NA	22.6	Lake	NO
HP301	NA	25.6	Lake	NO	HP311	3	7.3	Delta	VOC/TAL
HP301	11	28.7	Lake	VOC/TAL	HP311	4	10.4	Delta	VOC
HP302	1	7.3	Delta	VOC/TAL	HP311	5	13.4	Lake	VOC/TAL
HP302	2	10.4	Delta	VOC/TAL	HP311	6	16.5	Lake	VOC/TAL
HP302	3	13.4	Delta	VOC	HP311	NA	19.5	Lake	NO
HP302	NA	16.5	Lake	NO	HP311	8	22.6	Lake	VOC/TAL
HP302	5	19.5	Lake	VOC/TAL	HP311	NA	25.6	Lake	NO
HP302	6	22.6	Lake	VOC/TAL	HP311	10	28.7	Till	VOC/TAL
HP302	8	25.6	Till	VOC/TAL	HP312	1	7.3	Delta	VOC/TAL
HP303	1	7.3	Delta	VOC/TAL	HP312	2	10.4	Lake	VOC/TAL
HP303	2	10.4	Delta	VOC/TAL	HP312	3	13.4	Lake	VOC/TAL
HP303	3	13.4	Lake	VOC/TAL	HP312	4	16.5	Lake	VOC/TAL
HP303	4	16.5	Lake	VOC/TAL	HP312	5	19.5	Lake	VOC/TAL
HP304	1	7.3	Delta	VOC	HP312	7	22.6	Lake	VOC/TAL
HP304	NA	10.4	Lake	NO	HP312	8	25.6	Till	VOC/TAL
HP304	3	13.4	Lake	VOC	HP313	NA	7.3	Lake	NO
HP304	4	16.5	Lake	VOC/TAL	HP313	2	10.4	Lake	VOC/TAL
HP305	1	7.3	Delta	VOC/TAL	HP314	NA	7.3	Floodplain	NO
HP305	NA	10.4	Lake	NO	HP314	NA	10.4	Floodplain	NO
HP305	3	13.4	Lake	VOC	HP314	NA	13.4	Floodplain	NO
HP305	NA	16.5	Lake	NO	HP315	NA	7.3	Floodplain	NO
HP305	NA	19.5	Till	NO	HP315	NA	10.4	Floodplain	NO
HP305	NA	22.6	Till	NO	HP315	NA	13.4	Floodplain	NO
HP306	1	7.3	Delta	VOC/TAL	HP316	NA	7.3	Floodplain	NO
HP306	2	10.4	Lake	VOC/TAL	HP316	NA	10.4	Floodplain	NO
HP306	3	13.4	Lake	VOC/TAL	HP316	NA	13.4	Floodplain	NO
HP306	6	16.5	Lake	VOC/TAL	HP317	NA	7.3	Floodplain	NO
HP306	NA	19.5	Lake	NO	HP317	NA	10.4	Floodplain	NO
HP307	1	7.3	Delta	VOC/TAL	HP317	NA	13.4	Floodplain	NO
HP307	2	10.4	Delta	VOC/TAL	HP317	NA	16.5	Floodplain	NO
HP307	3	13.4	Lake	VOC	HP318	NA	7.3	Floodplain	NO
HP307	6	16.5	Lake	VOC	HP318	NA	10.4	Floodplain	NO
HP307	7	19.5	Lake	VOC/TAL	HP318	NA	13.4	Floodplain	NO
HP308	2	4.3	Delta	VOC	HP318	NA	16.5	Floodplain	NO
HP308	11	7.3	Delta	VOC	HP318	5	19.5	Floodplain	VOC/TAL
HP309	1	7.3	Delta	VOC/TAL	HP319	NA	7.3	Floodplain	NO
HP309	2	10.4	Lake	VOC	HP319	1	10.4	Floodplain	VOC/TAL
HP309	3	13.4	Lake	VOC/TAL	HP319	3	13.4	Floodplain	VOC/TAL
HP309	4	16.5	Lake	VOC/TAL	HP320	1	7.3	Floodplain	VOC/TAL
HP309	5	19.5	Lake	VOC/TAL	HP320	NA	10.4	Floodplain	NO
HP309	6	22.6	Lake	VOC	HP320	NA	13.4	Floodplain	NO
HP309	8	25.3	Till	VOC/TAL					

NA = Not Applicable; no sample collected.

TABLE 2--Hydropunch Sampling Analytical Summary

LOCATION	HP301	HP301	HP301	HP301	HP301	HP301	HP302	HP302	HP302	HP302
SAMPLE NO.	1	3	6	8	9	11	1	2	3	5
COMPOUND ($\mu g/l$)										
DCE	79	110	38	...	...	...	180	7	...	...
TCA	330	620	120	6	10	3	120	5	7	8
TCE	5 200	2 600	1 400	2	2		260	8	4	...
PCE	590	680	150	...	...	...	530	15	6	...
CCL4	640	120	120	...	...	...	...	...	...	...
Chromium	11.7	30.6	3	...	...	...	155	6.1	NS	...

LOCATION	HP302	HP302	HP303	HP303	HP303	HP303	HP304	HP304	HP304	HP305
SAMPLE NO.	6	8	1	2	3	4	1	3	4	1
COMPOUND ($\mu g/l$)										
DCE	...	...	...	2	...	...	...	...	...	34
TCA	4	...	16	2	2	...	9	2	...	30
TCE	...	...	18	160	...	...	...	...	...	49
PCE	...	...	...	1	...	...	...	...	...	46
CCL4	...	...	...	1	...	...	...	...	...	...
Chromium	...	...	...	3.1	...	...	NS	NS	...	...

LOCATION	HP305	HP306	HP306	HP306	HP306	HP307	HP307	HP307	HP307	HP307
SAMPLE NO.	3	1	2	3	6	1	2	3	6	7
COMPOUND ($\mu g/l$)										
DCE	...	130	...	...	...	120	3	2	...	...
TCA	3	37	5	...	...	29	2	2	...	1
TCE	...	340	3	...	...	880	19	11	...	...
PCE	...	150	...	...	...	100	...	16	...	...
CCL4	...	...	...	...	...	24	...	1	...	...
Chromium	NS	107	...	...	2.6	...	114	NS	NS	...

LOCATION	HP308	HP308	HP309	HP309	HP309	HP309	HP309	HP309	HP309	HP310
SAMPLE NO.	2	11	1	2	3	4	5	6	8	1
COMPOUND ($\mu g/l$)										
DCE	110	67	76	...	...	...	...	8	...	60
TCA	44	120	120	5	...	...	...	6	...	77
TCE	1 700	1 300	3 500	16	1	1	1	110	...	2 700
PCE	77	120	210	...	...	...	...	4	...	120
CCL4	260	200	430	...	...	...	...	3	...	180
Chromium	NS	NS	81	NS	...	...	...	NS	...	4.8

LOCATION	HP310	HP310	HP311	HP311	HP311	HP311	HP311	HP311	HP312	HP312
SAMPLE NO.	3	4	3	4	5	6	8	10	1	2
COMPOUND ($\mu g/l$)										
DCE	...	...	20	2	...	...	...	...	8	1
TCA	2	...	83	16	...	...	...	12	23	7
TCE	9	...	1 100	91	...	...	...	36	240	83
PCE	...	...	84	5	...	...	...	2	13	1
CCL4	...	...	39	...	...	...	...	...	7	2
Chromium	...	...	...	NS	...	...	...	2.6	...	...

LOCATION	HP312	HP312	HP312	HP312	HP312	HP313	HP318	HP319	HP319	HP320
SAMPLE NO.	3	4	5	7	8	2	5	1	3	1
COMPOUND ($\mu g/l$)										
DCE	...	...	...	...	...	...	...	...	...	...
TCA	1	...	...	2	...	2	...	2	...	2
TCE	...	...	...	...	...	...	...	...	...	...
PCE	1	...	...	...	...	...	...	...	...	...
CCL4	...	...	...	...	...	...	...	...	...	...
Chromium	...	...	...	...	...	...	...	...	...	...

... indicates not detected.
DCE = 1,2-Dichloroethene, TCA = 1,1,1-Trichloroethane
TCE = Trichloroethene, PCE = Tetrachloroethene
CCL4 = Carbon Tetrachloride, NS = Not Sampled

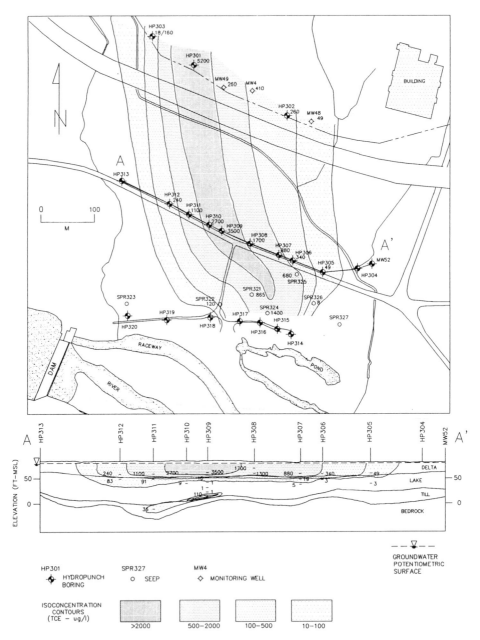

FIGURE 2 — Study area TCE plume map and cross—section.
Cross—section horizontal scale is 1.5X map scale.

190 ml), 58 to 70 percent of all sample attempts, while the floodplain
deposits had only a 17 percent successful target sample volume recovery.
The poorest sampling recovery, defined as insufficient volume for
minimal sample volumes (i.e., 80 ml), was observed in the floodplain (83
percent), till (33 percent), and lake deposits (29 percent). The delta
sands were the only geologic unit from which every sample attempt
collected at least the minimum sample volume required.

The delta sands, comprised principally of medium to fine sand with
varying percentages of silt, were the most permeable material, with
hydraulic conductivities of 5.5 to 23 m/day (18 to 75 ft/day), based on
previous aquifer hydraulic tests. This unit is sufficiently permeable
to transmit water to satisfy the minimal sampling volume requirement all
of the time and to provide the target sample volume over two-thirds of
the time.

The lake sediments, although comprised of finer-grained materials
than the delta sands, were sufficiently transmissive to fulfill the
target sample volume over half of the time, even though the hydraulic
conductivity of the lake deposits at the site averaged 0.15 m/day (0.5
ft/day). It is believed that the coarser-grained varves, typically
composed of very fine to fine sand, were sufficiently permeable to
provide the target sample volume. However, when the sample volumes were
not sufficient for even the minimal sample volume (29 percent of the
time), it is believed that finer-grained materials (silts and clays)
comprised both of the varve couplets at the sampling interval.

The till materials that were sampled with the HydroPunch were
typically the uppermost portion of the till deposit. The uppermost till
was generally more sandy and friable than the underlying, dense, compact
till, although the upper till was also characterized as very
heterogeneous and silty at some locations. The limited number of till
HydroPunch sample attempts do not permit a more thorough analysis.
Sample recoveries in the till appear to be strongly influenced by the
local permeability. Hydraulic conductivity values for the upper till,
measured from aquifer hydraulic tests, averaged 1.5 m/day (5 ft/day),
but ranged from 0.03 to 8 m/day (0.1 to 26 ft/day).

The floodplain deposits were minimally transmissive due to their
predominant silt and clay composition. In addition, no distinct bedding
features were observed to indicate preferential pathways for groundwater
flow. No hydraulic conductivity data were available for the floodplain
deposits, but representative horizontal hydraulic conductivity
literature values for silts and clays are 0.08 and 0.0002 m/day (0.26
and 0.0007 ft/day), respectively [4].

The HydroPunch sampling program met the project objective of
delineating the horizontal and vertical extent of off-site groundwater
contamination. Figure 2 depicts the horizontal and vertical extent of
the off-site groundwater TCE plume. The lateral extent of the plume is
well defined at the middle transect, with concentrations increasing from
"not detected" at the program's sample boundaries (HP304 to the east and
HP313 to the west) toward the plume's main axis area (HP309). This
general trend in concentration was observed for the other VOCs presented
in Table 2, while the limited chromium data indicated a more eastward
plume axis near HP306/HP307. Chromium data were limited compared to VOC
data due to the limited sample volumes at some sampling locations.

Similarly, the vertical extent of the TCE was well-defined. In all
cases, except for two (HP308 and HP311), no TCE was detected in the
bottom-most sample. At HP308, sampling at depth was suspended because
gross fuel oil contamination was observed in the drill cuttings. At
HP311, the dense till prevented further vertical profiling. TCE
concentrations dramatically decrease at the delta/lake geologic contact.

This is shown on Figure 2 for the middle transect and in Table 2 for the northern and middle transects. Monitoring well data at the facility's southern boundary also confirms this observation at the delta/lake geologic contact. However, the HydroPunch sampling did indicate the limited presence of TCE below the delta/lake sediment contact in the off-site area (HP309 and HP311).

The HydroPunch sampling data also correlated with historical screening data. Soil gas work, performed in 1989 in the area of the middle transect, approximated the lateral extent of the plume but did not provide the detailed concentration data, in particular vertical extent, that was generated by the HydroPunch sampling. Similarly, a TCE groundwater plume depiction from the early 1980s, constructed from sampling points in unspecified geologic horizons, was similar to the HydroPunch TCE plume in both concentration and lateral extent.

CONCLUSIONS AND RECOMMENDATIONS

For this particular case study, the HydroPunch tool was used as a field screening tool to provide the data necessary to meet the field investigation objective (i.e., delineate the vertical and horizontal extent of the groundwater plume for the subsequent placement of permanent monitoring wells). The TCE results, shown on Figure 2, clearly define the edge and center of the plume along the middle transect. An insufficient number of samples were collected along the southern transect to further define the plume. However, the analytical results from the seeps confirmed that the plume was discharging into the surface water in that area.

A comparison of the number of samples collected from the delta sands and lake sediments to the number of samples collected from the floodplain deposits shows that the HydroPunch was considerably less effective in low yielding aquifer materials. The delta sands have the highest horizontal hydraulic conductivity, the varved lake sediments have the next highest, and the floodplain deposits have the lowest horizontal hydraulic conductivity. Correspondingly, the highest sample collection rate (i.e., providing at least the minimal volume of 80 ml required for VOC analysis) was 100 percent (20 out of 20 sample attempts) from the delta sands, the second highest was 71 percent (32 out of 45 sample attempts) from the varved lake sediments, and the lowest sample collection rate was 17 percent (4 out of 24 samples) from the floodplain deposits. Generally, varved lake sediments have a hydraulic conductivity similar to floodplain deposits, however, the individual varve thickness at the site ranged between 0.15 and 0.75 m (0.5 and 2.5 feet). Therefore, based on the length of the sampler intake, the HydroPunch was usually in hydraulic communication with at least one of the fine sand varves with a higher hydraulic conductivity. This explains why the HydroPunch was relatively successful in the varved lake sediments.

The HydroPunch itself does not help to geologically characterize the materials in which sampling takes place. During this case study, the cuttings from the hollow stem augers were used to log the geologic materials with depth. This was acceptable for this particular investigation because the site geology was already characterized and well understood. Drilling methods which generate cuttings disturb the geologic materials brought to the surface making them difficult to characterize.

The use of hollow stem augers to advance the boring for HydroPunch sampling provided access to a specific depth, minimizing the contact of the sampling device with subsurface contaminants above the specified depth, and thereby minimizes any cross contamination. Therefore, the resulting HydroPunch sample is representative of groundwater conditions only at that specific depth. This is clearly shown by data presented in Table 2. TCE concentrations in the uppermost samples in HP309, HP310, and HP311 were 3 500 μg/l, 2 700 μg/l, and 1 100 μg/l, respectively. The corresponding TCE concentrations from the samples collected just 3 m (10 ft) below were only 16 μg/l, 9 μg/l, and 91 μg/l, respectively. The reduction in concentration was approximately two orders of magnitude showing that the hollow stem augers allowed the HydroPunch samples to be collected beneath significantly higher concentrations without creating cross contamination. However, drilling methods generate drill cuttings, which most often require additional sampling and handling for proper environmental management.

The authors make the following recommendations regarding the use of the HydroPunch.

1)   Use the HydroPunch tool as a field screening device to collect discrete interval groundwater samples from unconsolidated aquifer materials with sufficient horizontal hydraulic conductivity.

2)   Where there is little or no information regarding the geology, use another approach to collect geologic material samples in conjunction with the HydroPunch. This information will help define the geologic conditions from which the groundwater samples were collected.

3)   Consider the use of a drilling method which does not generate drill cuttings (i.e., cone penetrometer method). The proper environmental management of drill cuttings is a cost burden generally sought to be avoided during field screening programs. This should be evaluated against the possibility that the HydroPunch samples, collected with a drilling method which does not generate drill cuttings, has the potential for some amount of cross contamination as the sampler is pushed (through contaminated material) down to the discrete sampling interval.

REFERENCES

[1]   Colton, R. B., Surficial geologic map of the Windsor Locks quadrangle, Hartford County, Connecticut: U. S. Geological Survey Map GQ-137, 1953.

[2]   Stone, J. R., Shafer, J. P., and London, E. H., The surficial geological maps of Connecticut, illustrated by a field trip in central Connecticut: in Joesten, R. L. and Quarrier, S. S., eds.: New England Intercollegiate Geological Conference Guidebook for Field Trips in Connecticut and south-central Massachusetts, 1982, Storrs, Connecticut, p. 5-25.

[3]   Gustavson, T.C., and Ashley, G.M., Depositional sequences in glaciolacustrine deltas: Society of Economic Paleontologists and Mineralogist, Special Publication No. 53, 1975.

[4]   Todd, D. K., Groundwater Hydrology, John Wiley & Sons, New York, 1959.

Steven H. Edelman[1] and Andrew R. Holguin[2]

CONE PENETROMETER TESTING FOR CHARACTERIZATION AND SAMPLING OF SOIL AND GROUNDWATER

REFERENCE: Edelman, S. H. and Holguin, A. R., ''Cone Penetrometer Testing for Characterization and Sampling of Soil and Groundwater,'' Sampling Environmental Media, ASTM STP 1282, James Howard Morgan, Ed., American Society for Testing and Materials, 1996.

ABSTRACT: Cone penetrometer testing (CPT) is an alternative method to drilling for subsurface characterization of hazardous materials release sites. CPT provides higher quality data at lower cost and with fewer health and safety concerns than conventional drilling. CPT basically consists of pushing a cone-tipped, steel rod into the subsurface soils at a constant velocity and measuring the stresses used for automatic determination of soil types. CPT with concurrent measurement of pore pressure is used for determining hydraulic head and other parameters.

Several methods for in situ screening of subsurface contaminants have been added to CPT equipment, including vapor sampling, laser induced fluorescence (LIF), and pore water resistivity. CPT is also used for "direct push" sampling of soil and groundwater and for installation of small diameter "well points."

CPT can be used with an assessment strategy that parallels that of conventional drilling; however, the in situ testing capabilities of CPT lend themselves to a more comprehensive assessment strategy that minimizes soil and groundwater sampling. A case study illustrates this comprehensive assessment strategy.

KEYWORDS: Cone penetrometer testing, direct-push sampling, in situ analysis, subsurface characterization, soil sampling, groundwater sampling, vapor sampling.

---

[1]Senior Geologist, Holguin, Fahan & Associates, Inc., 143 South Figueroa Street, Ventura, California, 93001, and Instructor, Hazardous Materials Management, University of California, Santa Barbara, California

[2]President, Holguin, Fahan & Associates, Inc., 143 South Figueroa Street, Ventura, California, 93001

INTRODUCTION

Characterization of subsurface conditions at hazardous materials release sites is usually conducted using a combination of in situ testing and collection of subsurface samples for analysis. In situ measurements include water level measurements in wells to determine hydraulic head and the standard penetration test to determine soil strength. Subsurface samples include soil and groundwater.

In situ tests and sample collection are typically performed using a hollow-stem drill rig; in some cases air rotary or mud rotary drill rigs are used. However, "direct push" technologies, including all methods in which probes and/or samplers are attached to rigid steel rods and pushed or driven into the subsurface, have several advantages over drilling, including higher quality data, lower cost, fewer health and safety concerns, and virtually no production of waste (soil cuttings).

Cone penetrometer testing (CPT) encompasses all of the in situ testing and sampling methods available in direct-push equipment. This paper reviews the in situ testing and sampling capabilities of CPT as they apply to hydrogeological and environmental site characterizations, and presents strategies for conducting site investigations using CPT in a cost-effective manner. A case study is described to illustrate the data quality and cost effectiveness of properly designed CPT investigations.

CONE PENETROMETER TESTING

The capabilities of CPT as they relate to data collection for subsurface environmental site characterizations are summarized below (Table 1).

TABLE 1.--Summary of CPT data collection capabilities for environmental site characterizations

Mode	CPT Capability	Data Produced
In situ testing	CPT	• Soil behavior type
	Pore pressure dissipation test	• Hydraulic head • Permeability
	In situ chemical analysis	• In situ qualitative (screening) or quantitative analysis of concentrations of contaminants
Sampling	Soil sampling	• Concentrations of contaminants in soil
	Groundwater sampling	• Concentrations of contaminants in groundwater

CPT is, strictly speaking, the measurement of stress parameters at the rod tip during penetration to determine geotechnical properties and soil type. The hardware, procedures, and theory of CPT have been discussed extensively in the geotechnical literature.

Cone Penetrometer Testing Hardware and Procedures

CPT basically consists of pushing a cone-tipped, steel rod into the subsurface soils at a constant velocity and measuring the stresses on the cone tip during advancement (Figure 1). The basic components of a CPT system are a hydraulic ram, a steel rod that is pushed into the ground with the ram, and a cone-tipped probe at the tip of the rod. A standard cone has a 10 cm^2 area with an apical angle of 60 degrees, as specified by ASTM Standard Test Method for Deep, Quasi-Static, Cone and Friction-Cone Penetration Tests of Soil (D 3441-86). The hydraulic ram advances the rod and cone at a constant rate of 2 cm/sec.

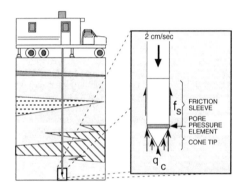

FIG. 1--Schematic diagram of CPT (q_C = tip stress, f_S = friction stress)

The CPT probe is mechanically separated into two parts, the cone tip and the friction sleeve (Figure 1). The cone tip consists of the entire conical part of the probe and the friction sleeve consists of the cylindrical area above the tip. A standard friction sleeve has a surface area of 150 cm^2. The tip and sleeve are each connected to strain-gauged load cells mounted within the probe body to measure the vertical components of stress as the probe is advanced downward through the soil. The vertical components of stress on the tip and sleeve are a bearing stress (q_C) on the cone, and a frictional stress (f_S) along the sleeve (Figure 1). The stress values measured by the load cells are transmitted electronically to an on-board computer, which logs these values at specified intervals, typically 2 to 5 cm.

The cone is called a "piezocone," or "CPTu," when outfitted to measure dynamic pore pressure, u, which is the pore water pressure on the cone surface during cone advancement in saturated soils. The measurement is made using a CPT probe equipped with a saturated, porous element (Figure 1) and an internal, saturated chamber with a pressure transducer.

Other probes and samplers can be placed at the tip of the rod and hydraulically driven into the subsurface using the same equipment. These other probes and samplers constitute the range of technologies

called "direct push," including vapor samplers, soil samplers, and
groundwater samplers.

In Situ Testing Capabilities for Environmental Site Characterization

CPT for soil type determination--Advancement of the cone into soil
induces a complex field of stresses and strains. Theoretical
relationships between q_c and soil rheology have to date proven to be
intractable. Therefore, soil classifications using CPT are empirically
based [1] [2] [3]. Empirical classification systems have been
presented by many workers; the system presented by Robertson and
Campanella [2] [3] and Robertson [4] is commonly used. This
classification differentiates gravel, sand, silt, clay, some
intermediate soil types, in a graph of q_c versus f_s/q_c (Figure 2). A
typical, modern CPT automatically prints soil types at 2-cm to 5-cm
intervals using this method. The empirical classification schemes have
been established through numerous side-by-side CPT and conventional
borehole samples. The basic soil behavior observations are that sands
display relatively high q_c (10-1,000 bars) and low f_s/q_c (0-5 bars),
whereas clays display relatively low q_c (10-50 bars) and high f_s/q_c(1-8
bars).

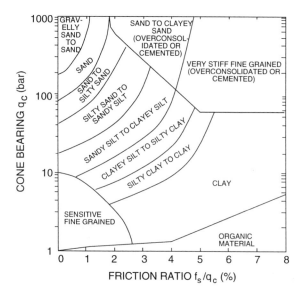

FIG. 2--Simplified Soil Behavior Type Classification for Standard Cone

CPT with dynamic pore pressure measurement (CPTu)--In saturated
soil, penetration of the cone causes shearing and compression in the
adjacent soil, which causes a change in the pore water pressure as
measured with CPTu. The dynamic pore pressure measurement, u, is an
important geotechnical parameter. It has several uses for
environmental site characterizations, including independent soil
classifications and detailed stratigraphic analysis [5] [6] [7] [8]
[9]. The most useful aspect of pore pressure measurements for

environmental investigations is determination of hydraulic head using dissipation tests.

   Dissipation tests--When the probe advance is halted in saturated soil, dynamic pore pressure dissipates and u approaches u_o (hydrostatic pressure) (Figure 3a). The value of u_o, when expressed in meters of water, is the hydrostatic pressure head at the location of the probe. The addition of u_o to the elevation of the measurement location yields a value of true hydraulic head or potentiometric surface elevation (Figure 3b).

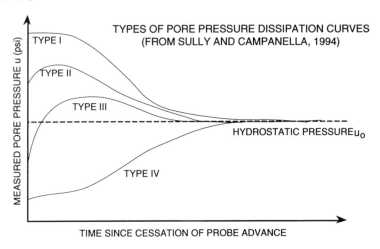

FIG. 3a--Dissipation test curves

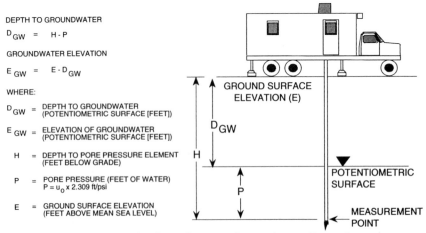

FIG 3b--Calculation of potentiometric surface elevation

Utilization of the dissipation test data and surveyed ground surface
elevations at three or more locations over a site can be used to
calculate the horizontal hydraulic gradient and direction of
groundwater flow. Because dissipation tests measure hydraulic head at
discrete points, similar to a piezometer, measurements at several
depths at a single location allows identification of perched and/or
confined aquifer zones and vertical hydraulic gradients.

The rate at which pore pressure dissipates depends on the horizontal
coefficient of consolidation, which is a function of soil
compressibility and permeability [8]. With the use of a Modified Time
Factor to plot theoretical dissipation curves, estimates of the
horizontal coefficient of permeability can be made [9]. This
capability is still in the developmental stage.

   In situ chemical analysis--Several methods for in situ qualitative
(screening) and quantitative analysis of chemical compounds are in
various stages of research, development, and commercial application.
Some of these technologies may revolutionize subsurface
characterization of hazardous materials release sites by producing
rapid, inexpensive screening and quantitative analytical data. In situ
analysis with which the authors are familiar include vapor analysis for
screening, in situ quantitative analysis (ISQA) for volatile organic
compounds, laser induced fluorescence (LIF), and resistivity.

   Vapor analysis--Although vapor sampling is, strictly speaking, not
an in situ measurement in that vapor samples are usually pumped to the
surface for analysis, it is considered here an in situ measurement
because soil or groundwater is not sampled; this is a semantic point.
Vapor sampling is a powerful CPT capability because it provides a rapid
and inexpensive screening method for soils impacted with volatile
organic compounds (VOCs) such as gasoline constituents and chlorinated
solvents. Vapors can be analyzed with conventional field analyzers
such as photoionization detectors, flame ionization detectors, and
field gas chromatographs. This capability is commercially available
and is often used in CPT investigations.

Several designs exist for downhole vapor sampling. A major technical
problem with vapor sampling has been contamination of the sampling
line, which requires removal and decontamination or replacement of the
vapor sampling line after collecting a contaminated vapor sample. One
solution that has worked well for the authors is a soil vapor sampler
that allows for purging and decontamination of the vapor collection
line between vapor samples (Figure 4a). Purging of the line between
samples is accomplished by a second line that sends ultrapure air
through the sampling line, allowing continual collection of vapor
samples with depth without having to withdraw the line for replacement
or purging at the surface.

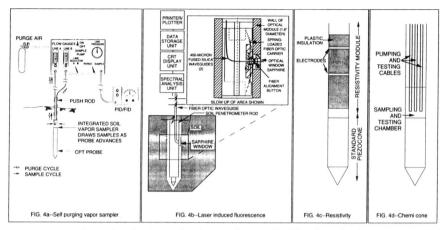

FIG 4.--In situ chemical analysis tools available for deployment on CPT rigs

ISQA--Because CPT is capable of collecting representative, reproducible samples of soil vapors, the potential exists to determine accurate total soil concentrations of VOCs via headspace analysis. This method is currently being developed. Major components of the method include laboratory headspace calibration using site-specific soil partitioning factors, a headspace volume correction, and a temperature correction.

LIF--LIF utilizes either Raman spectroscopy or fluorescence. The system fires solid-state lasers coupled into fiber optics that pass down through the center of a penetrometer rod (Figure 4b). The fiber ends at a lens installed at the side of the rod that focuses the light onto the soil surface adjacent to the lens. The return signal, which is a response to the laser excitation of the compounds, is collected by a second fiber optic cable and is carried back through the penetrometer rod to a spectral analysis unit that utilizes reference spectra to determine specificity and concentration level in the case of Raman spectroscopy or intensity in the case of fluorescence. This technology is currently moving from development to commercialization.

Resistivity--Two resistivity designs are commercially available. The more commonly used resistivity probe has electrodes mounted above the CPT cone to measure bulk soil resistivity (Figure 4c)[10] [11]. Another design, called the "chemi cone," involves pumping groundwater into a testing chamber where the water resistivity is measured (Figure 4d) [11]. The chemi cone is also capable of in situ pH and temperature measurements. The bulk soil resistivity is a continuous measurement, whereas the chemi cone is a discrete measurement. The resistivity probe is useful for screening the extent of a plume containing one or more compounds that measurably change the pore water resistivity.

Samplers

CPT equipment is capable of direct-push soil and groundwater sampling, similar to other direct-push technologies and drilling technologies.

   Soil sampling--For direct-push soil sampling, soil samples are retrieved using a probe with a retractable tip and standard brass sampling rings (Figure 5a). Undisturbed soil samples are collected just as they would be with a hollow-stem, flight auger drill rig.

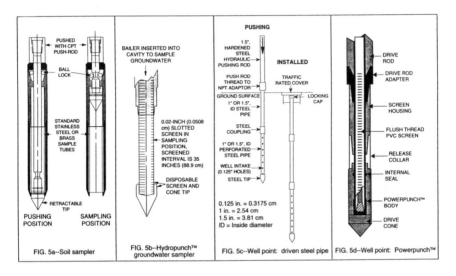

FIG. 5--Direct-push sampling tools

   Groundwater Sampling--Groundwater sampling technologies used with CPT equipment include one-time, Hydropunch™-type samplers, and well point installations for one-time or long-term sampling.

   One-time sampling--Groundwater samples are typically collected by means of Hydropunch™[3] or similar samplers (Figure 5b). These probes all operate by being pushed in a closed position, i.e., with the screen protected. When the correct depth is reached, the rod is pulled back, exposing the screen, and groundwater samples are collected. These probes collect groundwater samples at various depths within an aquifer, allowing for vertical profiling as well as lateral delineation of contaminant plume boundaries. The authors have used the BAT™[4] and Batch™[5] samplers and have found their performance to be poor compared to the simpler Hydropunch™ system (Figure 5b).

   Well point installation--Monitoring wells are traditionally installed with a hollow-stem auger or other drilling equipment.

[3]GeoInsight, Clayton, CA
[4]Hogentogler & Co., Inc., Columbia, MD
[5]Hogentogler & Co., Inc., Columbia, MD

Alternatively, narrow diameter (0.5 to 2 inches [≈1.27 to 5.08 cm]) monitoring wells can be hydraulically installed with direct-push equipment. Hydraulically installed "well points" differ from conventionally installed monitoring wells by having minimal or no gravel pack, i.e. the perforated casing is placed directly against the formation. Narrow diameter well points have lower well yields than conventional wells and are generally not useful as groundwater extraction wells. However, because of their low cost and ease of installation, they have advantages over conventional wells in several environmental applications. These applications include (1) monitoring wells (if accepted by the regulatory agency); (2) one-time screening sampling in low yielding formations; (3) one-time screening sampling to determine the extent of nonaqueous phase liquids; (4) wells and piezometers for hydrogeologic monitoring; (5) observation wells for pumping tests, air sparging tests, and vapor extraction tests; and (6) air sparging wells.

The authors are familiar with two well point designs. One is a simple driven steel rod using standard 1-inch or 1.5-inch (≈2.54 or 3.81 cm) diameter pipe[6] (Figure 5c). The other is the Powerpunch™[7], which is constructed of 0.75-inch (≈1.91 cm) diameter polyvinyl chloride (PVC) pipe that is driven within a 1.9-inch (≈4.83 cm) diameter steel rod that is removed after installation (Figure 5d). The annular space above the perforated interval may be grouted in the Powerpunch™ design.

COMPARISON TO CONVENTIONAL DRILLING

CPT is capable of performing all the same data collection functions as conventional drilling (Table 2). CPT should be the tool of choice for environmental site assessments due to its higher quality data, higher speed, lower cost, minimal site disturbance, and minimal health and safety risks compared to drilling. The principal disadvantage of CPT, compared to conventional drilling, is shallower refusal. CPT will usually encounter refusal in gravelly soils, particularly if gravel clasts are more than 2 inches (≈5.08 cm) in diameter, and in heavily overconsolidated soils. In normally consolidated soils, a standard 20-ton (≈18,000 kg) CPT rig can generally penetrate to depths of approximately 100 feet below grade (fbg) (≈30.48 meters below grade [mbg]).

---

[6]Holguin, Fahan & Associates, Inc., Ventura, CA
[7]GeoInsight, Clayton, CA

Table 2--Comparison of conventional drilling and CPT

Function	Conventional Drilling	CPT	Comparison
Stratigraphic delineation	Visual inspection of soil samples at 5-foot intervals and of drill cuttings by a field geologist	Automatic soil identification at 5-cm (0.16 foot) intervals	CPT stratigraphic delineation has higher resolution and is more objective than conventional drilling. Inspection of soil samples can be used for site-specific verification of soil types and detailed determination of grading, plasticity, and structure.
Screening of contaminants	Field testing of soil samples with hand held detectors, immunoassay, or other technique	In-situ testing of vapors, LIF, or resistivity	CPT screening methods are fast, inexpensive, and do not require soil sample collection
Collection of soil samples for laboratory analysis	Collection of undisturbed soil samples in brass or stainless steel sample tubes	Collection of undisturbed soil samples in brass or stainless steel sample tubes	Similar
Collection of groundwater samples for groundwater plume screening	Various techniques available utilizing Hydropunch™ samplers or similar methods	Various techniques available utilizing Hydropunch™ samplers or similar methods	Similar
Determination of water table depth and direction of groundwater flow	No satisfactory methods available to measure depth to water table without well installation	Dissipation test using CPTu	CPT can determine groundwater flow direction without monitoring well installation
Monitoring well installation	Efficient installation of wells for long-term plume and hydrogeologic monitoring	Installation of small diameter wells only	Conventional drilling is needed for conventional monitoring well installation, if required
Worker safety	Drillers and consultant working behind the drill rig are exposed to contaminated drill cuttings and heavy equipment. Heavy work is conducted outdoors with potential for heat stress. Work is difficult or impossible in inclement weather.	No cuttings are produced. No heavy equipment. CPT operators and consultant work inside air conditioned/heated rig.	CPT is intrinsically safe and personal protective equipment is virtually never needed
Waste disposal and site disturbance	Generation of soil cuttings	No generation of soil cuttings	CPT eliminates cost of handling, transporting, and disposal of soil cuttings. Minimal site disturbance compared to conventional drilling.

SITE CHARACTERIZATION STRATEGIES UTILIZING CPT

Because CPT uses different data collection techniques than drilling, different strategies can be used with CPT that have significant advantages over conventional drilling in terms of data quality, cost, and speed. The basis for utilizing different assessment strategies is the in situ testing capabilities of CPT, including in situ soil type identification, in situ hydrostatic head measurement, and in situ contaminant screening (e.g., vapor sampling or LIF analysis). Using in situ soil type identification and in situ contaminant screening, soil samples need only be collected for laboratory analysis. For example, if soils must be investigated to a depth of 50 fbg (≈15.24 mbg) at a given location, CPTu with a screening tool and dissipation tests produces a vertical profile of soil types, a profile of hydrostatic head values in saturated soils, and relative contaminant concentrations in the vadose zone. This data is then used to determine the depths at which soil samples should be collected for analysis based on screening concentrations, changes in soil type, and state of saturation.

Because CPT can produce all the same types of data produced with
conventional drilling, CPT can be used with an assessment strategy that
parallels that of conventional drilling. Thus, the investigator can
collect soil and groundwater samples at each investigation location
using CPT. Progressive "stepping out" could be performed until the
entire soil and groundwater plume is defined, just as with conventional
drilling.

However, the unique in situ testing capabilities of CPT lend themselves
to a more comprehensive assessment strategy. Basically, the strategy
consists of using CPTu with a screening tool to determine site
stratigraphy, the site hydrogeology, and the horizontal and vertical
extent of the contaminant plume prior to collecting soil and
groundwater samples. If groundwater is impacted and the contaminant is
one for which no in situ screening tool is available, Hydropunch™ or
similar groundwater samples may be collected during this stage of the
investigation as a screening tool. This is followed by collection of
soil samples for analysis in order to quantify soil contaminant
concentrations and verify the limits of impacted soil, and by
installation of monitoring wells for long-term hydrogeologic and water
quality monitoring. The well field would be designed based on the
known groundwater plume geometry and the known hydrogeologic conditions
such as perched water conditions and groundwater flow direction.

This strategy for site assessment can be implemented in the following
steps (Figure 6):

Step 1: Conduct CPTu with screening until the horizontal and vertical
        extent of the soil and groundwater plumes are defined
        (Figure 6a). The site stratigraphy, three-dimensional
        contaminant distribution, and hydrogeology (including
        groundwater depth and flow direction) should all be defined at
        the completion of this stage.

Step 2: Collect soil samples where the highest screening concentrations
        were found, at significant lithologic interfaces, low
        permeability zones, and around the contaminant plume to
        quantitate contaminant concentrations and to verify the
        horizontal and vertical extent of the soil plume. If
        groundwater has been found to be impacted, use a conventional
        drill rig (for conventional wells) or a CPT rig (for well
        points) to install permanent groundwater monitoring wells at
        the origins and boundaries of the groundwater plume for
        long-term monitoring (Figure 6b). The well field would be
        designed based on the known groundwater plume geometry and the
        known hydrogeologic conditions such as groundwater flow
        direction.

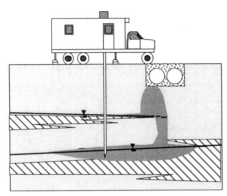 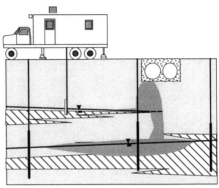

FIG. 6a--Step 1: Stratigraphy, hydrogeology, and extent of plume defined using in situ testing and screening tools

FIG. 6b--Step 2: Confirmatory soil samples collected for laboratory analysis (not illustrated), and monitoring wells installed for long-term monitoring

FIG. 6--Schematic of steps used for contaminant plume and hydrogeologic characterization

CASE STUDY OF GASOLINE SERVICE STATION

Three underground gasoline storage tanks were removed from the site during station remodeling, and impacted soil was found beneath a dispenser. A regulatory work plan was prepared showing a minimal number of CPT locations and optional on-site and off-site locations that would be implemented if needed to define the extent of the plume. Although soil sampling at 5-foot (≈1.5-meter) intervals was a regulatory requirement, the oversight agency waived this requirement because the soil type logging at 5-cm intervals and vapor sampling at 5-foot (≈1.5-meter) intervals produced the same data. The number of soil samples collected for laboratory analysis would need to be sufficient to convince the regulatory agency that likely areas of maximum soil concentrations were represented, and that the horizontal and vertical extent of the soil plume was defined. The agency required that impacted groundwater be defined by sampling from conventional 4-inch (≈10.16 cm) diameter monitoring wells.

This assessment closely followed the ideal two-step procedure described previously. Beginning at the known dispenser leak, a survey using CPTu outfitted with a soil vapor sampler was performed. A total of 16 soundings to a depth of approximately 60 fbg was performed in 3 days at a cost of about $10,000.

CPT soundings indicated that the site is underlain by interbedded sandy and clayey soils to a depth of 65 fbg (≈19.81 mbg) (Figures 7 and 8). The horizontal and vertical extent of the gasoline vapor plume as defined by this survey is shown in Figure 7a. A pore pressure dissipation test at two depths (45 and 50 fbg [≈13.72 and 15.24 mbg]) indicated that the depth to groundwater is 40 fbg with no significant vertical gradient. An elevation survey of the grade surface at all

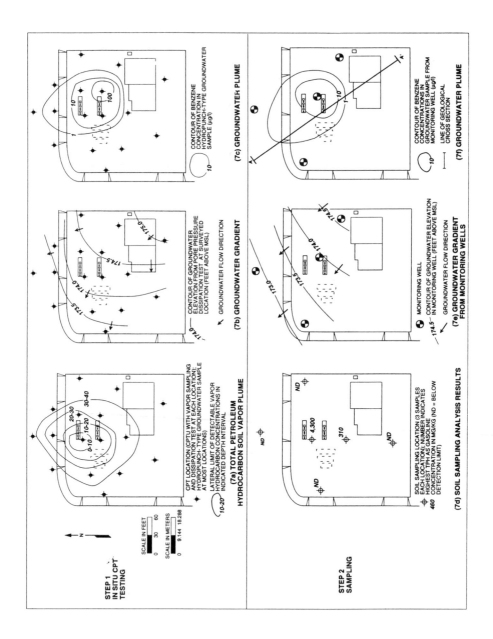

FIG. 7--Maps of service station case study site

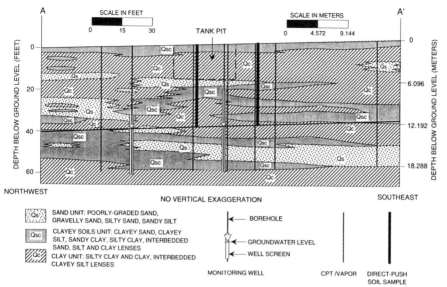

FIG. 8--Cross section of service station case study site

dissipation test locations indicated that the horizontal gradient is 0.01 with flow toward the northwest (Figure 7b). Hydropunch™ samples were collected at about half of the CPT locations, which delineated the center and limits of the groundwater plume (Figure 7c). The Hydropunch™ sampling took an additional 2 days at a cost of approximately $6,000. This concluded Step 1, and resulted in detailed delineation of site geology, site hydrogeology, and the extent of soil and groundwater plumes.

Step 2 began with collection of soil samples with the CPT rig at two locations within the soil vapor plume and at four locations around the vapor plume, and resulted in quantification of soil concentrations and confirmed the definition of the limits of the soil plume (Figure 7d). Soil sampling was completed in 1 day at a cost of about $3,000.

An optimal monitoring well field was designed using the hydrogeologic and contaminant information determined from Step 1. Two downgradient monitoring wells, one upgradient monitoring well, and one well at the center of the plume were installed with a hollow-stem drill rig and tested. These wells confirmed the groundwater gradient (Figure 7e) and quantified compound concentrations in groundwater (Figure 7f) to meet regulatory requirements and for long-term monitoring.

This case study illustrates the comprehensive CPT assessment strategy applied to a simple site. The authors have extensive experience assessing similar sites in which 6 to 12 wells were installed during "stepping out" in order to adequately characterize the plume. The costs of these installations and monitoring over a period of years is much more expensive than the total project cost of the case study site. The cost savings increase dramatically for sites that are more complex.

CONCLUSIONS

CPT is a superior investigative tool for subsurface environmental site investigations. The primary limitation of CPT compared to conventional drilling is shallower refusal. Compared to conventional drilling, CPT produces higher quality data at lower total cost with minimal site disturbance, no generation of soil cuttings, greater speed, and minimal health and safety risks.

REFERENCES

[1]   Douglas, B.J. and Olsen, R.S., "Soil Classification Using Electric Cone Penetrometer," Symposium on Cone Penetration Testing and Experience, Geotechnical Engineering Division, ASCE, pp. 209-227, 1981.

[2]   Robertson, P.K. and Campanella, R.G., "Interpretation of Cone Penetration Tests. Part I: Sand," Canadian Geotechnical Journal, v. 20, pp. 718-733, 1983a.

[3]   Robertson, P.K. and Campanella, R.G., "Interpretation of Cone Penetration Tests. Part II: Clay," Canadian Geotechnical Journal, v. 20, pp. 734-745, 1983b.

[4]   Robertson, P.K., "Soil Classification using the Cone Penetration Test," Canadian Geotechnical Journal, v. 27, pp. 151-158, 1990.

[5]   Campanella, R.G. and Robertson, P.K., "Current Status of the Piezocone Test," First International Symposium on Penetration Testing, ISOPT-I, pp. 1-24, 1988.

[6]   Senneset, K. and Janbu, N., "Shear Strength Parameters Obtained from Static Cone Penetration Tests," American Society for Testing and Materials, Special Publication 883, 1984.

[7]   Jones, G.A. and Rust, E.A., "Piezometer Penetration Testing CUPT," Proceedings, 2nd European Symposium on Penetration Testing in the U.K., Birmingham, 1982.

[8]   Robertson, P.K., Sully, J.P., Woeller, D.J., Lunne, T., Powell, J.J.M., and Gillespie, D.J., (in press), "Estimating Coefficient of Consolidation from Piezocone Tests," Canadian Geotechnical Journal, 1992.

[9]   Teh, C.I. and Houlsby, G.T., "An Analytical Study of the Cone Penetration Test in Clay," Geotechnique, v. 41, pp. 17-34, 1991.

[10] Campanella, R.G. and Weemees, I., "Development and Use of an Electrical Resistivity Cone for Groundwater Contamination Studies," Canadian Geotechnical Journal, v. 27, pp. 557-567, 1990.

[11] Woeller, D.J., Weemees, I., Kokan, M., Jolly, G., and Robertson, P.K., "Penetration Testing for Groundwater Contaminants," Geotechnical Engineering Congress, v. 1, 1991.

Daniel A. Leavell[1], Philip G. Malone[2], and Landris T. Lee[1]

THE MULTIPORT SAMPLER: AN INNOVATIVE SAMPLING TECHNOLOGY

REFERENCE: Leavell, D. A., Malone, P. G., and Lee, L. T., "The Multiport Sampler: An Innovative Sampling Technology," Sampling Environmental Media, ASTM STP 1282, James Howard Morgan, Ed., American Society for Testing and Materials, 1996.

ABSTRACT: The Multiport Sampler (MPS) was developed to overcome current cone penetrometer sampling limitations. The MPS was specifically designed to obtain multiple fluid (gas, vapor, and/or liquid) samples during a single penetration while minimizing sample cross-contamination from each sampling cycle. The MPS consists of a series of vertically stacked sampling modules that are independently operated from the surface. As the cone penetrometer is advanced to the desired sampling depth, a module is selectively opened to allow soil pore fluid to be drawn through a sidewall port. Fluid samples may be either analyzed during the penetration event using equipment on the surface or collected inside the penetrometer for subsequent analysis. The MPS has been used only to sample vapors from soils contaminated with chlorinated organic solvents to determine the relative concentration of the contaminants in different soil strata.

KEY WORDS: penetrometer, electric cone, sampler, contaminant, ion trap mass spectrometer, contaminant trap

The Multiport Sampler (MPS) was designed and developed to provide a more effective means of obtaining soil pore fluid (gases, vapor, and liquid) samples using a cone

[1]Research Civil Engineer, Geotechnical Laboratory and [2]Geophysicist, Structures Laboratory, U.S. Army Engineer Waterways Experiment Station, Vicksburg, MS 39180.

penetrometer. Current technology typically uses one sampling port to obtain one or more samples during each penetration event. A sample is collected and either stored in the sampler or brought to the surface before subsequent samples are taken. For multiple sampling events during a single penetration, sample cross-contamination is kept to a minimum by purging the sampling port (with water) between each successive sampling cycle. However, even low-level residual contamination in the sampler may adversely affect the analysis when the analyte concentrations are in parts per million (ppm) or less.

A variety of samplers have been designed for retrieving liquid and gas samples from soils or other geologic media. The challenge is to design a sampler that remains clean and sealed, can be constantly verified as sealed, and opens on demand unfailingly. A variety of different solutions to the problem have been proposed. A device patented by Raugust [1] in 1967 uses a sample barrel with a side wall port and a shear-pin that holds a plug in the port. A plunger activated from the surface knocks the shear pin free and allows the pressure outside the sampler to force the port open. This approach is useful but the mechanical operation limits this device to an oil-well tool application.

An alternate approach proposed by Peterson [2] used a sampler with side-wall ports that open when the outer barrel of the sampler is rotated so that a port in the outer barrel is aligned with an opening to a central cylinder that receives the sample. This design presents potential problems in maintaining a seal and requires that all of the ports be opened at one time.

A pneumatically-controlled sampler device was proposed by Knight and Miller [3] in 1953, but this device was developed as a stab-type soil sampler. The point of the sampler was maintained at the end of the sampling tube by filling the interior of the sampler with compressed air. The pressure of the air behind the point was controlled at the ground surface. Releasing the air pressure allowed the point and the soil immediately in front of it to move into the sampler and be brought to the surface in the sampler tube.

The MPS was designed and developed by Peters et al. [4] to obtain multiple vapor and liquid samples from soil strata during any single penetration without compromising the integrity of the analyte due to cross-contamination between sampling events. Each MPS sampling port is dedicated to obtaining one sample during a single penetration event. As the penetrometer is advanced to any given depth (within the cone penetrometer depth capability) each sampling port is kept tightly closed under positive gas pressure. At the sampling depth, any selected port may be independently opened to allow entrance of a soil pore fluid. The fluid may then be analyzed during the sampling event or collected for subsequent analysis.

EQUIPMENT DESCRIPTION

The MPS consists of three components: the probe, umbilical cable, and auxiliary equipment. The probe contains the sampling modules, the umbilical cable connects the probe to the surface equipment, and the auxiliary equipment consists of support items required for operation, data collection, and sample analysis. Figure 1 shows the MPS components.

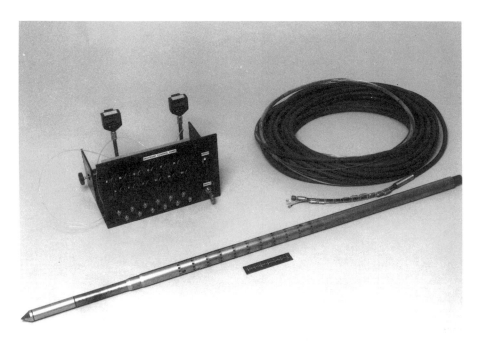

Fig. 1--Photo of the Multiport Sampler (MPS) system showing a six-port control console (upper left), the umbilical cable (upper right), and the probe with sampling modules installed (bottom). The scale shown is 6.0 in. in length (1.0 in. = 25.4 mm).

Probe

The probe contains up to twelve (12) sampling modules or ports, and is designed for attachment to a standard electric cone penetrometer.

Electric cone--The electric cone penetrometer is used to measure the penetration resistance of the soil as described in ASTM Test Method for Deep, Quasi-Static, Cone and Friction-Cone Penetration Tests for Soil (D 3441). The measured data are used to estimate the soil type as the penetrometer advances, enabling rapid delineation of the soil stratigraphy [5 and 6].

Sampling Modules--The port of each sampling module is independently opened to allow fluids (gas, vapor, and/or liquid) to enter the module. Figures 2 and 3 indicate the sectional view and top plan view, respectively. All sampling module parts are fabricated

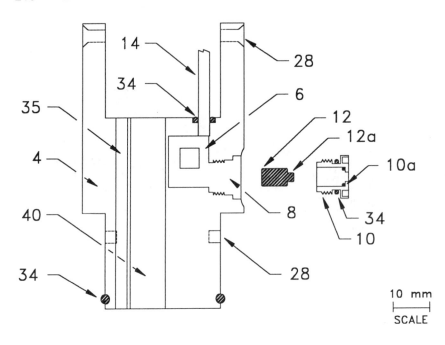

Fig. 2--Sectional view of a MPS sampling module.

from D-2 steel and vacuum heat treated to minimize abrasion from soils. The parts of the sampling module are numbered in Figures 2 and 3 and described below.

Item	Description
4	Steel housing
6	Sampling cavity
8	Transverse opening
10	Threaded insert (port)
10a	Inner lip and o-ring
12	Piston
12a	Piston tip
14	3.18 mm (0.125 in.) nylon tubing or contaminant trap
28	Connecting screw holes
34	O-ring
35	3.18 mm (0.125 in.) nylon tubing passageway
40	Grout tube and electrical wire passageway

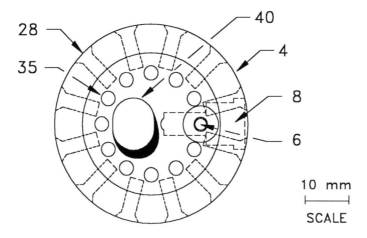

Fig. 3--Plan view of the MPS sampling module.

The sampling module is a 44.5-mm (1.75-in.)-diameter housing (4) containing a sampling cavity (6). The cavity includes a transverse passage or opening (8) which extends to the exterior of the module. An insert (10) is threaded into the housing within the transverse opening (8). The insert (also referred to as the port) contains a cylindrical chamber configured to receive a piston (12) which slides outward to close the port or slides inward to open the port (Figure 4a and 4b). Movement of the piston is controlled

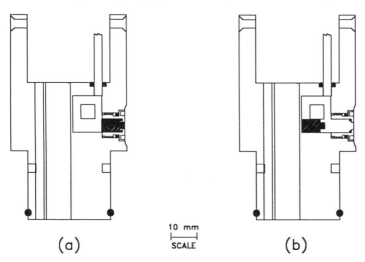

Fig. 4--Cross-section of the MPS Sampling Module showing (a) piston in the closed position and (b) piston in the open position.

by varying the gas pressure within the sampling cavity. The gas (clean compressed air or inert gas) is supplied via the umbilical tubes. The insert (10) has an inner lip portion (10a) which prevents movement of the piston (12) outside of the opening (8) and beyond the exterior surface of the housing (4). When the piston is in the closed position, the piston tip (12a) is flush with the exterior surface of the housing. When sampling, a sample is drawn through the port (10), into the sampling cavity (6), and either into 1/8-in nylon tubing (14) to the surface for direct analysis or into a contaminant trap (14) for a post analysis after the MPS is withdrawn.

Umbilical Cable

        The umbilical cable (Figure 5) connects the MPS probe to the surface (above ground) equipment. It consists of up to twelve 3.18 mm (0.125 in.) nylon tubes concentrically bundled around a 9.53-mm (0.375-in)-diameter grout delivery tube and electrical wires for collecting data from the strain gauges in the cone. Each tube end has an o-ring-sealing snap-type connector to allow easy connection to the probe assembly. A single tube connects each module to the equipment on the surface. The grout delivery tube allows the open penetration hole to be filled with grout as the penetrometer rod is withdrawn to the surface. The grout tube connects to the grout pump at the surface and terminates at the sacrificial penetrometer tip. The outside sheath of the umbilical cable is a heat-shrunk neoprene jacket. The length of the umbilical cable is approximately 45 m (150 ft) (cable lengths of 90 m (300 ft) are possible) with the overall outside diameter being approximately 17 mm (0.675 in).

Auxiliary Equipment

        A control console provides the capability for independent operation of each sampling module. Figure 6 shows a schematic drawing of the control console. The tubes

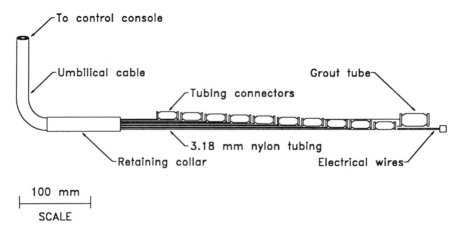

Fig. 5--Umbilical cable that connects MPS Probe to surface instrumentation.

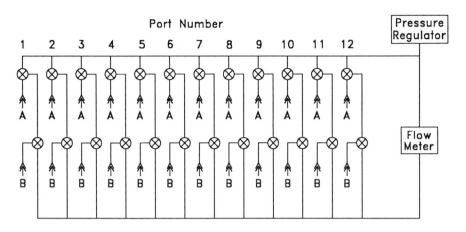

A — Connection point for Ion Trap Mass Spectrometer (ITMS) or vacuum source

B — Connection point for MPS tube from umbilical cable

⊗ — Three—way valve

Fig. 6--Schematic drawing of a twelve-port control console.

in the umbilical cable attach to the console at point B. Instrumentation for direct sample analysis or a vacuum source for alternate sampling requirements is connected at point A. An ion trap mass spectrometer (ITMS) has been used for direct analysis of contaminated soil pore vapor. Each pair of three-way valves control sampling and airflow monitoring of each MPS sampling module. A non-bleeding pressure regulator provides pressure control, and a digital flowmeter allows the airflow within each tube to be monitored. A compressed gas (nitrogen) cylinder is typically connected to the pressure regulator and a small portable vacuum pump is used for an alternate vacuum source.

OPERATION AND MAINTENANCE

Assembly

Prior to assembly of the MPS, the individual modules are thoroughly cleaned in an ultrasonic cleaner using either organic solvents or water. All O-rings are lubricated with an inert silicone lubricant and installed in their respective locations. The pistons are inserted into the threaded ports, and the piston-port assemblies are installed in the MPS modules. Each module is sequentially aligned and attached to an adjoining module by twelve hardened screws. An alternative design that replaces the twelve screws with three steel dowel pins is currently being tested. After all modules are assembled, the electrical wires from the electric cone are threaded through the grout tube passageway prior to attachment of the electric cone. The grout tube and all individual 3.18 mm (0.125 in.) nylon tubes that attach the MPS to the umbilical cable are cut to size and installed. The final step in the assembly is to attach the MPS to the umbilical cable using o-ring-sealing snap-type connectors.

After all connections are complete, the umbilical cable is connected to the control console and the tubing pressurized to 3.5 bar (50 psi). The valves are sequentially opened to allow each module to be pressurized. Observing the pressure gauge on the console for any pressure drop and listening will determine if leaks are present. A flowmeter can be used to determine if there is a pressure leak (no flow means there are no leaks). At this point, the modules are individually checked for pressure leaks and the piston is cycled (opened and closed). Any leakage should be investigated and corrected.

Operation

Prior to deployment of the MPS, the sampling ports are sealed at the surface using a positive gas pressure. The positive gas pressure forces the port piston against the o-ring and lip of the port, sealing the port and restricting fluids from entering the sampling chamber. The gas pressure also forces the piston flush against the exterior surface of the housing minimizing cross-contamination as the MPS is pushed to the sampling depth. Positive gas pressures of 8.6 bar (125 psi) or more are used for pressurizing the interior of the sampling chamber. High purity gas is typically used to guarantee that contaminants are not introduced into the sampling system. The port sealing piston is the only moving part in the sampling module making its operation simple; there are no springs or check valves.

When the sampling port is at the desired depth, a sample is obtained by releasing the port sealing pressure and applying a slight vacuum, approximately -0.35 bar (-5 psi), to slide the piston back into the sampling chamber. The soil pore fluid sample enters the chamber and is either pulled through a single tube to the surface for analysis or pulled into a contaminant trap (CT) at the sampling chamber for subsequent analysis upon withdrawal of the probe. The CT is a specially designed tube filled with chemical absorbents (such as activated carbon or chemically treated silica gel) that absorb contaminants.

Disassembly

After sampling is complete and the MPS probe retracted to the surface, it is disassembled by reversing the sequence for assembly. First, disconnect the electrical connections, grout tube, and each MPS sampling module tube from the umbilical cable. Carefully disconnect the MPS sampling modules from the electric cone, ensuring that the electrical wiring does not twist. Disconnect the grout tube from the electric cone and remove it from the MPS sampling modules. Finally, pull the electrical wire through the grout tube passageway. The piston and port assemblies can now be removed. The individual modules can be separated for cleaning, if necessary.

<u>Cleaning</u>

The MPS is typically washed and/or wiped clean as it is pulled from the soil. The screw heads and open ports are typically filled with soil and will require cleaning prior to their removal. If the screws are removed they are discarded. All O-rings are always inspected and cleaned before they are reused. An ultrasonic cleaner is used to clean the major parts of the MPS probe after it has been disassembled. The cleaner is of sufficient length to immerse either the assembled MPS modules or the disassembled modules after disconnecting them from the umbilical cable and electric cone. The MPS parts are placed in the ultrasonic cleaner for approximately one-half to one hour. Other utensils (squirt bottles filled with methyl alcohol, Q-tips, and tweezers) can be used for detail cleaning after the parts are removed from the ultrasonic cleaner.

The cleaning fluid may be distilled water, organic solvents, or a mixture of distilled water and solvents. Methyl alcohol is a useful solvent which is soluble in water or can be used in an undiluted form. Other solvents (i.e. ethyl alcohol, acetone, etc) may also be used. In all cases precautions are needed to prevent undue personnel exposure and accidental ignition. No cleaning agents containing chlorinated hydrocarbons or other target contaminants may be used due to the possibility of contaminating the sampler prior to sample collection.

OPERATIONAL TESTS

The MPS probe was subjected to laboratory and field testing to determine its structural integrity and operational capabilities. During the laboratory tests, the probe was hydraulically pushed into a large-scale stress chamber filled with sand to a simulated *in situ* soil stresses equivalent to a depth of 61 m (200 ft). The laboratory tests were successfully completed.

A field test integrating the probe and the Site Characterization and Analysis Penetrometer System (SCAPS) truck (Figure 7) was conducted at the U.S. Army Engineer Waterways Experiment Station (WES), Vicksburg, Mississippi. A site containing homogeneous silt deposits (water table at 10 m (33 ft)) was chosen, and the probe was hydraulically pushed to 8.5 m (28 ft) and retracted. During the push, the ports were held in the closed position under compressed gas (nitrogen) pressure; no pressure leakage was detected. At a depth of 8.5 m (28 ft) several ports were opened to ensure their operation under field conditions.

LABORATORY SAMPLING TEST

The laboratory test was used to evaluate the MPS probe's capability to retrieve samples from the vadose zone for quantitative measurement of contaminant concentrations. The test bed consisted of a 305-mm (12-in.)-diameter stainless steel cylinder, 508 mm (20 in.) deep, filled with sand. The sand was given a 10 percent water content using a 5 ppm aqueous solution of trichloroethylene (TCE), causing the bulk concentration of

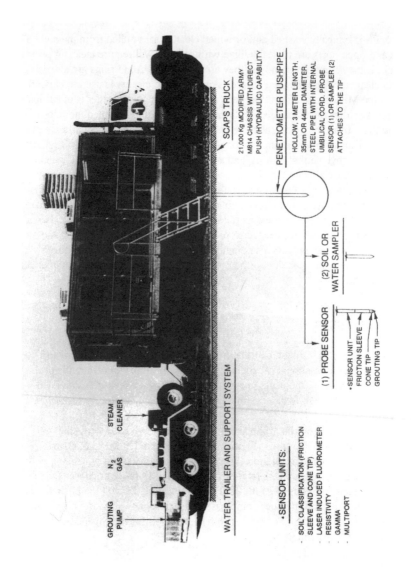

Fig. 7.--The SCAPS truck.

TCE to be 0.5 ppm. Therefore, the vapor phase concentration was between 0.5 and 5 ppm. The cylinder was filled with the sand and TCE mixture to within 50 mm (2 in.) of its top, covered, and then allowed to equilibrate over a 24 hour period.

The probe was configured with three modules. The ports were spaced 50.8 mm (2 in.) apart with port 1 being 267 mm (10.5 in.) from the MPS tip. The MPS tip was pushed until it was in contact with the bottom of the test bed. Sampling of the pore vapor was conducted after instrument calibration was completed. The sampling procedure was: (1) sample ambient air through all ports and tubing, using an ITMS for direct analysis, to ensure the modules and tubing are free of contamination, (2) sample through a port opened at depth (in test bed) for direct analysis with the ITMS, (3) sample through two ports at different flow rates into CTs, and (4) sample through the ports with a 45 m (150 ft) of 3.18 mm (0.125 in.) nylon tubing for direct ITMS and CT analyses.

During the first test (ambient air sampling), a trace of hydrocarbon contaminant was found in one port. However, the trace amount was not significant enough to mask the ITMS identification of the targeted contaminant, TCE. The second and third tests (*in situ* sampling) were conducted using tube lengths of approximately 1.6 m (5 ft). Both direct analysis (nylon tube connection to the ITMS) and post-test analysis (CT inside probe) were performed. The results were:

Port 1 - TCE = 2.52 ppm, collected using a 2.2 ml/min mass flow rate (11 ml total volume drawn), data from a post-test analysis of the CT.

Port 2 - TCE = 3.33 ppm, collected using a 2.2 ml/min mass flow rate (5.5 ml total volume drawn), data from a post-test analysis of the CT.

Port 3 - TCE = 3.30 ppm, collected using a 30 ml/min mass flow rate, data from a direct ITMS analysis.

The fourth test was conducted for both ITMS and CT analyses through approximately 45 m (150 ft) of clean tubing. The results were:

Port 1 - TCE = 2.46 ppm, collected using a 3.3 ml/min mass flow rate (19.8 ml total volume drawn), data from a post-test analysis of the CT.

Port 3 - TCE = 2.82 ppm, collected using a 16 ml/min mass flow rate, data from a direct ITMS analysis.

The laboratory tests clearly demonstrate the ability to obtain quantitative concentration measurements using both CT and the ITMS measurement methods. The ITMS tests using port 3 showed that the nylon tubing absorbed some of the TCE contaminant (3.30 ppm versus 2.82 ppm for 1.6 m (5 ft) and 45 m (150 ft) lengths of tubing, respectively). The tests also showed that the vapor phase was not homogeneous throughout the test bed (higher concentrations were measured near the top of the test bed).

FIELD SAMPLING TEST

A field demonstration of the MPS was conducted at the Savannah River Integrated Demonstration Site (SRS), located near Aiken, South Carolina. The purpose of the field demonstration was to complete integration of the SCAPS truck and *in situ* sampling methods for measuring TCE and perchloroethylene (PCE) contamination. The field test was based on the capability of determining discrete contamination levels of TCE and PCE by drawing the analyte to the surface for direct analysis with the ITMS and using a CT to collect the contaminant for post analysis with the ITMS after penetrometer withdrawal. The demonstration was conducted at SRS because of its known areas of subsurface TCE and PCE contamination.

The MPS was configured with ten sampling modules of which eight were used. The odd numbered ports, beginning above the electric cone, were connected to the ITMS for direct measurements and CTs were connected to the even numbered ports. The ports connected to the ITMS were used for qualitative and quantitative screening while the CTs were used for quantitative measurements. Figure 8 depicts the field operation of the MPS.

Prior to pushing the MPS, the instrumentation was calibrated through the probe for measuring TCE and PCE concentration levels less than 500 ppm. All ports were pressurized to 8.6 bar (125 psi) and checked for leakage prior to the push. Figure 9 shows the soil type classification with depth (interpreted from electric cone data), type of sample obtained, contaminant, and contaminant concentration.

The probe was advanced to 7 m (23 ft) and ITMS (port 1) and CT (port 2) samples were taken. The ITMS sample showed 0.26 and 1.2 ppm of TCE and PCE, respectively. The CT sample showed 1.7 and 7.3 ppm of TCE and PCE, respectively. The ITMS data showed lower contamination levels than did the CT. The reduction in the measured concentration may be caused by the absorption of contaminant into the walls of the nylon tubing that links the subsurface port with the surface analytical equipment (ITMS). The laboratory tests at ORNL also showed that TCE is absorbed into the walls of the nylon tubing.

Data for sampling at 8.5 m (28 ft) is only available for the ITMS sample (port 3). The data file for the desorbed CT sample (port 4) was corrupted by a computer malfunction and could not be retrieved. The ITMS sample showed the concentration of both TCE and PCE increased with depth to 9.2 and 73 ppm, respectively.

The ITMS sample (port 7) showed 7.4 ppm of TCE and 58 ppm PCE at 9.5 m (31 ft). The CT sample (port 6) showed 8.5 and 87.5 ppm of TCE and PCE, respectively. The correlation between the ITMS and CT samples was good with the CT sample giving higher concentration levels than the ITMS (same response at 7 m (23 ft)). Also, the contaminant concentration levels decreased, as indicated by the ITMS samples, after 8.5 m (28 ft) indicating an uneven distribution of the contaminant with depth.

The probe was advanced into a clay layer at 11.1 m (36.5 ft) and could not be advanced any further. An ITMS sample (port 5) was obtained at this depth even though there was low fluid flow (1 ml/min under a full vacuum). The sample showed TCE and PCE concentrations in this layer at 4.3 ppm and 20 ppm, respectively. The drop in contaminate concentration levels after 8.5 m (28 ft) indicates that contamination is not being dragged by the MPS to lower levels in the strata. CT sampling was not performed

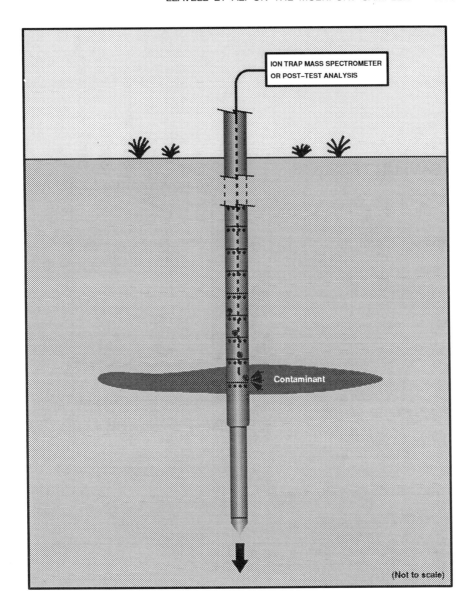

Fig. 8--Field operation of the MPS.

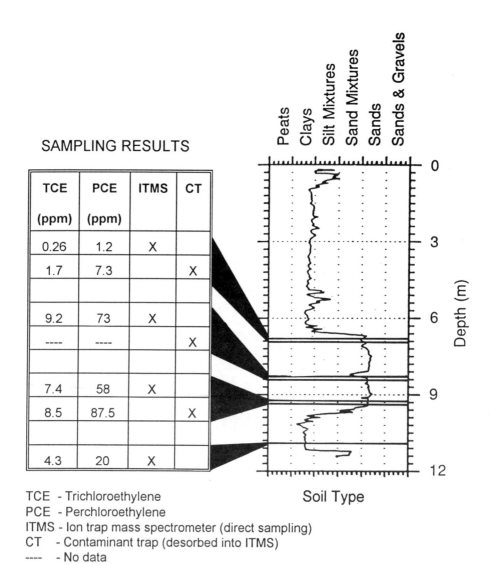

Fig. 9--The soil type classification with depth (derived from cone penetrometer correlations), type of sample obtained, contaminant, and contaminant concentration.

at this depth because the impermeable clay layer prevented the development of a steady state flow rate which is necessary for CT sampling. The probe was withdrawn.

CONCLUSION

Laboratory and field testing has shown the MPS to be a viable method for sampling soil pore gases for contaminant analyses. The MPS is capable of collecting samples for on-site analyses or for analyses in a field or analytical laboratory after the site investigation is complete. The CT and ITMS interface well with the MPS and promise to add to its usefulness and versatility as a site characterization and fluid sampling tool.

ACKNOWLEDGMENTS

The design, development, and testing of the MPS was performed at WES with additional laboratory testing at ORNL and field testing at SRS. The research was sponsored by the Army Environment Center (AEC) under their project officer, Mr. George E. Robitaille. The CT and direct measuring ITMS technologies were developed and adapted to the MPS by ORNL personnel. Currently, development of the MPS is being funded jointly by the AEC and the Strategic Environmental Research and Development Program (SERDP). Permission to publish this information was granted by the sponsoring agency and the Chief of Engineers.

REFERENCES

[1]   U.S. Patent Office. (1967). Method and Apparatus for Obtaining a Fluid Sample From an Earth Formation." Patent No. 3,356,137 issued December 5, 1967.

[2]   U.S. Patent Office. (1982). "Variable-length Sampling Device." Patent No. 4,359,110 issued November 16, 1982.

[3]   U.S. Patent Office. (1953). "Pneumatically-controlled Soil Sampler." Patent No. 2,664,269 issued December 29, 1953.

[4]   U.S. Patent Office. (1994). "A Modular Device for collecting Multiple Fluid Samples from Soil Using a Cone Penetrometer." Patent No. 5,358,057 issued October 25, 1994.

[5]   Campanella, R. and Robertson, P. (1982). "State of the Art in In-Situ Testing of Soil: Developments Since 1978," Department of Civil Engineering, Vancouver: University of British Columbia.

[6]   Olsen, R. S. and Farr, J. V. (1986). "Site Characterization Using the Cone Penetrometer Test," Proceedings of the ASCE Conference on Use of In-Situ Testing in Geotechnical Engineering. American Society of Civil Engineers, New York, NY.

Sampling Strategies

Charles W. Attebery[1]

ACCELERATING REMEDIAL INVESTIGATIONS AT CLOSING AIR FORCE INSTALLATIONS

REFERENCE: Attebery, C. W., ''Accelerating Remedial Investigations at Closing Air Force Installations,'' Sampling Environmental Media, ASTM STP 1282, James Howard Morgan, Ed., American Society for Testing and Materials, 1996.

ABSTRACT: The Air Force, in coordination with the Environmental Protection Agency and state environmental regulatory agencies, is attempting to expedite the transfer of federal land to the public. The Air Force has implemented four programmatic and technical initiatives focused on streamlining environmental sampling strategies and decision-making processes to accelerate remedial investigations at a limited number of closing Air Force installations. The success of these initiatives may result in standardization of remedial investigation environmental sampling strategies and decision-making processes for the Air Force.

KEYWORDS: Air Force, remedial investigation, closing Air Force installations

INTRODUCTION

At the direction of the President and Congress, the Air Force, in coordination with the Environmental Protection Agency (EPA) and state regulatory agencies, is working to expedite remedial investigations (RIs) at closing Air Force installations. The Air Force has implemented four initiatives that have affected environmental sampling strategies and the use of environmental samples to expedite decision making. Three of the initiatives are focused on facilitating communication and decision-making processes, while the fourth initiative addresses the collection and analysis of environmental samples. This paper presents (1) a summary Presidential and Congressional directive concerning the "fast-tracking" of environmental projects, (2) a brief description of the Air Force environmental investigation sampling strategy, (3) four Air Force initiatives intended to accelerate environmental projects, (4) conclusions concerning the effectiveness of the initiatives and their effect on environmental sampling, and (5) recommendations.

[1]Member of the Technical Staff, The MITRE Corporation, Brooks Air Force Base, San Antonio, Texas 78235-5357.

BACKGROUND

Government Directive

The United States military is currently reducing its basing and staffing requirements to match a force-structure plan that was created in response to the Defense Authorization Amendments and Base Realignment and Closure (BRAC) Act of 1988 (BRAC 88), Public Law 100-526, 10 United States Code (USC) 2687 note, and the Defense Base Closure and Realignment Act of 1990 (DBCRA 90), Pubic Law 101-510, 10 USC 2687 note. They are currently closing military installations or realigning military missions throughout the United States.

To ease the military installation closure and realignment economic effects on the communities surrounding these installations, the President and Congress agreed that environmental cleanup was a top priority and that government agencies and departments should work together to expedite the transfer of military real property for beneficial economic reuse by executing "fast-track environmental cleanup."

The Air Force responded to the call for cooperation by attempting to streamline the integration of the myriad of federal real property and environmental laws and regulations that are applicable to property transfer at closing military installations. The relationship of the various laws is complex, and the time lines of their satisfaction is largely dependent on the Comprehensive Environmental Response, Compensation and Liability Act (CERCLA) as amended and the Resource Conservation and Recovery Act (RCRA) as amended [1,2,3].

CERCLA and RCRA Investigation and Cleanup Processes

The CERCLA and RCRA investigation and cleanup processes are basically linear (Fig. 1). Perhaps the most significant difference between CERCLA and RCRA investigation and cleanup processes, aside from CERCLA addressing old contaminated sites and RCRA focusing on newly contaminated sites, is that CERCLA cleanup criteria are negotiated based on established risk assessment procedures, and RCRA cleanup criteria are generally preestablished. Environmental sampling requirements for the programs are basically the same.

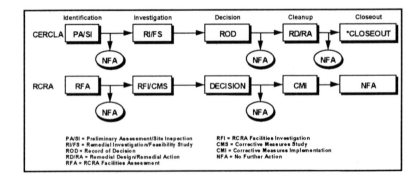

FIG. 1--CERCLA and RCRA investigation and cleanup processes

Air Force Iterative Investigation Process and Environmental Sampling Strategy

The Air Force employs an iterative process to establish environmental sampling strategies for RIs (Fig. 2). The process, which includes development of a conceptual site model (CSM) and establishment of data quality objectives (DQOs), historically relied on the experience of environmental contractors to establish a cost-effective environmental sampling strategy even though some quality oversight was provided by the Air Force. The Air Force provided guidance concerning quality assurance procedures for the collection of environmental samples and analytical laboratory quality control requirements to support the collection of scientifically defensible environmental data [4].

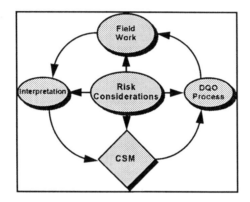

Fig. 2--Air Force iterative investigation process

With the formation of the Air Force Center for Environmental Excellence (AFCEE) in 1991, the Air Force began providing more detailed technical direction to environmental contractors concerning the creation of environmental sampling strategies. Emphasis was placed on the use of lower-cost soil gas analysis and geophysical technologies to define environmental contamination. More expensive soil and groundwater sampling and analysis was reserved for later iterations of the iterative RI process. The intent of the Air Force iterative investigation process was to collect the minimum amount of information necessary to make a decision concerning the remediation of a contaminated site. The amount of data necessary to make a decision is subjective and appears to vary between EPA regions, states, and regulatory representatives. "How good is good" seems to depend on the personality of the persons evaluating the site-specific data. Subsequently, RI and cleanup sampling strategies were not standardized.

AIR FORCE INITIATIVES TO EXPEDITE ENVIRONMENTAL INVESTIGATIONS

To meet the Presidential and Congressional direction to expedite the transfer of property at closing installations, the Air Force has attempted to expedite their iterative investigation process by (1) facilitating communication and empowering installation-level personnel by forming BRAC Cleanup Teams (BCTs), (2) developing innovative work execution strategies by developing nontraditional work plan (WP) formats that exploit computerized data management and geographical information systems, (3) incorporating presumptive remedy requirements into RI sampling strategies, and (4) emphasizing the use of "push-technology" and field screening of environmental samples with on-site mobile laboratories.

Base Realignment and Closure Cleanup Teams

BCTs are an attempt by the Air Force to expedite the RI and cleanup process through increased communication and shared decision-making authority and responsibility with regulatory agencies. Implemented at all closing Air Force installations, a BCT is made up of Air Force, EPA, and state environmental regulatory agency representatives and has the authority, responsibility, and accountability for environmental restoration and closure-related environmental compliance programs at closing installations [3]. BCT members typically manage only the closing installation project and are responsible for preparing the BRAC Cleanup Plan (BCP), which is a comprehensive planning document that contains status and life cycle planning information for all environmental programs at a closing installation. The Air Force provides the EPA and state regulatory agencies with funds to pay for the extra project support.

The BCT is advised on technical issues by a project team that is made up of regulatory and Air Force technical experts and environmental contractor personnel. At the request of the BCT, project team personnel organize and present economic and technical site-specific information to the BCT to facilitate BCT decisions. Site-specific environmental sampling strategies are typically established by the BCT before site-specific WP documents are prepared by the environmental contractor. Environmental contractor quality oversight is still provided by AFCEE.

The BCTs minimize normal Air Force funding bureaucracy and delays by receiving funds directly from the Air Force Base Conversion Agency (AFBCA), which is responsible to the Air Force for the successful closure of designated Air Force installations. The BCT membership will include an installation-level AFBCA representative.

Nontraditional Work Plan Formats

The Air Force developed a nontraditional RI WP format to streamline iterative RIs by capitalizing on increased regulatory support, computerized data management, and geographical information system capabilities. The format incorporated abbreviated site-specific WPs and a comprehensive investigation decision tree, which required that the BCT agree on the decision tree once and then be kept informed concerning the application of the decision tree as site-specific investigations progressed. Site-specific WP reviews were accelerated because of their abbreviated format, which referenced the generic decision tree for supporting investigation rationale.

Traditional WPs describe one iteration of data collection. Collected data are summarized in an RI report where recommendations for additional RIs are included. Subsequently, another WP or WP addendum is prepared, reviewed, and revised, and another iteration of data collection (field work) is accomplished. No data collection is accomplished during the preparation and review of the RI report and WP addendum.

The Air Force tried to circumvent the RI report/WP addendum cycle by implementing the nontraditional WP format at a closing installation [5]. The RI WP incorporated a comprehensive investigation decision tree that contained proposed investigation scenarios (including statistical sampling) for numerous generic site conditions. As new site contaminant data were collected, investigation strategies could be adjusted by the contractor (according to the appendix) without formal comment by BCT members. The BCT was briefed on investigation results and the application of the investigation decision tree for each site as investigations progressed.

Several iterations of investigation data collection were accomplished without the formal preparation of iterative WP addenda. WP documentation for the investigation was reduced to updates of standardized forms that (1) identified site conditions based on newly collected data, (2) referenced the decision tree appendix, and (3) contained a brief summary of the proposed field work for the next iteration. The BCT members were able to stagger site-specific field work completion dates to allow for continuous field work for several iterations of data collection. Field work was discontinued when the contractor felt that enough data had been collected to characterize a site.

Aside from the continuous BCT review of ongoing investigations, the use of a computerized data management system combined with a geographical information system was critical to the successful implementation of the nontraditional WP format. The time necessary to manipulate collected data had to be minimized to allow continuous field work to continue. For this project, geologic and site-specific lithologic data were put into electronic format by scientists in the field with laptop computers, and data were fed directly into the database by disk or modem. Analytical data, which are put in electronic format by the analytical laboratory, were also loaded directly into the database. Access to the data is almost immediate, and this allowed scientists to begin interpreting the new data almost immediately. Briefings to the decision-making BCT were accelerated correspondingly.

The three-dimensional presentation of data, using a geographical information system also facilitated the BCT's understanding of site conditions. Geologic and lithologic data correlation, typically done by hand, was done on the geographical information system. The resulting cross sections were sliced, rotated, and moved in three-dimensions, which gave scientists greater insight into complex geologic and contaminant migration scenarios and allowed the BCT to make faster decisions.

AFCEE requires its environmental contractors to submit analytical and geologic data in a predetermined electronic format to correspond with the Air Force's Installation Restoration Program (IRP) Information Management System (IRPIMS) requirements. The Air Force uses this database as a permanent record of its environmental investigations. The database and geographical information systems used by the contractor at this installation were programmed to organize project data into the IRPIMS format for submittal to AFCEE.

Incorporation of Presumptive Remedy Requirements Into Remedial Investigations

The selection of a remedial technology for a contaminated site and the inclusion of strategies to collect additional data to support the design and implementation of that remedial technology during an RI can expedite the environmental investigation and cleanup process. Implemented at a minimum of one closing installation, the collection of remedial technology design and implementation data during an RI process will potentially remove the need to remobilize equipment for collecting data after an appropriate remedial technology has been selected through a feasibility study process.

AFCEE's restoration technology function (AFCEE/ERT) has published a tool to help investigators identify and prioritize technologies that the Air Force is interested in implementing at contaminated sites. The AFCEE remediation matrix [6] is not intended to limit the selection of remedial technology to the prioritized choices, but is intended to focus remedial investigators on the design and implementation requirements of a remedial technology early in the investigation process.

Expanded Use of Push Environmental Sampling Technology and Mobile Laboratories

AFCEE is emphasizing the use of push-technology sampling devices, such as the stratoprobe and HydroPunch, as efficient alternatives to traditional drilling and soil sampling methods. Push-technology typically employs hydraulic presses or hammers to drive sampling devices into the ground to collect soil and groundwater samples. Advantages of push-technology over traditional drilling are (1) more efficient soil and groundwater sample collection, (2) no drill cuttings and groundwater monitor well development water, and (3) increased safety concerning encounters with underground piping and conduits. The technology is effective up to 80 feet below ground surface in some geologic settings.

Push-technology is generally used in conjunction with an on-site mobile laboratory. With a mobile laboratory, environmental samples can be analyzed on site on the same day that they were taken, which gives investigators an opportunity to adjust iterative investigations without interruption of their field effort.

CONCLUSIONS

The initiatives employed by the Air force to expedite the RI and cleanup process can be grouped into two categories: programmatic and technical. Programmatic initiatives include the (1) formation of BCTs, (2) development of a nontraditional WP format, and (3) incorporation of presumptive remedy requirements into RI strategies. Technical initiatives include the (1) expanded use of push-technology and (2) field screening of environmental samples with on-site mobile laboratories.

The apparent Air Force emphasis on programmatic versus technical initiatives suggests that they believe changes in how environmental sampling strategies are established will expedite the environmental investigations more than changes in technical approach. By minimizing the number of BCT decisions made during an RI and adding structure to those decisions that are made, it appears that the Air Force can avoid lengthy reviews and discussions of investigation alternatives. These programmatic initiatives seem to be moving environmental investigations and cleanups at closing installations forward faster than those being conducted with more traditional investigation approaches.

Among programmatic initiatives, the formation of the BCT appears to be the key to making the other initiatives identified in this paper effective with regard to expediting the investigation and cleanup process. The formation of the BCT seems to have facilitated an increased communication and understanding among its members, and the structured BCT communication process, created as part of the BCP preparation, appears to have facilitated the cross-pollination of innovative ideas from environmental contractors into the Air Force IRP. The formation of the BCT also meets the Air Force and regulatory agencies' common goal of working together to expedite the RI and cleanup process on closing installations.

The use of computerized data management systems appears to be critical to the success of the nontraditional WP format described in this paper for expediting the investigation and cleanup process. The nontraditional WP format takes advantage of the communicative environment established by the formation of the BCT and the speed with which data management systems manipulate data. The nontraditional WP format focuses on getting RI and cleanup information to BCT decision makers in the fastest and simplest format possible and predetermining decisions, where appropriate. The use of the nontraditional WP format without the BCT or data management system may not be effective in expediting the overall investigation process.

The incorporation of presumptive remedy investigation requirements into environmental sampling strategies gives the BCT flexibility to move quickly to remedial technology design and implementation if supporting RI data are collected. It also minimizes the cost of additional data collection. Even though unnecessary data may be collected, this initiative appears effective in expediting the investigation and cleanup process, and it is being implemented at closing Air Force installations nationwide.

The use of push-technology and on-site mobile laboratories has increased the speed of environmental sample collection and analysis and has become a standard tool of BCTs at closing installations. These tools have been used with other initiatives described in this paper. The net effect is an accelerated BCT decision-making process that relies on the quick collection and analysis of site-specific data, computerized data manipulation, and presentation. Another benefit of the use of push-technology is that there are no investigation-derived wastes (i.e., drill cuttings and monitor well development water).

RECOMMENDATIONS

The BCT forum facilitates the open transfer of environmental experience that is critical for establishing time and cost-effective environmental sampling strategies. Although the BCT has lessened the effect of the experience of environmental contractors in favor of an Air Force and regulatory team, environmental contractors are encouraged to embrace their project team role with the BCT and participate actively in the BCP development process at closing installations. The cooperative nature of BCTs and the structured BCP development process allows cross-pollination of new ideas among environmental professionals.

The Air Force and regulatory agencies are encouraged to work toward a more standardized environmental investigation and cleanup process across EPA regions, states, and Air Force installations. The initiatives described in this paper should be evaluated as part of their standardization.

REFERENCES

[1]   Department of Defense, *Base Realignment and Closure (BRAC) Cleanup Plan (BCP) Guidebook*, Fall 1993.

[2]   Department of Defense, *Draft Base Conversion Handbook*, Fall 1994.

[3]   Air Force Base Conversion Agency, *Base Conversion Primer*, Spring 1994.

[4]   Air Force Center for Environmental Excellence (AFCEE), *Handbook for the Installation Restoration Program (IRP) Remedial Investigations and Feasibility Studies (RI/FS)*, September 1993.

[5]   United States Air Force, *Draft Remedial Investigation Work Plan for Source Control Operable Unit at Castle Air Force Base*, April 1994.

[6]   Air Force Center for Environmental Excellence Technology Transfer (AFCEE/ERT), "AFCEE Remediation Matrix - Hierarchy of Preferred Alternatives," March 1994.

Eric A. Weinstock, CPG, CGWP, PG [1]

Methods for the Collection of Subsurface Samples During Environmental Site Assessments

---

REFERENCE: Weinstock, E. A., "Methods for the Collection of Subsurface Samples During Environmental Site Assessments," Sampling Environmental Media, ASTM STP 1282, James Howard Morgan, Ed. American Society for Testing and Materials, 1996.

ABSTRACT: With the market for Environmental Site Assessments growing in response to the requirements of lending institutions, the need to collect subsurface samples quickly and inexpensively is greater today than ever before. ASTM, through committee E-50, has developed standardized procedures for the performance of a Pre-Transaction Environmental Screening and for a Phase I Environmental Site Assessment. Standards for the performance of subsurface sample collection during a Phase II Environmental Site Assessment are currently being drafted by committee E-50.

This paper discusses numerous sample collection techniques that have been successfully employed during Phase II Assessments and presents case histories of their application.

The collection of shallow soil samples is described using commercially available hand augers and hand-driven core samplers. These devices are modified with extensions to collect deeper samples from storm drains and leaching pools.

The performance of soil gas surveys are described using both hand-driven sample probes and vehicle-mounted, hydraulically driven vapor probes. Once the soil vapor is collected at the ground surface, a sample of the media is either analyzed on-site using a field-operated detection device or delivered to a laboratory for analysis.

Application and case histories of the Geoprobe(TM) sampling system, a form of "direct push" technology, are described. This device uses vehicle-mounted, hydraulically-driven sample probes. The probe can be advanced to depths as great as 100 feet below grade and can retrieve soil, soil gas and groundwater samples.

---

[1]Senior Hydrogeologist, CA RICH Consultants, Inc., 404 Glen Cove Avenue, Sea Cliff, New York 11579

Use of standard truck-mounted drill rigs are also considered. Collection of soil samples are discussed using standard techniques such as a split-barrel core sampler (ASTM Method D1586-84, Standard Method for Penetration Test and Split-Barrel Sampling of Soils) and installation of monitoring wells. Applications and case histories of the Hydropunch(™) in-situ groundwater sampler are also described. This device allows for the collection of a groundwater sample during the process of performing a soil boring without the installation of a monitoring well.

The method selected for a given project will vary depending on site-specific criteria such as the depth to groundwater and the subsurface geology. This paper discusses the advantages and disadvantages of each method. It also evaluates the criteria that should be considered in selecting in-situ sampling devices, such as a driven sampling probe, versus a permanent monitoring well.

KEY WORDS: Phase II, Environmental Site Assessment, in-situ, sampling, soil, soil gas, groundwater, direct push.

INTRODUCTION

The need to collect subsurface samples quickly and inexpensively is greater today than ever before. Phase I environmental site assessments for real estate transactions typically are limited to file searches, site walk-throughs and limited sample collection. During a follow-up or Phase II environmental assessment, there are currently a variety of methods that can be applied to collect subsurface soil, soil gas and groundwater samples quickly and inexpensively.

This article will explore several techniques currently available to collect in-situ soil, soil gas and groundwater samples without the need to install a permanent well. As these samples are often collected on property not owned by the party requesting the sampling, it may be beneficial to collect samples using a technique that does not require the expense and future liability of a permanent monitoring well [1]. The method selected for a given project will depend on site-specific criteria, such as the depth to groundwater and the subsurface geology at the property. It is not the intention of the author to endorse any one particular product or manufacturer. The purpose of this article is to serve as an overview of the methods available to perform this work including some of the advantages and disadvantages of each technique. Also, the use of trade names is for identification purposes only. There may be other similar and equivalent equipment manufactured under different trade names that are not mentioned in this article.

The equipment available to collect subsurface samples has been divided into four general categories: hand-operated augers; hand-driven coring devices and sampling probes; hydraulically-driven coring devices and sampling probes (also known as "direct push" technology); and mobile drilling rigs. The four categories are summarized on Table 1 of this paper.

Table 1
Categories of Available Equipment for the Collection of
Subsurface Soil, Soil Gas and Groundwater Samples

Category	Soil	Soil Gas	Ground Water	Advantages	Disadvantages
Hand-Operated Soil Augers	X			Inexpensive, quick, easy use, ideal for sampling leaching pool sediments	Depth limitations usually 3-1/2 to 7 meters (5 to10 feet) in most soils, borings in sandy soils below the water table may collapse.
Hand-Driven Coring Devices and Sampling Probes	X	X	X	Inexpensive, versatile mobility, effective in areas with limited head room such as inside buildings or basements.	Depth limitations usually 7 meters (approximately 20 feet), may be difficult to obtain soil samples from a discrete depth, ground water sampling limited by suction lift of equipment
Hydraulically-Driven Coring Devices and Sampling Probes	X	X	X	Fast set-up time, no generation of drill cuttings, equipped with a pneumatic hammer for pavements.	Soil sample volumes may be limited, maximum depth typically 33 meters (100 feet)
Conventional Drill Rigs & Associated Sampling Equipment	X	X	X	Depth limited only by the capacity of the drill rig, wide variety of soil samplers available, option to install a permanent well may be selected.	Most expensive method, drill cuttings produced that may require disposal, water samples below the water table are limited to 500 mils.

HAND OPERATED AUGERS

The common hand auger is by far the least expensive tool for collecting subsurface soil samples. An assortment of hand augering tools are available from numerous equipment vendors. This tool consists of a tee shaped handle that is attached to a 8.9 centimeter (3-1/2 inch) diameter bucket using steel pipe, as shown on Figure 1.

The advantages of this method are its inexpensive cost, the portable nature of the equipment, and the ease in cleaning the sampler between samples.

The depth that this technique can be employed is directly proportional to the strength of the person turning the auger handle and the local geologic conditions. Generally, this tool can be used to achieve a depth of about 3-1/2 meters (10 feet) in most soils, with as much as about 7 meters (20 feet) possible in areas of soft, loosely-packed soil. Also, the presence of saturated sands can limit the effective working depth of this tool. If the area is covered by concrete or asphalt, a portable jack hammer or some other means must be available to open a hole in the pavement. The hand auger method is not very effective in frozen or very gravely soils. This tool is extremely effective for the collection bottom sediments from storm drains and leaching pools.

HAND-DRIVEN CORING DEVICES, SOIL GAS PROBES AND GROUNDWATER SAMPLERS

A number of products are currently on the market that allow you to drive a soil coring device or a sample probe by hand using a slide hammer or other manual driving device (see Figure 2). These also come with attachments which allow for the use of a hand held electric-pneumatic hammer to assist in the installation of a probe. Numerous vendors manufacture this type of equipment for commercial resale. In addition to collecting soil samples, this equipment can also be used to set sampling probes either above the water table for the collection of soil gas samples or below the water table for the collection of shallow groundwater samples.

The system typically consists of flush-threaded, small-diameter pipe that attaches to a variety of sampling tools. A soil sampling attachment can be attached to obtain soil samples at selected depths. Also, slotted probe tips can be attached to collect soil gas samples at a desired depth with the use of a vacuum pump. These probes can be used as a temporary sampling device or left in the ground as permanent sampling ports. If groundwater is within 7 to 8 meters (20 to 25 feet) of the ground surface, a sampling device is available that will allow for the collection of groundwater samples from this same type of probe.

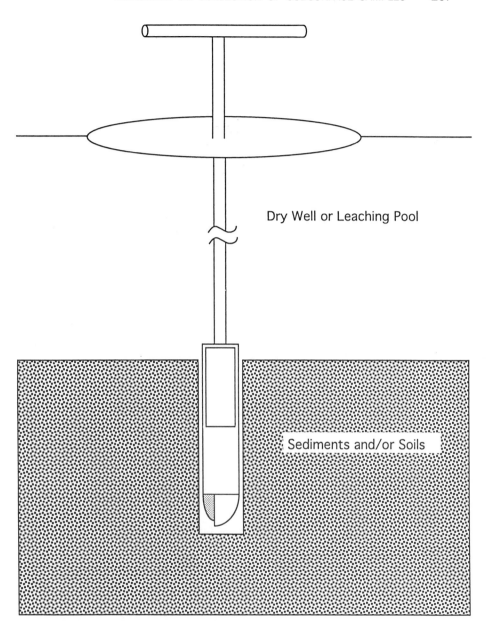

Fig. 1--Hand Auger

Prepared by CA RICH Consultants, Inc.

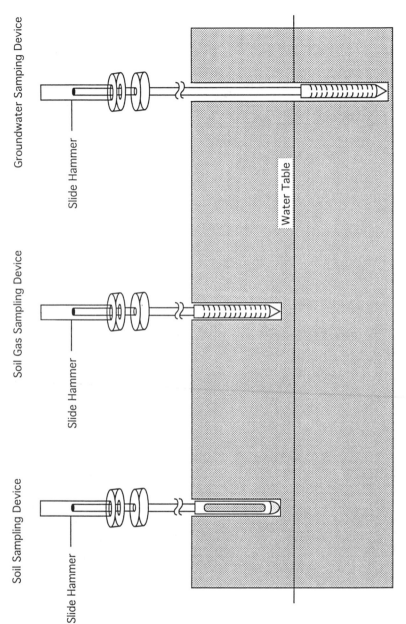

Fig. 2-- Hand-Driven Probes

Prepared by CA RICH Consultants, Inc.

Although more expensive than a hand auger, the hand driven devices are still relatively economical to purchase and fairly quick to operate. They are easy to mobilize to a job site and clean between samples. This type of equipment is particularly useful inside buildings where ceiling height may be a factor in the selection of sampling methodology.

This type of sampling equipment generally has an effective working depth of approximately 7 to 8 meters (20 to 25 feet) below grade, depending on soil conditions. The device is very effective for obtaining soil gas samples using a portable vacuum pump coupled with a field-operated meter, sample bags or sorbent media tubes.

The soil core sampler generally retrieves limited volumes of material. If samples will be sent out for analysis, the sample volume required by the laboratory should be confirmed before selecting this technique.

HYDRAULICALLY DRIVEN CORING DEVICES, SOIL GAS PROBES AND GROUNDWATER SAMPLERS

A technique somewhat similar to the hand-driven system but operated by a hydraulic ram which is typically mounted to the rear of a vehicle is commonly refereed to as "direct push" technology. One popular system manufactured by Geoprobe(TM) of Salina, Kansas consists of 3.175 cm. (1-1/4 inch) diameter hollow steel drive points which are hydraulically driven down into the ground (see Figure 3). A rotating impact hammer is included to quickly penetrate concrete or asphalt pavement. Once the underlying soil is encountered, the drive point is hydraulically driven to the desired sampling depth. The solid plug at the bottom of the point is then pulled up and a hollow soil sampling tube is driven into the soil. The effective working depth of this tool is approximately 33 meters (100 feet) below grade.

Soil gas samples are collected by advancing the sampling rods to the desired depth and then retracting the rods several inches to expose the sample port. New polyethylene tubing is then lowered down the rods and set in place. A vacuum pump is used to purge the tubing. Soil gas samples can either be analyzed on-site with mobile equipment or collected for analysis at a laboratory.

Groundwater samples are collected by advancing a 0.33 meter (1 foot) temporary sample screen attached to the bottom of the sample rods. Once the desired depth is achieved, groundwater samples are collected by lowering 0.95 cm. (3/8-inch) outside diameter tubing fitted with a bottom check valve into the screened section and pumping a sample to the ground surface by gently lifting and lowering the tubing. A 1.1 cm. (7/16-inch)outside diameter mini bailer is also available for the collection of groundwater samples. If the depth to water is within approximately 9 meters (27 feet) a peristaltic pump may also be utilized. Groundwater samples will typically be quite turbid. If sampling for dissolved metals, filtering is advised.

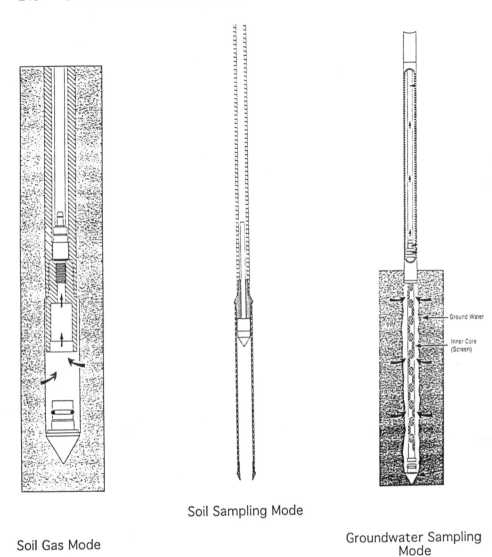

Soil Sampling Mode

Soil Gas Mode

Groundwater Sampling
Mode

Fig. 3--Hydraulically Driven Probes or "Direct Push" devices in the Soil
Gas, Soil Sampling and Groundwater Sampling Modes

(Source: Geoprobe Systems, 1993-1994 General Equipment and Tools Catalog)

The purchase price of this system is significantly greater than the less sophisticated, hand-driven methods. However, it is usually possible to subcontract the services of a firm that operates this equipment. A contractor will typically charge a set daily rate and an additional per sample charge for materials.

The advantage of this method is speed. Set-up time is minimal and the probe can be advanced and retrieved very quickly, especially when concrete or asphalt pavements are encountered.

The cost of using this technique is more than the hand operated methods but less than the services of a conventional drill rig. Also, since the probe is pushed into the ground, no drill cuttings are produced which may have to be drummed and potentially disposed of as a hazardous waste.

Similar to the hand driven technique, the amount of soil contained in the sampler is limited. The laboratory performing your analysis should be consulted regarding required sample volumes before using this method. Also the equipment is typically mounted to a vehicle. Four-wheel drive golf car style vehicles are available for sites with difficult access.

CONVENTIONAL DRILL RIGS WITH SPLIT-BARREL CORE SAMPLERS AND IN-SITU GROUNDWATER SAMPLERS

The conventional method for collecting deeper soil samples - in excess of 3-1/2 to 7 meters (10 to 20 feet) below grade - is through the use of a hollow-stem auger drilling rig. An auger with a hollow center is advanced into the soil to a desired depth. At a selected point, a split-barrel core sampler is driven into the soil to collect a soil sample (ASTM Method D1586-84, Standard Method for Penetration Test and Split-Barrel Sampling of Soils). Core barrels ranging in diameter from to 5 to 14 cm.(2 to 5-1/2 inches) are available to obtain larger quantities of soil for a wider range of laboratory analysis. Thin-walled tube samplers can also be used to sample soft clayey soils.

At selected depths, in-situ soil gas or groundwater samples can be collected using either a temporary soil gas probe or a Hydropunch(TM) groundwater sampler [2]. The Hydropunch(TM), manufactured by QED, Ann Arbor, Michigan, is driven ahead of the hollow stem auger, similar to a split barrel core sampler. This system is illustrated on Figure 4. This device can be advanced into unconsolidated strata using either a drill rig or hydraulically driven using a cone penetrometer arrangement [3].

The Hydropunch(TM) operates in two modes. In the water table interface mode, the sampler acts as a temporary well point and a small-diameter bailer is lowered down to collect a sample. This allows for the detection and sampling of an immiscible floating layer on the groundwater surface. The bailer can be used to collect a sufficient sample volume to fill the required sample bottles.

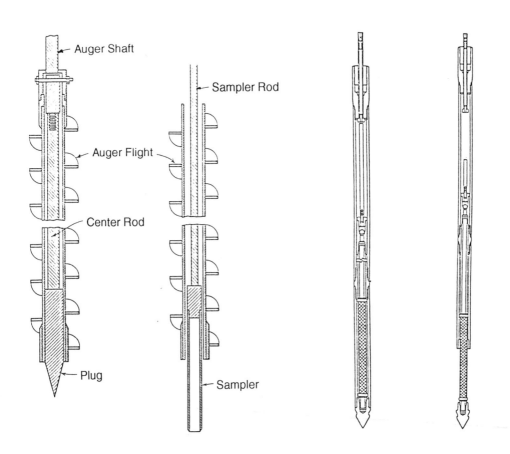

Hollow Stem Auger Drilling
Method with a Standard
Split Barrel Soil Sampler
(Source: Terzaghi and Peck, 1967) [5]

HydroPunch™ Insitu Groundwater
Sampler in the Closed and Open
Positions
(Source: QED-Technical Information and
Applications Guidlines)

Fig. 4--Conventional Drilling Methods Using a Standard Split Barrel Soil
Sampler and HydroPunch™ Insitu Groundwater Sampler

At depths below the water table surface, an internal chamber within the sampling device allows for the collection of approximately 500 milliliters of groundwater at a selected depth. In a comparison of monitoring well samples and Hydropunch(™) samples collected using this mode of operation at three different geographic locations, Edge and Cordry [4] obtained fairly similar results in analyses for volatile organic compounds. As with the direct push method, groundwater samples will typically be quite turbid. If sampling for dissolved metals, filtering is advised.

The depth that an in-situ water sample can be collected in unconsolidated material using this method is essentially limited only by the ability of the drill rig to advance the boring. With the drilling rigs available on the market today, this device can be effectively used to depths of 33 to 66 meters (100 to 200 feet).

The main advantage of using a hollow-stem auger drilling rig is versatility. This method allows you to take advantage of a wider variety of samplers and to achieve a greater sampling depth than the other methods mentioned in this article. Also if something unexpected is discovered during the drilling operation, such as a layer of immiscible, floating product, the boring can easily be completed as a small diameter monitoring well for long-term monitoring of the problem.

This is, however, the most expensive of the methods discussed in this article. Use of a hollow-stem auger drilling rig, quite often produces drill cuttings that may need to be containerized and disposed of as a regulated waste material. The initial capital cost, maintenance and operation of a drilling rig and crew typically limits ownership of this type of equipment to either drilling contracting companies or larger engineering firms.

CASE HISTORIES

Case History 1 - A Phase I/II environmental site assessment of an existing dry cleaner was requested by a mortgage holder prior to foreclosure. The building was constructed over sandy soil with a water table at approximately 7 meters (21 feet) below grade. The main concerns with respect to subsurface conditions were possible releases of perchloroethene (PCE or dry cleaning fluid) to the underlying soils & groundwater, and possible releases of No. 2 heating oil from an on-site underground storage tank (UST).

A soil gas survey was performed outside the rear and side doors of the Facility to check for historic releases of PCE. A hand-driven soil gas probe was driven approximately 0.66 meters (2 feet) into the soil at several locations by each door. The probe was purged using a portable vacuum pump and a field reading was recorded using a portable photo ionization detector (PID) to measure volatile organic vapors. A surfical soil sample at each location was also collected using a stainless steel hand-operated auger.

To check the integrity of the UST, direct push soil samples were collected at 5 meters (15 feet) below grade along all four sides of the tank using a Geoprobe (™) device and analyzed for semivolatile organic hydrocarbons.

Groundwater samples were also collected from the upgraident and downgradient property boundaries using the Geoprobe (TM) device equipped with a 1.1 cm.(7/16-inch) diameter mini bailer.

As shown on Figure 5, the results of this investigation are quite significant. A release of PCE to the underlying aquifer is evident by groundwater concentrations in excess of 30,000 ppb. Also, two of the soil samples from around the UST indicate evidence of either leakage from the tank or accidental overfilling and spillage of petroleum.

As a result of this investigation, the mortgage holder decided not to foreclose on the property. It was estimated that the clean-up costs -- believed to be on the order of $250,000 to $500,000 -- out weighted the value of the property.

Case History 2 - The owner of a commercial/industrial property was informed by a tenant of more than 40 years that they planned to move to a new location once their lease expired. Knowing that this tenant performed automotive maintenance/repairs, and operated gasoline, diesel fuel and heating oil underground storage tanks on the property, the owner retained the services of an environmental consultant prior to the termination of the lease. The terms of the lease stated that the tenant was responsible for all maintenance and repair of the building, including the underground storage tanks.

Initially, a Phase I inspection and file search was performed for this property. As a result of this effort, it was discovered that only one of the two heating USTs on the property was registered and both of the USTs had not been tightness tested for more than five years. Using a hollow-stem auger drill rig, in-situ groundwater samples were collected at five location around the property using the Hydropunch (TM) method operated in the interface mode. Soil cores were also collected just above the water table which was encountered at 1.7 meters (5 feet) below grade. Both the Hydropunch (TM) samples and the soil cores displayed a strong petroleum odor and elevated PID readings in the area of one of the heating oil tanks. The results of the laboratory analyses are displayed on Figure 6.

Based on the observed petroleum product in the soil cores and groundwater samples, it was decided to complete the five borings as 5.1-cm. (2-inch) diameter monitoring wells. This was performed so that the thickness of petroleum floating on the groundwater could be measured and so that the occurrence of spilled oil could be demonstrated to the tenant.

Based on these initial tests, a spill report was filled with the State regulatory agency. Five additional monitoring wells were installed downgradient of the property by the tenant's consultant. After the boundaries of dissolved hydrocarbons in the groundwater were determined, two groundwater recovery wells and a 560 liter per minute (140 gallon per minute) groundwater treatment system consisting of an air stripping tower were installed at the property. As it was clearly demonstrated that the release was caused during the tenant's occupancy of the property and the lease stated that the tenant was responsible for

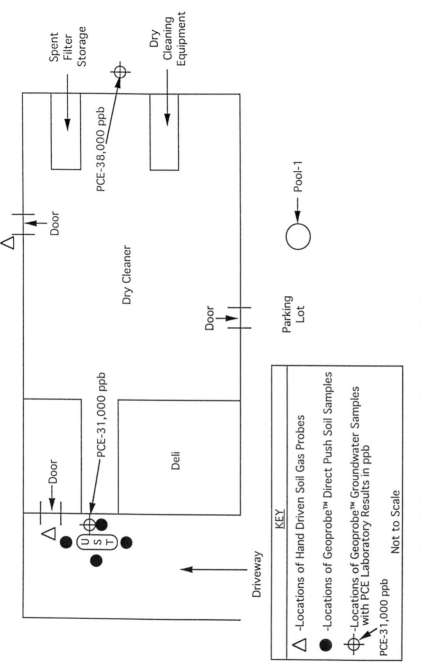

Fig. 5--Case History 1 - Schematic Site Plan and Sample Locations

Prepared by CA RICH Consultants, Inc.

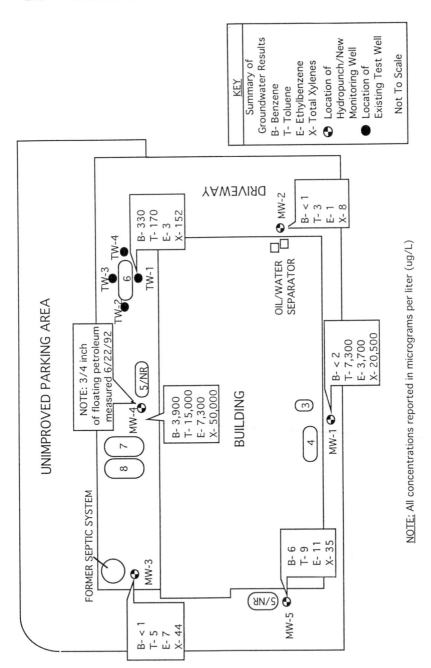

NOTE: All concentrations reported in micrograms per liter (ug/L)

Fig. 6--Case History 2 - Schematic Site Plan, Sample Locations and BTEX Concentrations

Prepared by CA RICH Consultants, Inc.

all building maintenance and repairs, the cost for installation and operation of the system was borne by the tenant.

Case History 3 - A developer was considering purchasing a suburban shopping center located over glacial outwash deposits. A Phase I Environmental Site Assessment revealed that one of the previous tenants, a dry cleaner, released perchloroethylene (PCE) to the underlying soil and groundwater. Five groundwater monitoring wells existed on the property indicting a depth to water of approximately 5 meters (15 feet). The property was serviced by municipal wells and on-site septic fields.

The prospective buyer was interested in further defining the extent of PCE contamination and in developing remediation cost estimates for the property. A series of soil gas probes were performed around the exterior of the former dry cleaner, as shown on Figure 7, using the Geoprobe (TM) device operated in the soil gas mode. Soil gas samples were collected at 1.7, 3.3 and 5 meters (5, 10 and 15 feet) below grade at each location. The width of the doorways and the ceiling height precluded the use of the vehicle-mounted Geoprobe (TM) device indoors. To determine the extent of subsurface PCE vapors below the building, hand-driven soil gas probes were performed at several locations inside the structure.

Historical releases to the on-site septic systems were also an item of concern. Using a hand-operated soil auger, sediment samples were collected from all of the primary leaching pools on the property. In addition, three addition "micro wells" consisting of 2.54 cm. (1-inch) diameter slotted PVC pipe were installed at the site using the Geoprobe(TM) device. Groundwater samples were obtained by lowering 0.95 cm. (3/8-inch) outside diameter tubing fitted with a bottom check valve into the screened section of the well.

The results of this investigation are included on Figures 8 and 9. Based on these test results, cost estimates for a soil vapor extraction system and a groundwater treatment system were provided for the prospective buyer.

SUMMARY

The need to collect representative soil, soil gas and groundwater samples quickly and inexpensively presents a growing market, particularly in the performance of Phase II environmental assessments for real estate transactions. Quite often a method that does not require the cost and liability of a permanent monitoring well installation is preferred in this type of investigation.

This article presents four types of techniques that can be employed for these purposes. It is up to the professionals practicing in this field to evaluate and apply these techniques as the developing procedures for performing Phase II Environmental Site Assessments evolve and become more standardized with time. Three case histories of the application of these techniques are included.

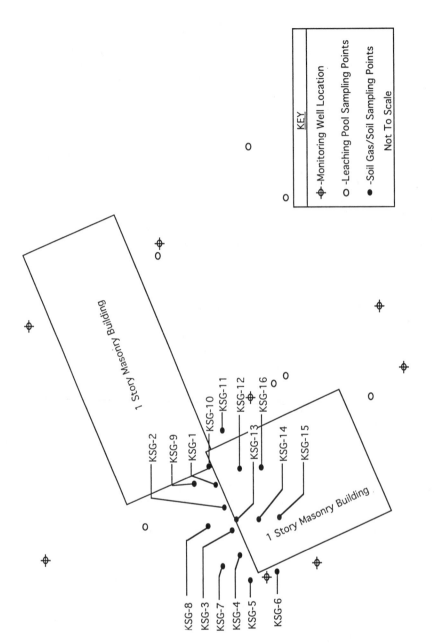

Fig. 7--Case History 3- Schematic Site Plan and Sample Locations

Prepared by CA RICH Consultants, Inc.

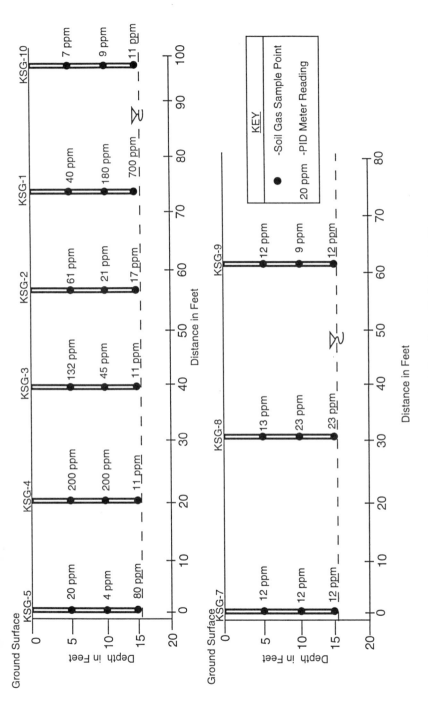

Fig. 8--Case History 3-Subsurface Soil Gas Profile

Prepared by CA RICH Consultants, Inc.

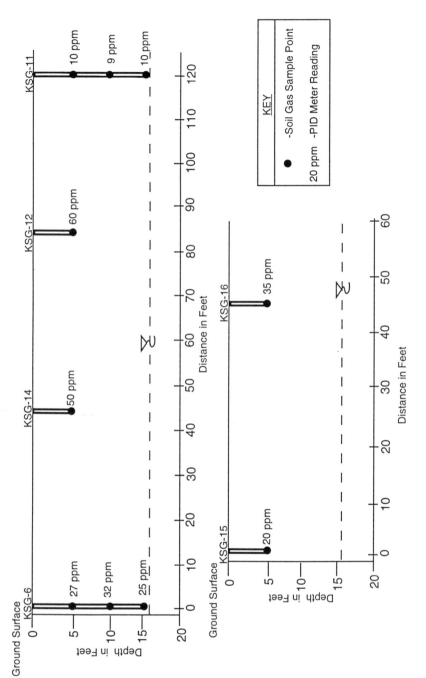

Fig. 9--Case History 3-Subsurface Soil Gas Profile

Prepared by CA RICH Consultants, Inc.

REFERENCES

[1] Weinstock, Eric A., "Phase II Environmental Assessments", Water Well Journal, April, 1991

[2] Chalfant, Chris, "HydroPunch and Soil Gas Probes Simplify Groundwater Assessments", The National Environmental Journal, May/June 1992.

[3] Smolly, Mark, and Kappmeyer, Janet C., "Cone Penetrometer Tests and Hydropunch (R) Sampling: A Screening Technique for Plume Definition", Ground Water Monitoring Review, Spring 1991, Vol. 11, No. 2.

[4] Edge, Russel W., and Cordy, Kent, "The Hydropunch (TM): An In Situ Sampling Tool for Collecting Ground Water from Unconsolidated Sediments", Ground Water Monitoring Review, Summer 1989, Vol. 9, No. 3.

[5] Terzaghi, K., and Peck, R.B., Soil Mechanics in Engineering Practice,2nd Ed. John Wiley & Sons, New York, 1967

Note: Portions of this paper we adapted from a earlier article printed in the November / December 1992 issue of The National Environmental Journal

Soil and Soil Gas Sampling

Robert K. Sextro[1]

ESTIMATION OF VOLATILE ORGANIC COMPOUND CONCENTRATIONS IN THE
VADOSE ZONE: A CASE STUDY USING SOIL GAS AND SOIL SAMPLE RESULTS

REFERENCE: Sextro, R. K., "Estimation of Volatile Organic Compound
Concentrations in the Vadose Zone: A Case Study using Soil Gas and Soil Sample
Results," Sampling Environmental Media, ASTM STP 1282, James Howard
Morgan, Ed., American Society for Testing and Materials, 1996.

ABSTRACT: Obtaining defensible and conservative estimates of the nature, extent and
concentration of volatile organic compounds (VOCs) in the vadose zone is extremely important
when formulating the conceptual model of the site, when performing risk assessments, for
estimating contaminant mass, for assessing remedial alternatives, for selecting target areas for
cleanup, and/or for making no further action/investigation decisions. Studies have shown that
soil gas analytical results provide both a more complete indication of the VOCs present and a
higher estimate of their respective concentrations in the vadose zone than the analysis of soil
samples alone. For the past several years deep downhole (to 30+ meters) soil gas sampling and
analysis has been performed by various consultants during remedial investigations (RIs) and
remedial actions (RAs) of the vadose zone at McClellan Air Force Base. A number of these soil
gas results have been confirmed by the concurrent collection and analysis (for VOCs) of soil
samples (preserved by either refrigeration to 4 degrees centigrade or refrigeration combined with
methanol preservation). The use of this VOC sampling and analysis strategy has resulted in the
optimization of VOC sampling and analysis procedures, in a better understanding of the
relationship between the concentration of VOCs in soil gas and in the soil, and in a more
accurate and comprehensive conceptual model for VOC contamination in the vadose zone. The
paper will present the methodologies used by the various consultants for sample collection,
preservation, and in the analysis of soil gas and soil samples, present the results of a focused QC
study on soil gas sampling and analysis, and discuss the correlation between the soil gas and soil
matrix analytical results. The current and future strategies for the sampling, analysis, and
estimation of VOCs in the vadose zone during RIs and RAs will also be presented.

KEYWORDS: volatile organic compounds, VOCs, soil gas, soil, sampling, analysis,
methanol preservation

---

[1] Quality Assurance Manager, Northern California Environmental Operations, Jacobs
Engineering Group, 2525 Natomas Park Drive, Suite 370, Sacramento, California 95833.

HISTORIC PERSPECTIVE

At many hazardous waste sites on Air Force (AF) bases being investigated throughout the United States (US), shallow [1.5 to 3 meters (m)] soil gas sampling and analysis has been performed to target (1) hot spots, (2) potential release locations, and (3) sites for further investigation. This sampling strategy has shown mixed results, as some results show good agreement between soil and soil gas VOC findings while other results have shown virtually no agreement (detected soil gas VOCs but no detected [ND] soil VOCs). In this author's experience the usefulness of the results appears to be dependent not only on the reliability of the analytical method used, but on the nature of the VOC source and on the sub-surface geology. At AF bases located in the far western United States, and particularly those in semi-arid areas with declining water tables (or saturated zones) at depths of 15 to 30 meters and those with thick vadose (unsaturated) zones, shallow soil gas sampling and analysis has proven unsuccessful, as indicated by detects of soil gas VOCs and ND for soil VOCs, (in this author's opinion) in reliably targeting VOC sources or delineating soil and/or groundwater VOC plumes. However, studies conducted (started in 1991) at a private industrial waste site [1] in northern California have shown that deep (6 to 25+ m) soil gas sampling and analysis can provide a more complete characterization of the VOCs present and result in higher estimates of contaminant concentrations than sampling and analysis of soil samples alone. These studies were conducted in similar conditions, and with sampling and analytical methodologies comparable (hence comparable accuracy and precision) to those described in this paper.

The following factors about defining the nature and extent of VOC contamination and use of VOC data were considered by McClellan Air Force Base (McAFB). Shallow soil gas VOC results were very erratic when compared to soil VOC results (e.g. detected soil gas VOCs and ND soil VOCs); soil VOC results were also very erratic and often near the lower limit of reliable quantitation; soil results were suspected to be biased low due to losses of VOCs during sample collection (as documented by Hewitt [2], and losses during shipment, storage and subsequent sub-sampling in the laboratory (also as documented by others [3]); and eventual remediation was presumed to be soil vapor extraction (SVE) for vadose zone contamination. Therefore, soil gas results were considered to be more appropriate for decision making and soil gas measurements were thought to be more sensitive in delineating the nature and extent of VOC contamination of the vadose zone. Also, results obtained from a comparison study of measured and soil gas estimated concentrations at the aforementioned private industrial waste site concluded (and received regulatory agency concurrence) that estimating soil VOC contamination using soil gas results was more conservative than using direct soil measurements alone.

Using this information and at the encouragement of the agency representatives from the Environmental Protection Agency (EPA) Region IX and California EPA working with McAFB on their Installation Restoration Program (IRP), the strategy for determining the nature, extent, and concentration of VOCs in the vadose zone at McAFB was altered in late 1992-early 1993. The shift in strategy was from the more traditional shallow soil gas screening methodology followed by discrete soil sample collection and VOC analysis to one of using primarily deep downhole soil gas sampling and analysis augmented by infrequent soil sampling and analysis of VOCs to determine the nature, extent and concentration of VOCs in the vadose zone. Initially this strategy received a boost in support by the material presented and discussed at the EPA National Symposium in Las Vegas in January 1993 on Measuring and Interpreting VOCs in Soils: State of the Art and Research Needs, which supported the notion that soil sampling alone underestimates the extent and concentration of VOCs. The McAFB basewide Quality Assurance Project Plan (QAPP)[4], developed in late 1992, provided the technical support for the IRP consultants working on McAFB in the form of some elementary standard operating procedures (SOPs) for the techniques of sampling the soil gas from depth and some preliminary SOPs for the analysis of samples by gas chromatography (GC) either in on-site or off-site laboratories.

This paper takes a retrospective look at the sampling and analyses performed by the various consultants and the data they collected (as reported to the base), introduces some data evaluation procedures and guidelines, and discusses the results.

METHODOLOGY

Field and Analytical Procedures

Downhole soil gas sampling, done in conjunction with hollow-stem augering, push-probe, or sonication penetration methods, is really an extension of shallow soil gas sampling. The gas sampling probe used in conjunction with the hollow-stem auger was a hollow chromium/molybdenum steel alloy tube [20 centimeters (cm) outside diameter]. Attached at the bottom of the probe was a perforated, retractable, stainless-steel tip to which a length of polytetrafluoroethylene (PTFE) tubing was attached by air-tight stainless-steel fittings. When the desired depth was reached with the auger, the probe with PTFE tubing attached, was lowered down the hollow-stem and driven about 50 to 100 cm beyond the auger head with the downhole hammer or sonic unit. The probe was then retracted approximately 5 centimeters exposing the perforations to the surrounding lithology and soil gas (Figure 1). Using a vacuum pump operating between - 250 and - 750 millimeter (mm) of mercury, the sample probe and tubing was purged for 3 to 5 probe/tubing volumes. The soil gas sample was routinely collected in either a glass gas-tight syringe, a expandable gas bag made of relatively inert material, or a passivated stainless-steel canister. Samples were not preserved in any way except to minimize exposure to the sun by storing them in the dark (typically in sample coolers without ice). Appropriate chain-of-custody documentation reportedly was used for each sampling episode. The sample holding times varied by sample container type and the location of analysis. Analysis of bag and syringe samples on-site had holding times of several hours for the syringe samples and up to 24 hours for the samples in the bags. Samples analyzed off-site had holding times of up to 24 hours for the bag samples and 72 hours for the canister samples.

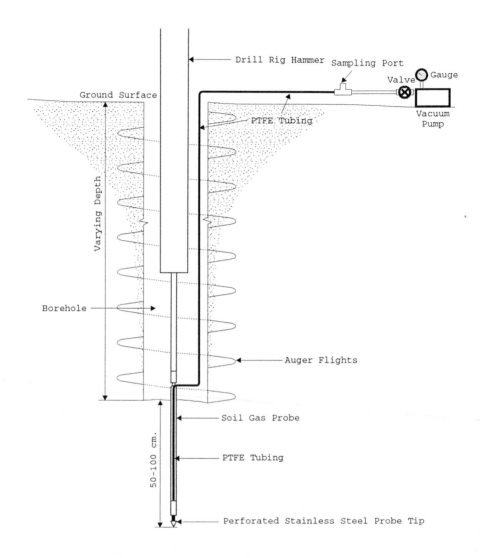

Figure 1. Downhole Soil Gas Sampling Apparatus

12/28/94 JG f:data\cadd\05k69904\sgsamp.cdr

For soil gas samples designated for on-site analysis, routine analytical procedures included the use of modified EPA method 18 (E-18) [5] or one of several different methods (together referred to as screening methods in the remainder of this paper) described as follows:

1. On-site analysis using GC;

   - multi-detector arrays consisting of flame ionization detection (FID), photo-ionization detection (PID) and/or electrolytic conductivity detection (HECD),
   - column characteristics similar to those used in EPA Solid Waste (SW) methods 8010 and SW8020 [6],
   - on-column injection of the gas sample taken directly from the syringe or the gas bag,
   - prior multi-standard, multi-point calibration with single point daily calibration verification,
   - a 11 to 13 compound target analyte list that varied slightly (by addition or deletion of 1 or 2 compounds) but routinely measured vinyl chloride (VC), cis- and trans-1,2-dichloroethylene (cis-1,2-DCE and trans-1,2-DCE), trichloroethylene (TCE), tetrachloroethylene (PCE), 1,1,1-trichloroethane (1,1,1-TCA), chloroform, benzene, ethylbenzene, toluene, xylenes, and chlorofluoro-methanes and -ethanes (Freons 11, 12, and 113),
   - detection limits that ranged from 10 to 50 parts per billion by volume (ppbv),
   - a precision criterion of 30 percent relative percent difference (RPD) as measured by sample replicates and laboratory duplicates, and an accuracy criterion of +/- 30 percent as measured by the use of control samples and/or blank spikes.

2. Off-site analysis using GC or GC/mass spectrometry described as follows:

   - GC analysis was routinely performed using a modification to E-18, consisting of a PID and a electron capture detector (ECD); on-column injection of the gas sample taken directly from the gas bag; 15 target analytes including those previously mentioned plus 1,1-dichloroethylene, and carbon tetrachloride; detection limits at 5 ppbv; and an accuracy acceptance criterion of +/- 30 percent.
   - GC/mass spectrometry analysis was performed on confirmation samples collected in passivated stainless-steel canisters using EPA method TO-14 [7] without modification.

For the soil sampling performed in conjunction with the soil gas sampling the method of sampling (following the SOPs outlined in the McAFB QAPP) is described as follows:

   - A split-spoon sampler with previously decontaminated brass or stainless steel rings (sleeves) was used in conjunction with the hollow-stem auger. The sampler was driven about 0.5 to 1 m beyond the auger head (and beyond the location of soil gas sampling point) with a downhole hammer to collect undisturbed soil in the rings. After the sampler was removed from the hole and auger, the split spoon was opened and the rings were carefully removed. Those rings selected for analyses were separated, capped with PTFE sheets and covered with plastic end caps. Sample holding times ranged up to 14 days prior to analysis (the soil remained in the sleeve but in contact with the PTFE sheets).

- For most samples preservation followed EPA protocol, which was refrigeration at 4 degrees Centigrade (^{0}C). However, for a limited number of samples preservation was in a methanol solution followed by refrigeration to 4^0C. All subsampling for analysis was performed in the laboratory using undocumented procedures, which could have attributed to the observed biases as noted by other authors [3].

- Methanol preservation (as described in the 1994 version of the McAFB QAPP) was accomplished in the field by selecting the ring [the standard ring size being 10 cm in diameter and 5 cm in length, containing approximately 150 to 200 grams (g) of sample] to be preserved from the split-spoon and placing it directly (without covering or capping with PTFE sheets) in a previously prepared and weighed (tared) 500 milliliter (mL), amber, wide-mouth, glass sample container. Preparation included adding (either in the field or laboratory and prior to sample collection) 250 mLs of pesticide-grade methanol, chemical surrogates at measured concentrations appropriate for use with the SW8240 analytical technique (including 1,2-dichloroethane-d4, toluene-d8, and 1-bromo-4-fluorobenzene), weighing the bottle to the nearest 0.1 g, followed by refrigerated storage of the bottle. The pre-prepared bottles containing methanol were stored in the field for up to seven days with refrigeration and without the degradation of the surrogates or methanol (although minor absorbances of methylene chloride were noted).

- All soil sample analyses, regardless of the method of preservation, were performed using SW8240, GC/mass spectrometry.

At approximately 10 percent of the sample locations a second soil gas sample was collected sequentially in the aforementioned stainless- steel canister for off-site confirmational analysis by EPA method TO-14. The confirmation was aimed at corroborating the identity of the speciated VOCs found and, to the extent possible, quantitatively confirming the concentrations of the VOCs.

Field replicates (collocated samples) were collected and analyzed for all media at a frequency of approximately 10 percent of the total number of samples for each media. Laboratory duplicates were normally analyzed at a frequency of one per analytical batch (up to 20 samples of similar media). Matrix and surrogate spiking, while done on soil samples, was not routinely performed on the soil gas samples.

After sample collection and withdrawal of the probe from the hole the PTFE sample line was either purged with high-purity nitrogen until VOC readings from on-site screening instruments were lower than 1 000 ppbv, or replaced with new line. Gas sampling syringes were also purged with the high-purity nitrogen and new gas sampling bags were used for each sampling event. The down-hole probe and retractable tip (and any downhole soil sampling equipment) were decontaminated as follows; (1) scrub with water and laboratory-grade detergent for removal of dirt, grease, etc., (2) rinse with copious amounts of potable water, and (3) final rinse with analyte-free, distilled water. An additional step for all sample sleeves (rings), soil gas probe tips, and gas sampling syringes was baking overnight in an oven at 160 °C.

A special study was conducted at McAFB on 2 sites in late 1993. The emphasis of the study was to obtain analytical data that could be used to help resolve some of the observed concentration differences for analytes determined by both modified method E-18 and TO-14. For this study the gas sampling probe was emplaced in the designated bore hole at the desired depth and not moved until after each set of soil gas samples were collected. The sample set collected consisted of (1) simultaneous duplicate gas bags for E-18 analysis, followed by (2) simultaneous canister samples for TO-14 analysis, and lastly (3) a third canister collected sequentially for TO-14 analysis. A total of 9 sets of 5 samples each were collected (from 2 boreholes) at the 2 sites. The results of this study are presented in Table 2 and are discussed later in this paper.

Data Use Procedures

For the evaluation of analytical methods the results from EPA method TO-14 using GC/MS techniques were considered to be the most accurate for both analyte identification and quantitation. This method's results from field replicate samples were compared against those of the screening results by plotting (using scatter plots as illustrated in Figure 2) actual results over the entire measured range of concentrations (about 6 orders of magnitude), and by calculating percent differences and plotting the percentages versus the concentrations in ppbv from TO-14 (as shown in Figure 3).

To compare direct soil VOC measurements to results obtained from soil gas it was necessary to calculate an equivalent soil concentration using soil gas concentrations. This equivalent or total concentration is the sum of the VOCs in the three phases present in the soil: vapor (soil gas), aqueous (dissolved) and soil solids (adsorbed)[1]. If nonaqueous phase liquids [(NAPLs), neat or free product] are absent, contaminant mass can be estimated for individual compounds using the soil gas concentration of an analyte. The calculation was based on the assumption that in-situ equilibrium exists among the gas, liquid, and solid phases. If NAPLs are present this approach will underestimate the concentration of contaminants. The equation used to estimate soil concentrations, as cited in [1], is presented as follows together with required parameters identified in Table 1.

Equilibrium Equation:
$$C_T = 0.001 \times C_G[0.01 \times \rho_B \times K_{OC} \times TOC \times 1/H + (\phi_w \times 1/H) + \phi_a] \, 1/\rho_B$$

Several parameters are specific to the type of soil from which the soil gas was collected. However, due to the variability of soils at McAFB several simplifications were needed. Silt and sand are the two most dominant soil types observed at McAFB and generally constitute more than 95 percent of the total soil. Silt, as used here, includes all sandy and clayey silts and sand includes silty, clayey, well and poorly sorted sands. The total organic carbon (TOC) concentrations observed at McAFB are generally less than 0.1 percent (the detection limit for TOC by EPA method SW9060) and therefore, the values were set arbitrarily at 0.10 for silt and 0.03 percent for sand. Bulk density, total porosity, and water saturated porosity values were based on local measurements.

TABLE 1: Parameters Used in the Equilibrium Equation

Parameter	Soil Type	
	Silt	Sand
C_T = Soil VOC Concentration, mg/kg		
C_G = Soil Gas VOC Concentration, ng/mL		
ρ_B = Bulk Density of Soil, g/cc	1.32	1.39
K_{OC} = Soil Partitioning Coefficient mL/g (for TCE)	126	126
TOC = Total Organic Carbon , Percent	0.10	0.03
H = Henry's Law Constant, Dimensionless (for TCE)	0.297	0.297
ϕ_w = Fractional Water Saturated Soil Porosity	0.39	0.24
ϕ_a = Fractional Air Saturated Soil Porosity	0.11	0.23

DISCUSSION OF RESULTS

The compound TCE was chosen for evaluation because of its wide-spread presence at McAFB in the groundwater, vadose zone soil and soil gas, and chromatographically it was subject to less interfering and coeluting compounds (its identification was probably the most reliable of the dozen or so aromatic and halogenated VOCs suspected to be present). Note: No attempt has been made by the author to perform the type of data collection previously discussed under more controlled conditions (thus eliminating or reducing variables), but rather this discussion is aimed at evaluating the findings and information as reported by the various consultants.

Data Quality Evaluation

Because the data used for evaluation were obtained from the various consultants working at McAFB and it had been previously evaluated by them in accordance with the criteria provided in the McAFB basewide QAPP, no further evaluation of the quality was performed by the author. All usable data, as reported to the base by the end of 1994, were considered in the preparation of this paper as these same data have been used to make decisions at the base regarding further investigation and/or remedial actions.

Analytical Method Evaluation

In most cases the initial analytical results from both screening and TO-14 analyses were not evaluated until some months (and in one case over a year) had passed. This lack of timely evaluation of the screening results and their comparison to those from method TO-14 lead to several important oversights: (1) the recognition of total petroleum hydrocarbon (TPH) VOC contamination mixed with halogenated VOC contaminants and its implication to definitive interpretation of the chromatograms for halogenated VOCs; (2) the recognition of the potential for a high quantitative and perhaps false positive bias from the screening methods because of the TPH concentrations (resulting in pattern-peak chromatograms typical of jet fuel, making identification and quantitation of individual analytes difficult); and (3) little discussion about the severity of these analytical problems, or the potential effects on the use and interpretation of the results.

Comparison of soil gas results from field replicates, one analyzed by the screening methods and the other analyzed by TO-14, for TCE are presented in Figure 2. Although visually it may appear to be a reasonable correlation between the screening results and those by TO-14, the scales used for data ranging over 6 orders of magnitude distorts the visual appearance of the curve. As it is not very meaningful to statistically correlate data over this range of values, those data at concentrations below approximately 20 000 ppbv were evaluated separately from those above that concentration. The coefficient of determination for the data set below 20 000 ppbv is $r^2=0.007$ with a regression formula of C1(screening results)=8 517 + C2(TO-14 data) indicating random variation. The coefficient of determination for the data set above 20 000 ppbv is $r^2=0.931$ with a regression formula of C1=49 822 + 1.04C2 indicating good correlation at the 99% upper confidence level.

For the data comparison presented in Figure 3 it was assumed that method TO-14 results are the more accurate (as previously stated, at McAFB TO-14 was used for confirmation of both the on- and off-site screening methods). In Figure 3, a positive percent difference indicates that the screening results were higher than the corresponding TO-14 results. Results plotted as + 1 000 percent difference are truncated and may represent outliers. Results plotted at -100 percent represent screening results that in some

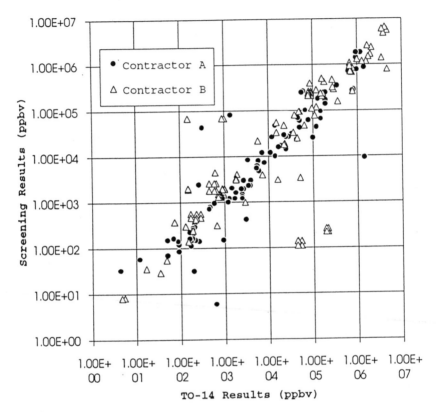

Figure 2. Screening versus TO-14 Results for TCE

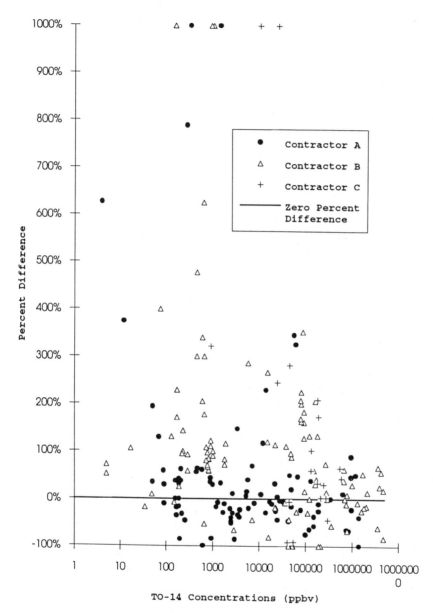

Figure 3. Percent Difference for TCE Between
Screening and TO-14 Results

cases were reported as not detected at a specified concentration which was significantly lower than the reported concentration obtained using method TO-14. Twenty five percent of the screening results are biased high by more than 100 percent [equivalent to an RPD of 67 percent] and the remaining results are nearly equally divided between + 100 and - 100 percent difference. For TCE, an overall high bias for screening results of about 90 percent is indicated, but an increasing tendency toward higher bias below concentrations of about 20 000 ppbv is evident (and supported by the statistics from Figure 2). Similar high biased results are observed for several of the commonly occurring halogenated VOCs including 1,1,1-TCA, *cis* 1,2-DCE, Freon 113, and 1,1-dichloroethane.

A closer inspection of these results indicates that these data can be grouped into three concentration ranges. At concentrations below 2 000 ppbv the average bias is approximately +180 percent, between 2 000 and 200 000 ppbv the average bias is about +55 percent, and at concentrations greater than 200 000 ppbv the average bias is negligible (-2 percent). Thus, at concentrations below 200 000 ppbv there is a tendency for screening results to be biased high. This indicates that the accuracy of screening results is affected by systematic factors in performing the methodology and/or in the composition of the soil gas. Thus it is likely that the observed biases are analytical method and gas composition related and not attributable to the type of sample container used (results from the passivated canisters are typically lower than those from the gas bags and/or glass syringes).

Data Use Evaluation

For the comparison study of calculated versus measured soil TCE concentrations (using the screening results only), only samples that were collected within about 50 to 100 cm of each other were used to assure that the soil gas was actually extracted from the same soil type used for the calculation. Nominally, the soil gas sample was collected at a depth less than the soil sample. Figure 4 illustrates the relationship between calculated and directly measured concentrations. The coefficient of determination for these data (ranging over about 4 orders of magnitude) is $r^2=0.231$ with a regression formula of C1(calculated soil concentration)=2.93 + C2(measured soil concentration), indicating little correlation and more of a random variation. Another obvious conclusion that can be reached is that the calculated concentrations are very conservative (much higher) when compared to the direct measured concentrations. Of a total of 109 samples only about 16 percent of calculated TCE values have lower concentrations than the direct measured TCE in soil. Also, the calculated estimates are as much as 3 orders of magnitude higher than direct measured results. Greater differences are more commonly observed at direct measured concentrations below 0.1 mg/kg.

The reasons for the differences may be related to losses of VOCs from soil samples due to the method of sample collection, shipment, storage and laboratory sub-sampling [2 and 3]. It may also be related to uncertainties about equilibrium assumptions made for the soil gas-based calculations, or the lack of correlation may be due to problems with both analytical methods. Based on comparisons of soil gas results from

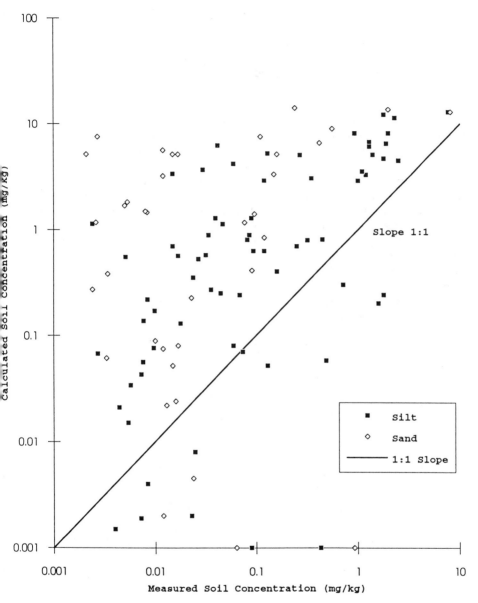

Figure 4. Measured Versus Calculated Soil Concentrations
(mg/kg) in Silts and Sands

duplicate samples analyzed by on- and off-site screening methods and method TO-14, TCE soil gas concentrations are generally overestimated using the screening methods by an average of 50 to 200 percent relative to TO-14 results. Screening results tend to be significantly higher (several 100 percent) at lower concentrations (less than about 2 000 ppbv) than at higher concentrations. Therefore, calculated estimates based on screening results (used exclusively for this study) could be biased high but not enough to account for the order of magnitude differences illustrated in Figure 4.

The results for TCE from the special study at McAFB given in Table 2 can be summarized as follows: (1) regardless of analytical method was used, the correlation between duplicate (paired) samples results was good (within 30 percent difference); (2) comparison of results from the paired canisters and the third canister were also good (within 30 percent difference); (3) the differences between the method E-18 and TO-14 results often were significant (approximately a factor of 2), but indistinguishable from those sequentially collected routine and confirmation samples previously discussed in this paper; and (4) the E-18 method bias can not be attributed to high TPH VOC concentrations because for these samples TPH was, for the most part ND at a detection limit of 10 ppbv. This study showed that the soil gas concentrations did not change dramatically whether samples were collected sequentially (the routine procedure) or simultaneously (in other words, sequential gas sample appear to be representative), and that analytical method differences appear responsible for the observed differences in results obtained by methods E-18 and TO-14.

TABLE 2: TCE Results (in ppbv) from Special Study

Sample Set # Type	Gas Bag 1	Gas Bag 2	Canister 1	Canister 2	Canister 3
1	110 000	140 000	52 000	53 000	54 000
2	260 000	220 000	190 000	190 000	210 000
3	460	560	230	270	290
4	1 300	1 500	810	760	820
5	1 900	2 000	850	900	1000
6	94 000	96 000	48 000	50 000	50 000
7	1 500	1 700	850	950	950
8	2 600	2 600	650	590	550
9	240 000	260 000	92 000	78 000	80 000

CURRENT APPROACH

Sampling

The revised McAFB QAPP (1994) does not substantially change the methods to be used for down-hole soil gas sampling or sub-surface soil sampling. However, it does now prescribe a SOP for collecting simultaneous duplicate soil gas samples by using an inlet splitter prior to entering the sample containers (that are arranged in parallel). Also promulgated is the SOP for methanol preservation of soil samples (which follows the method previously described in this paper) and the introduction of both the "flexible decision plan" (for sampling) and guidelines for data quality and data sufficiency (for analysis).

The introduction of what has been termed the "flexible decision plan" was done to encourage the IRP support contractors to make field decisions about (1) advancement or,, (2) termination of a boring, (3) to perform stepout drilling for further delineation of a source, or (4) to convert a boring to an SVE well. These decisions will be based on specific 'trigger concentrations' and will be guided by one of two conceptual models of the site (for details of the "flexible decision plan" and trigger concentrations refer to the McAFB QAPP [4]).

Analysis

The revised McAFB QAPP (1994) does offer guidelines for data quality and data sufficiency (how and when to choose an analytical approach), provides two additional SOPs for soil gas analytical methodologies and deletes the previous on-site GC procedure.

The data quality guidelines in the revised McAFB QAPP follow the approach established by the EPA in their revised guidance on data quality objectives [8]. The emphasis in the data quality guidelines is on the use of screening data with definitive confirmation (percentage of confirmation specified by the project). For the purposes of soil gas sampling this translates into on-site, or rapid turn-around off-site GC techniques ("calibrated" to the confirmation analysis) with confirmation by method TO-14.

This author cannot recommend too strongly the necessity to "calibrate" either the on- or off-site screening methods to the results of samples analyzed with either a normal or modified TO-14, and to do so in near real-time. This should be performed for initial samples collected for soil gas VOC analysis by selecting 20 to 30 samples representing a wide range of concentrations and a variety of analytes. The evaluation of the results from both methods should also be performed as rapidly as possible. "Calibration" means comparison of chromatography, analyte identification and quantitation. In this way, if modifications to analytical or sampling procedures are needed (such as modifying chromatographic procedures or taking less or more samples for duplicate analysis), changes can be made very early in the field program.

CONCLUSIONS

The high bias in the screening soil gas results represents only one aspect of the uncertainties in the relationship between estimated and directly measured soil VOC concentrations. Only the comparison of collocated soil gas, methanol preserved, and normally preserved soil results will provide some indication whether normally preserved soil samples yield low results or the assumptions about equilibrium conditions yield high results. The high bias in the screening results is, however, significant for making decisions about the nature and extent of soil gas contamination that may require remediation, about removal rates for SVE systems, or when to terminate SVE operations.

The greater sensitivity of soil gas sampling and analysis when defining the nature and extent of VOC contamination and the conservative results obtained from estimating VOC contamination in the vadose zone using soil gas concentrations continues to be sufficient justification at McAFB for the application of this methodology. Also, with the proper planning, selection of analytical methodology and early evaluation of results, this author believes it is possible to accurately sample and analyze soil gas of varying compositions and provide reliable estimates of potential vadose zone soil VOC contamination to decision makers.

REFERENCES

[1]   Aerojet General Corporation, Demonstration Project Report, prepared by ICF Technology, 1992.

[2]   Hewitt, A. D., "Comparison of Methods for Sampling Vadose Zone Soils for Determination of Trichloroethylene", Journal of AOAC International, Vol. 77, No. 2, 1994.

[3]   US Army Corps of Engineers, Losses of Trichloroethylene from Soil During Sample Collection, Storage and Laboratory Handling, Special Report 94-8, 1994.

[4]   McClellan Air Force Base, Basewide Remedial Investigation/Feasibility Study (RI/FS) Quality Assurance Project Plan, prepared by Radian Corporation, 1992 and updated 1994.

[5]   Code of Federal Regulations, Protection of the Environment, Title 40, Part 60, appendix A, 1994.

[6]   US Environmental Protection Agency, Test Methods for Evaluating Solid Waste, 3rd Edition, including revisions and updates of 1986, 1992, 1994.

[7]   US Environmental Protection Agency, Compendium of Methods for the Determination of Toxic Organic Compounds in Ambient Air, EPA/600/4-89/017, 1988.

[8]   US Environmental Protection Agency, Data Quality Objectives Process for Superfund: Interim Final Guidance, EPA 540/R-93-071, 1993.

Don J. DeGroot[1], Alan J. Lutenegger[1], James G. Panton[2], David W. Ostendorf[1], and Samuel J. Pollock[3]

METHODS OF DETERMINING IN SITU OXYGEN PROFILES IN THE VADOSE ZONE OF GRANULAR SOILS

REFERENCE: DeGroot, D. J., Lutenegger, A. J., Panton, J. G., Ostendorf, D. W., and Pollock, S. J., **"Methods of Determining in Situ Oxygen Profiles in the Vadose Zone of Granular Soils,"** Sampling Environmental Media, ASTM STP 1282, James Howard Morgan, Ed., American Society for Testing and Materials, 1996.

ABSTRACT: This paper describes seven different methods evaluated to determine in situ oxygen profiles in the vadose zone at a site consisting of granular soils. The techniques investigated included both permanent installations (four) and rapid deployment drive probes (three). The soil and ground water conditions at the site are first described followed by details of the design, deployment and sampling for each method. Typical oxygen profile results obtained from the different methods are presented and compared based on their ease of deployment, efficiency, repeatability, and reliability. Based on the results of this study, the most reliable and consistent sampler types tested were two of the permanent installation designs. The ease of construction and relatively low material costs of these particular methods, combined with their simplicity, relative quickness and accuracy of sampling make them superior to the others tested. The rapid deployment drive probe methods were prone to errors due to short circuiting, damage during deployment or sampling difficulties. However, when care was taken to minimize these factors, some of these methods also produced reliable and consistent results.

KEYWORDS: calcium magnesium acetate, deicer, drive probes, granular soils, oxygen profiles, soil gas, vadose zone

[1]Associate Professor, Department of Civil and Environmental Engineering, University of Massachusetts, Amherst, MA 01003

[2]Geotechnical Engineer, Shannon & Wilson, Inc., Boston, MA 02116

[3]Hydrogeologist, Massachusetts Highway Department, Boston, MA 02210

271

INTRODUCTION

Soil gas sampling can be performed using a variety of different techniques. ASTM Standard Guide for Soil Gas Monitoring in the Vadose Zone (D 5314) divides soil gas sampling techniques into six basic categories based upon the method of collection: (1) whole air-active, (2) sorbed contaminants-active, (3) whole air-passive, (4) sorbed contaminants-passive, (5) soil sampling for subsequent headspace atmosphere extraction sampling, and (6) soil pore liquid headspace gas. Active methods involve the forced movement of soil gas from the sampling horizon to a collection device whereas passive methods rely on the passive movement of soil gas to a collection device. For sorbed methods, the sampled soil gas passes through a collection device and collected by adsorption prior to analysis. Of these six methods, the whole air active approach is the most appropriate for measurement of oxygen concentrations. Oxygen does not lend itself to sampling by sorptive methods, and passive sampling techniques cannot produce the required results. Headspace sampling from soil or pore liquid samples is an indirect soil gas measurement since it is derived from the soil or pore liquid sample itself after it is sampled and retrieved to the ground surface.

The whole air-active method is typically performed with a drive probe or similar device. Once drawn to the surface and contained, soil gas samples can either be analyzed on site or transported to a laboratory for subsequent analysis. Alternatively, the sampling device can be directly coupled to a portable oxygen analyzer. A whole air-active sampler may be designed to either be installed in a predrilled borehole (permanent installation) or driven/pushed to the required sampling depth (rapid deployment drive/push probe). These two methods often serve two different purposes with the permanent installations being used when frequent measurements are required at a site and the rapid deployment drive probes being used when measurements are required infrequently. Most of the whole air-active samplers described in the literature were primarily developed for measuring concentrations of volatile organic compounds [e.g., 1, 2, 3, 4, and 5]. However, some samplers have been explicitly developed for sampling of oxygen and/or other soil gases [e.g., 6, 7 and 8].

The authors' interest in determining effective methods for rapid direct measurement of in situ oxygen profiles is a result of the development and use of an alterative highway deicer known as Calcium Magnesium Acetate (CMA). CMA was developed by the Federal Highway Administration to provide an effective yet environmentally safer deicer than rock salt [9]. Acetate is biodegradable in soil and consumes oxygen during aerobic decomposition in the subsurface of areas adjacent to highway shoulders. However, there is concern for oxygen depletion if an appreciable amount of acetate reaches underlying unconfined aquifers. To investigate this potential, the authors are studying the degradation of CMA in the vadose zone of a highway shoulder in southeastern Massachusetts. The site is ideally suited for this investigation because it has only been deiced with CMA since it opened in 1987. One of the major objectives of the project is to determine cost effective, efficient and reliable means for measuring spatial and temporal variations in subsurface oxygen concentrations at the site. Accurate temporal measurement of subsurface oxygen concentration profiles is considered essential to assessing the subsurface degradation, or attenuation of acetate through natural aerobic biodegradation processes.

For this project, seven different whole air-active methods of directly measuring oxygen profiles within the vadose zone at the site were investigated to determine the most effective method(s) considering ease of deployment, repeatability and reliability of measurements. Four different permanent installation samplers and three different rapid deployment drive probes were investigated. Samples of soil gas were drawn from the soil using the same oxygen measurement device attached to the different samplers and probes. By using the same protocol to analyze the soil gas samples, the measured results allowed direct comparison among different sampling devices to determine their relative attributes.

The paper gives a brief description of the soil and groundwater conditions at the site followed by details of the design, deployment and sampling for each of the seven different methods investigated. Typical oxygen concentration profiles obtained using the different methods are presented and compared. This is followed by a discussion of the results and identification of which method(s) proved to be most effective. The paper concludes with recommendations for the use of all of the methods investigated. Oxygen profiles such as those presented herein are being used along with other data (e.g., acetate concentrations, carbon dioxide, etc.) for a companion study of aerobic biodegradation process and transport.

TEST SITE

The test site is located in southeastern Massachusetts on the North shoulder of the westbound lane of State Route 25 in Plymouth County. The section of highway at the test site was constructed on a 3.5 m embankment of a controlled sand fill over what was formerly a portion of the Mann bog. The subsurface stratigraphy generally consists of top soil from 0 to 0.1 m, underlain by a poorly graded sand fill to a depth of approximately 3.0 m. At this depth, there exists a thin layer of organics which serves as a marker bed between the controlled fill and the poorly graded native sand soil that existed at the site prior to construction of State Route 25. Measurements of water levels over an eighteen month period in open standpipe piezometers installed at the site indicate that the water table varies between 3 to 4 m below grade.

The in situ physical characteristics of the soil were evaluated by conducting standard penetration tests (SPT) in general accordance with ASTM Standard Test Method for Penetration Test and Split-Barrel Sampling of Soils (D 1586). The SPT blow count values (N) are generally low near the surface and increase significantly up to a depth of approximately 2.5 to 3.0 m. Below this depth, which is coincident with the thin layer of organics and the controlled fill/native soil interface, the N values decrease. The N values in the fill layer are generally greater than 30 indicating that the soil is dense to very dense.

The gravimetric water content of the soil ranges from 1 to 5 % to a depth of approximately 3 meters (i.e., above the water table) and increases thereafter to values ranging from 10 to 20%. The grain size distribution of samples was determined in general accordance with ASTM Standard Test Method for Particle-Size Analysis of Soils (D422). Fifty-eight particle-size analysis tests were conducted on samples from four SPT borings. The results indicate that both the controlled sand fill and the native sand is poorly graded with some silt fines (1 to 12 % by mass) using ASTM Standard Test Method for Classification of Soils for Engineering Purposes (D2487). All of the grain size distribution curves are very similar and therefore the soil at the site is considered uniform throughout the depth of sampling (0 to 6.0 m).

METHODS OF INVESTIGATION

This section describes the different samplers investigated. It contains descriptions of the design and construction of the various samplers as well as details of their installation. Oxygen concentrations of the soil gas obtained using the different samplers (with the exception of one of the rapid deployment drive probes as noted below) were measured using a digital readout Bacharach Sniffer 302 portable oxygen meter. Gas samples were obtained either by directly drawing a sample from the soil through a sampling line using the vacuum pump contained within the Bacharach unit (direct read) or by a separate vacuum pump to draw a 2 liter sample into a 3 liter Tedlar bag [10] which was subsequently connected to the Bacharach unit for an oxygen reading. For both methods, the Bacharach unit was first set to an oxygen reading corresponding to ambient air (20.9%) prior to sampling. Reported oxygen concentrations for depth specific gas samples were taken as the steady state reading. For a soil with a constant air permeability, the time to reach a steady state reading is largely a function of the volume of gas contained within the sample lines (i.e., the sample lines must be purged before a representative depth specific value is obtained). During sampling the meter was periodically checked to ensure that it was still reading 20.9% for ambient air. Tests conducted comparing measured oxygen concentrations indicate that both the direct read and Tedlar bag procedures give the same values provided all connections are carefully inspected prior to sampling so as to minimize the possibility of ambient air leaking into the sampling stream.

Permanent Installation Samplers

Flexible Sampling Tubes--Two flexible sampling tube clusters were constructed and installed, one with 3 mm OD polypropelene sampling tubes and the other with 6 mm OD nylon sampling tubes. Both clusters were constructed by attaching the individual sampling tubes to the outside of a 3 m long, 13 mm ID schedule 40 polyvinylchloride (PVC) pipe using nylon cable ties (Fig. 1a). The ends of the individual tubes were located at 0.25 m increments along the length of the PVC pipe. The open end of the sampling tubes were covered with 20 μm nylon mesh to protect from clogging. The mesh was wrapped around the tube and PVC pipe and secured with nylon cable ties. The other end of the tubes were capped with Swagelok fittings to provide a means of attaching the surface sampling equipment.

Both clusters were installed using 0.11 m inside diameter (ID) hollow stem augers (HSA). The boring was advanced to a depth of 3.05 m and the assembled sampling cluster was placed inside the augers. The augers were withdrawn, taking care to ensure that the sampling ports remained at the correct elevation, and the borehole was backfilled with native soil cuttings. During backfilling, the cuttings were continuously tamped around the cluster using a capped 13 mm ID PVC pipe to prevent short circuiting. This method was selected over the use of bentonite because the shallow depths of installation and low water content of the uniform sand would likely result in significant desiccation cracking of the bentonite after initial installation. Both borings were backfilled to within approximately 0.25 m below the surface and finished with a cast iron well cover that was set in concrete.

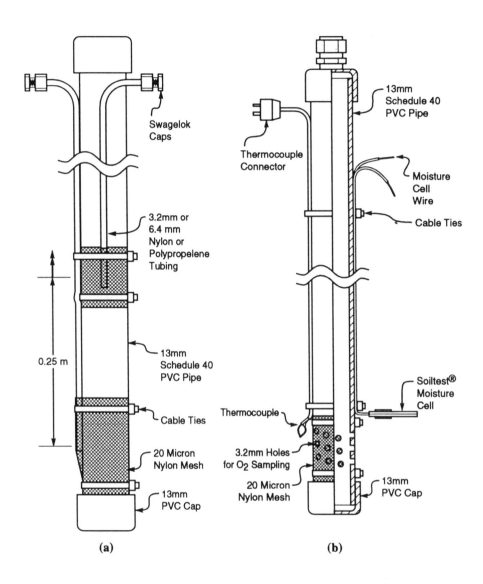

FIG. 1-- Schematic of (a) flexible sampling tube cluster and (b) rigid PVC sampler tube (not to scale).

Rigid PVC Sampling Tubes--The rigid PVC sampling tubes were constructed from 13 mm ID schedule 40 PVC pipe, and were equipped with a moisture cell, thermocouple, and ports for measuring oxygen (Fig. 1b). The top of each tube was slip fit with a PVC cap initially, but sampling difficulties (i.e., connecting and disconnecting surface sampling equipment) led to the retrofitting of each tube with 13 mm female national pipe thread (NPT) couplers and 13 mm male NPT by 6 mm tube Swagelok unions. All pipe connections were made using PVC cement. At the base of each tube, a series of 3 mm diameter holes were drilled to create sampling ports. The sampling ports were protected from clogging by wrapping 20 μm nylon mesh around the tube and securing it with nylon cable ties. Four PVC clusters consisting of three tubes each were installed using a drill rig. To begin the installation, each boring was advanced to the desired depth using 0.11 m ID HSA. A sampling tube was placed inside the HSA and held at the desired position while the HSA was carefully withdrawn. The sampling tubes were placed in clusters of three per borehole, with each tube based at a different depth below the ground surface. The minimum separation distance between sample tubes within a given cluster was set equal to 1 m to minimize cross-interference during sampling. The boring was backfilled and surface finished using the same procedure as described for the flexible sampling tube clusters.

Multi-Port Stainless Steel Sampling Tube--The multi-port stainless steel sampling tube was constructed from a 2.8 m long section of stainless steel tubing with an ID of 25 mm and an OD of 38 mm (Fig. 2a). One end of the tube was fitted with a stainless steel point that was machined to a 60° apex. The point was slipped in the end of the tubing and was attached with a 3 mm diameter pin. An adapter that had an AW female box thread and a 76 mm x 10 mm slot machined into it was welded to the other end of the tube. Sampling ports were installed along the length of the tube at 0.25 m increments. Initially, the sampling ports consisted of a 6 mm diameter, 3 mm thick, 20 μm sintered stainless steel filter installed in a 10 mm diameter stainless steel filter housing. The filters were installed slightly recessed and held in the filter housing with a small dab of inert silicone adhesive. A 3 mm diameter stainless steel tube, bent at 90° to form an elbow, was press fit into the housing. A 3 mm OD polypropelene riser tube was attached to this elbow using teflon heat shrink tubing. The entire assembly was installed into the sampler tube by feeding the riser tubing through a 10 mm hole drilled in the side of the sampler, through the inside of the sampler and then out the slot in the side of the adapter. The sampling port was then pressed into the sampler and the end of the riser was capped with a Swagelok cap fitting.

The sampler was installed by driving it into the ground with a 0.62 kN automatic SPT hammer. The free end of the individual riser tubes were temporarily taped to the sampler to protect them from damage during driving. However, during driving it was noted that the riser tubes were working their way out of the sampler through the slot. The riser to port connections appeared to have failed, so the sampler was removed from the ground. Once removed, it was noted that the filter elements in the sampling ports appeared to be clogged. When the sampler was disassembled, it was found that most of the elbow to housing connections had failed, as had a few of the riser to elbow connections.

The filter housing was redesigned (Fig. 2a) to allow the filters to be installed using a shoulder which helped position them flush with the outside surface of the sampler in an attempt to minimize clogging. The filters were held in place by crimping the edge of the

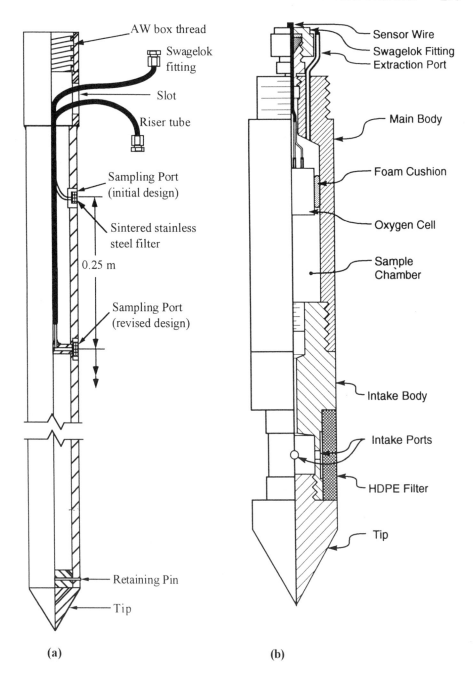

FIG. 2-- Schematic of (a) multi-port stainless steel sampling tube and
(b) flush filter probe with downhole sensor (not to scale).

housing and were fitted with a straight section of 2.3 mm OD hypodermic stainless steel tube attached with silver solder. The polypropelene riser was forced over this tubing, providing a snug connection that required no heat shrink tubing. The revised sampler was installed using the same technique and equipment as the initial installation until the bottom sampling port was 3 m below the ground surface. The ports were checked to see if a sample could be obtained and it was found that no soil gas could be drawn through a majority of the ports. The sampler was removed from the ground and the filters examined. It appeared that the filters had become clogged with fines and had also been damaged during driving. The ports were tested again while the sampler was lying on the ground, first with the filter elements in place and then with the elements removed, and it was determined that the filters were indeed clogged.

Rapid Deployment Drive Probe Samplers

All of the rapid deployment drive probes were attached to a drill string consisting of AW rods and driven in 0.23 m increments with a 0.62 kN automatic SPT hammer. An adapter section with a friction reducer connected each probe to the AW rod drill string. The friction reducer with its larger diameter serves to reduce friction forces along the drill string making it easier to drive the probes. For each probe, the soil gas sample tube lines were run inside the drill string rods and out through a slotted AW pin for connection to an oxygen meter at the surface.

Flush Filter Probe with Downhole Sensor--The flush filter drive probe with downhole sensor consisted of a tip, an intake body, and a main body containing a remote oxygen sensor (Fig. 2b). The probe was constructed of stainless steel with a 60° apex conical tip on one end and a 13 mm fine thread on the other end. The tip threaded onto an intake body which was machined to accept a High Density Polyethylene (HDPE) filter protecting a series of 3 mm diameter intake ports. This intake body was attached to the main body which contained a chamber housing a G.C. Industries model GC 33-200 electrochemical oxygen sensor. A 2 mm diameter stainless steel tube was soldered to the top of the chamber to create an extraction port, and a 2 mm OD Tygon tube connected this port to a hand vacuum pump at the surface. The oxygen cell was connected to a GC 501 Digital Oxygen Meter with a 2 conductor, 20 gauge sensor wire that was soldered to the terminals on the cell and fed through a Swagelok fitting at the top of the chamber. A 3 mm phono jack allowed the oxygen cell sensor lead to be disconnected in order to feed it through the drill string. The soil gas oxygen concentration at each sampling depth was determined by drawing a sample into the downhole chamber with the vacuum pump and then reading the digital meter. Ambient air was used as the control for calibration of the sensor.

The initial design of the probe allowed the oxygen cell to rest on a shoulder within the chamber, with foam padding surrounding the cell to help cushion it. After this arrangement proved to lead to sensor damage during driving, the sensor was further cushioned with a piece of HDPE filter material. When this design also resulted in a damaged sensor, a section of 13 mm schedule 80 PVC pipe was machined to create a reinforcing housing for the sensor and provide a secure support for the unit within the chamber.

Flush Filter Probe with Surface Sampling--The flush filter drive probe with surface sampling consisted of a tip and an intake section, with a tube attached to the filter section for sampling (Figure 3a). The probe was constructed of stainless steel with a conical tip machined to a 60° apex on one end and a 13 mm fine thread on the other end. The filter section was machined to accept an HDPE filter, and was tapped to accept an 3 mm NPT to 3 mm Swagelok fitting. Soil gas samples were drawn to the surface through a 3 mm diameter polypropelene line using a Bacharach Sniffer 302 portable oxygen meter.

Retractable Tip Probe with Surface Sampling--The retractable tip drive probe with surface sampling consisted of a sliding tip, a filter housing containing a sintered stainless steel filter and a tip housing (Fig. 3b). The tip was retained in the housing by a ring held in place by four set screws. The ring limited the travel of the tip when the probe was retracted to expose the filter. A series of 3 mm diameter holes were drilled around the circumference of the filter housing to allow a sample of soil gas to be drawn into the probe. The top of the filter housing was tapped to accept a 3 mm NPT to 3 mm Swagelok fitting. Soil gas samples were drawn to the surface through a 3 mm diameter polypropelene line using a Bacharach Sniffer 302 portable oxygen meter. Once the probe was driven to the desired sampling depth, the drill string was pulled back approximately 5 cm to expose the sampling ports. Once the sampling was complete, the probe was driven to the next sampling depth.

PRESENTATION OF RESULTS

Flexible Sampling Tubes

The flexible sampling tubes (Fig 1a) were sampled fourteen times between 05 October 1993 and 30 November 1994. Measured values for both the 3 mm diameter polypropelene and 6 mm diameter nylon flexible tubes display nearly identical spatial and temporal trends over the entire sixteen month sampling period. Fig. 4 plots oxygen concentrations measured for the 3 mm diameter polypropelene sampling tubes. The measured data shows the presence of relatively high oxygen concentrations near the ground surface followed by a gradual decrease in the concentrations with depth. The data also indicate a temporal variation with the lowest values being measured during the month of March. These results indicate the probable influence of CMA being used on the highway during January, February and March 1994. During these months there were numerous precipitation events that required the use of CMA on the highway. During the months of January and February there was considerable build-up of plowed snow adjacent to the highway shoulder. However, the ground was frozen during January and February thus preventing much infiltration of snow melt. As temperatures increased during March, the near surface soil thawed allowing infiltration of CMA laden snow melt and rain. This can result in a decrease in the oxygen concentration within the soil which was found to be the case for the two sampling dates in March. As the CMA gets aerobically degraded and/or passes through the unsaturated zone to the water table the oxygen concentrations should eventually return to the prewinter values. This was also found to be the case with the 1994 summer and fall readings being nearly identical to the prewinter values as shown in Fig. 4.

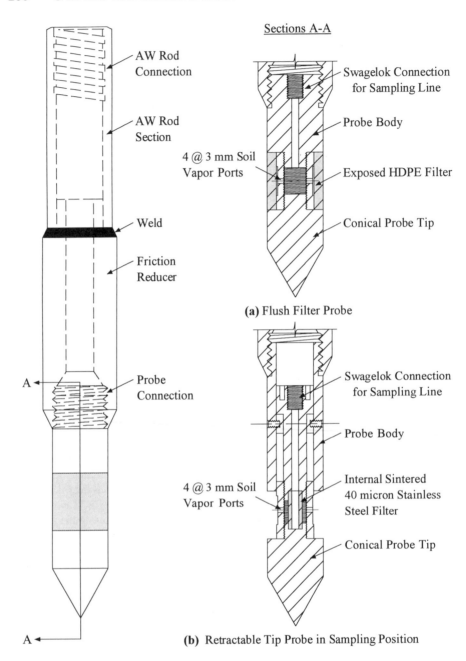

Sections A-A

AW Rod Connection

AW Rod Section

Swagelok Connection for Sampling Line

Probe Body

4 @ 3 mm Soil Vapor Ports

Exposed HDPE Filter

Weld

Conical Probe Tip

Friction Reducer

(a) Flush Filter Probe

A ◄

Probe Connection

Swagelok Connection for Sampling Line

Probe Body

4 @ 3 mm Soil Vapor Ports

Internal Sintered 40 micron Stainless Steel Filter

Conical Probe Tip

A ◄

(b) Retractable Tip Probe in Sampling Position

FIG. 3--Schematic of (a) flush filter probe with surface sampling and (b) retractable tip probe with surface sampling (not to scale).

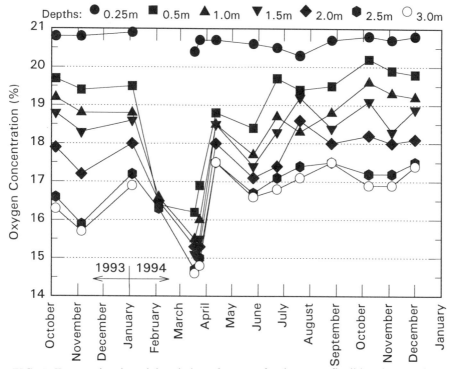

FIG. 4--Temporal and spatial variation of oxygen for the 3 mm flexible tube sampler.

Rigid PVC Sampling Tubes

When initially installed, the rigid PVC sampling tubes were fitted with slip-fit caps for protection. Soil gas sampling was performed by first removing the cap then forcing a rubber stopper fitted with a sampling line into the open PVC tube. However, it was discovered that this method presented sealing and disturbance problems because the caps were difficult to remove. They often required the use of a twisting motion during removal that would twist the entire PVC tube sampler causing disturbance of the surrounding soil seal. In addition to this, the stopper to PVC pipe seal would often leak allowing ambient air into the sampling stream. To correct these problems, all of the PVC cluster tubes were retrofit with a cap and Swagelok fitting (Fig. 1b) on 26 November 1993. These fittings proved to be much more effective providing a more convenient connection and also a positive seal during sampling.

The rigid PVC clusters were sampled sixteen times between 11 August 1993 and 30 November 1994. Fig. 5 presents plots of oxygen concentrations versus depth for five of the sampling dates. All of these data were obtained after the clusters were retrofit with a new top fitting on 26 November 1993. The data plotted in Fig. 5 represent the combined results of two clusters - one with three samplers with tube ports at 0.5, 1.5 and 2.5 m and the other with three samplers with tube ports at 1.0, 2.0 and 3.0 m. The trends in the measured data are very similar to those measured from the flexible sampling tubes with a general decrease in oxygen concentration with depth and also with the lowest oxygen values being measured during the month of March followed by an eventual return to prewinter values.

Flush Filter Probe with Downhole Sensor

The flush filter probe with downhole sensor (Fig. 2b) was deployed four times at the site between 01 July 1993 and 23 March 1994. On three of the four test dates, erratic readings were obtained from the probe resulting in only one complete and one partial profile. Although both profiles produced expected trends for the site with decreasing oxygen concentrations with depth, the magnitude of the measured values are lower than expected. This finding is based on comparison with values obtained from the rigid PVC and flexible tube samplers.

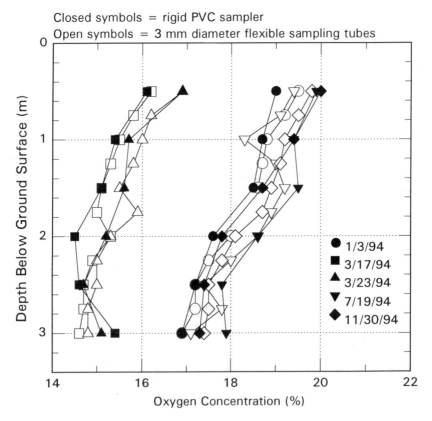

FIG. 5--Comparison of oxygen concentration profiles obtained using the rigid PVC sampling tubes and the 3 mm flexible sampling tubes.

Flush Filter Probe with Surface Sampling

The flush filter probe with surface sampling (Fig. 3a) was deployed twice at the site on 23 March 1994 and 25 May 1994. Fig. 6 plots oxygen concentrations versus depth for both sampling dates. Both profiles display expected trends with decreasing oxygen concentrations with depth and lower values at all depths for the March data relative to May.

Retractable Tip Probe with Surface Sampling

The retractable tip probe with surface sampling (Fig. 3b) was deployed three times at the site between 04 November 1993 and 25 May 1994. Fig. 7 plots oxygen concentrations versus depth for all three sampling dates. The profile for 04 November 1994 displays expected results with a gradual decrease in oxygen concentrations with depth. The other two profiles also display a general trend of decreasing oxygen concentrations with depth but are more erratic.

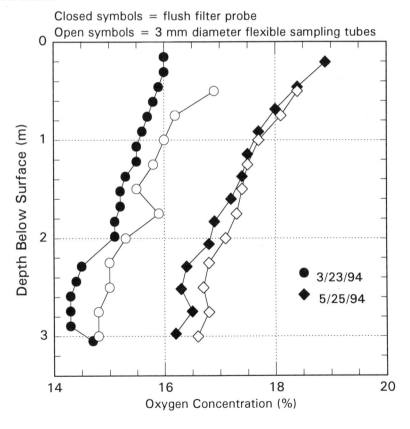

FIG. 6--Comparison of oxygen concentration profiles obtained using the flush filter drive probe with surface sampling and the 3 mm flexible sampling tubes.

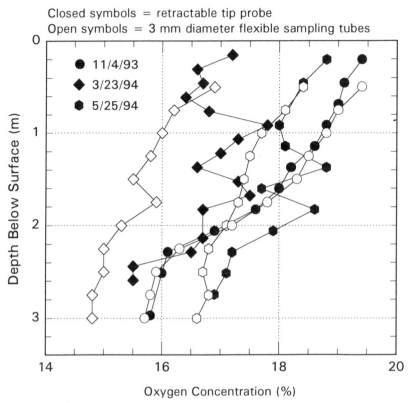

FIG. 7--Comparison of oxygen concentration profiles obtained using the retractable tip drive probe with surface sampling and the 3 mm flexible sampling tubes.

EVALUATION AND DISCUSSION OF RESULTS

Based on all of the data measured, the flexible sampling tubes appeared to produce the most complete and consistent set of measurements. The results obtained using this method are discussed first and then used as a standard for comparison for the other permanent installation and rapid deployment methods.

Permanent Installation Samplers

For all sampling dates, the oxygen concentrations measured at each depth for the 3 mm and 6 mm diameter flexible tubes were essentially the same. The values measured at 0.5 m show some discrepancies which are most likely a result of short circuiting between this shallow sampling depth and the surface due to ambient air being drawn along the soil-sampler interface. These results reflect the excellent repeatability of measurements obtained from the two different samplers and also suggest that the measured values were not dependent on the diameter of the flexible tubing. Sampling of the flexible tubing clusters was the easiest to perform of all the permanent installations designs. The volume of the individual tubes was very small, so the required purge volume and time prior to sampling was also small. This benefits sample quality because the in situ soil gas is disturbed less by a smaller total volume being withdrawn which in turn reduces the effects on the surrounding soil gas. This reduces the possibility of altering concentrations measured at adjacent depths. In addition, flexibility of the sampling tubes allowed test equipment to be connected with a minimum of disturbance to the installed sampler.

Once the rigid PVC clusters were retrofitted with the Swagelok fittings, they also produced very consistent results. Fig. 5 compares the values obtained from the rigid PVC clusters with those measured from the 3 mm diameter flexible sampling tubes for five sampling dates. The data from both samplers compare very well. Most of the readings are nearly identical with the exception of a few measurements that deviate up to a maximum value of approximately one percent oxygen.

Rapid Deployment Drive Probe Samplers

The flush filtered probe with surface sampling produced results that compared extremely well with those obtained from the flexible tube and rigid PVC samplers. As shown in Fig. 6, the probe values are within 0.5% oxygen of values obtained with the 3 mm diameter flexible sampling tubes at all depths except for two of the 25 May 1994 readings which are within 1% oxygen. These results clearly indicate that this probe design worked very well and did not suffer from any clogging problems in spite of having to be driven into the very dense soil at the site.

Fig. 7 compares oxygen concentration values obtained using the retractable tip probe with surface sampling on three sampling dates with those from the 3 mm diameter flexible sampling tubes. The measured data using the probe on 04 November 1994 compare extremely well with that measured in the flexible sampling tubes. Both profiles are essentially identical with a maximum deviation between the two methods of approximately 0.5% oxygen near 0.5 m. The data do not, however, compare well for the other two sampling dates. At some depths the readings are similar to those measured in the flexible sampling tubes, but a majority of the probe readings are greater. This deviation may be a result of the fact that this method requires the probe to be driven to a specific depth and then retracted a short distance to expose the filter section. This action may reduce the integrity of the seal between the sampler and the soil, leading to leakage and sampling problems.

Another problem which was noted during sampling, was that the filter was not consistently exposed by a set amount of drill string retraction. At some depths, retracting the drill string 5 cm was sufficient, while at other depths the drill string had to be retracted 10 cm or more before the oxygen level read by the sampler began to drop. When the Bacharach Sniffer 302 oxygen meter cannot draw a sample it defaults to a reading of 20.9%. This can be used as an indication of whether the drill string is sufficiently retracted or the filter element is clogged. During driving, with the retractable tip closed, the oxygen meter was monitored and if upon retraction of the drill string at a sampling depth the meter reading dropped from 20.9%, this indicated that the retraction was successful and the filter was exposed. Since most of the retractable tip probe readings are greater than those of the flexible sampling tubes, the error was most likely due to short circuiting. This problem is easily produced because of the presence of a friction reducer and also during retraction of the drill string which is required in order to expose the filter element. The friction reducer which is located just above the sample ports creates an open annulus between the native soil and drill string in soils that remain open during drilling. In addition, if the drill rods are pulled at an angle, the rods will no longer have intimate contact with the soil and will allow ambient air to travel down the annulus and be sampled. This problem can be minimized by being very careful to ensure that the drill string always remains vertical during driving and is not pulled back using the cable and hook but should be pulled using the hydraulic feed of the drill rig. Additionally, the friction reducer could be lengthened, moved to a position further away from the sample port (i.e., higher up on the drill string) or even completely removed.

SUMMARY AND CONCLUSIONS

Based on the results of this study, the most reliable and consistent sampler type tested is the flexible sampler tube design. The other styles were prone to errors due to short circuiting, damage during deployment or sampling difficulties. The ease of construction and relatively low material cost of this design, combined with the simplicity, relative quickness and accuracy of sampling make it superior to the others tested. The quality of the resulting data from this method can be attributed to several factors. Among these are the small purge volumes required prior to sampling and the lack of disturbance of the sampler caused during gas collection. The flexibility of the sampling tubes themselves allow the line to be readily manipulated in order to connect the sampling equipment. Since the tube flexes, the soil around the sampler is less likely to be disturbed during connection of the sampling equipment. This reduces the chances of short circuiting and sampling of ambient air. Both the 3 mm and 6 mm styles produced the same basic oxygen profiles despite having two different sample tube diameters.

Once the rigid PVC sampler tubes were retrofitted with Swagelok couplings attached to the top of the tubes, the measured data was very consistent and essentially identical to that obtained from the flexible tubes. These results suggest that this method can also be very effective. Although the samplers are as inexpensive and easy to construct as the flexible tube samplers they are not as efficient. Only a maximum of three can be installed in a borehole drilled with a 0.11 m ID HSA auger whereas the flexible samplers have sample ports every 0.25 m over the entire length of the borehole. Another disadvantage it that the purge volume during sampling is greater than that of the flexible tubes.

The multi-port stainless steel sampler was a difficult design to machine and assemble, and could not be installed in large part due to problems caused by driving it into the very dense soil at the site. Other problems included the fact that in order to sample several depths while maintaining a design that was practical to machine and preserved the integrity of the sampler body, the area of the filter had to be quite small. The small surface area was easily clogged, and the sintered material used for the filter was damaged during probe installation. A more durable filter material may eliminate this problem, but the surface area of the individual filters cannot be sufficiently increased without extensive machine work on the sampler body.

The flush filter probe with downhole sensor had limited success mainly due to the sensitivity of the sensor element to the dynamic forces encountered during driving into the dense soil at the site. This design may be more successful if a more sturdy sensor element and support housing can be designed. The probe design used for this project may have the potential to work at a site consisting of looser soils.

The flush filter probe with surface sampling produced very good results. The measured data compared very well with those obtained from the flexible tube and rigid PVC samplers. This method was the most successful of all the rapid deployment drive probe methods investigated. Relative to the permanent installation designs it has the advantage that it can be repeatedly used to rapidly obtain oxygen profiles but it does require the use of a drill rig or other method of deployment each time it is used.

The accuracy of the results produced by the retractable tip probe with surface sampling was affected in part by the design and by the method of deployment. This design produced results that were almost exactly the same as the flexible sampling tubes on one date but showed considerable deviation on other dates. This inconsistency is attributed to the potential for short circuits to develop during driving and retraction of the drill string. Care needs to be taken to ensure that this sampler, and in fact all rapid deployment drive probes, be driven vertically. Lengthening the friction reducer, moving it to a position higher on the drill string or eliminating it completely may minimize the potential for short circuiting.

ACKNOWLEDGMENTS

This research was funded in part by the Federal Highway Administration (FHWA) and the Massachusetts Highway Department (MHD). The authors thank Neal Plante of MHD District 5 and UMass graduate students Matthew Bonus, Paul Cheever, Charlene Howell, Jennifer Jordan, Lannae Long, and Heather Vandewalker for their assistance during field sampling. The opinions and findings contained herein are those of the authors and do not necessarily reflect that of FHWA and MHD.

REFERENCES

[1] Marrin, D.L. and Thompson, G.M., "Remote Detection of Volatile Organic Compounds in Groundwater Via Shallow Soil Gas Sampling," Proc. Conf. Petroleum Hydrocarbons and Organic Chemicals in Groundwater, National Water Well Association, Dublin, Ohio, 1984, pp. 172-187.

[2] Spittler, T.M., Clifford, W.S. and Fitch, L.G., "A New Method for Detection of Organic Vapors in the Vadose Zone," National Water Well Association Meeting, Baltimore, Maryland, 1985, pp. 236-246.

[3] Nadeau, R.J., Stone, T.S., and Klinger, G.S., "Sampling Soil Vapors to Detect Subsurface Contamination; A Technique and Case Study," USEPA, Region 8 Environmental Response Team, 1985, pp. 215-226.

[4] Thompson, G.M. and Marrin, D.L., "Soil Gas Contaminant Investigations: A Dynamic Approach," Ground Water Monitoring Review, Vol. 7, No. 3, 1987, pp. 88-93.

[5] Heins, T.R., Kerfoot, H.B., Miller, D.J., and Peterson, D.A., "Quality Assurance in Soil Gas Surveys: False Positive Acetone Identifications at a Codisposal Site," Current Practices in Ground Water and Vadose Zone Investigations, ASTM STP 1118, David M. Neilsen and Martin N. Sara, Eds., American Society for Testing and Materials, Philadelphia, 1992, pp. 151-165.

[6] Deyo, B.G., Robbins, G.A. and Binkhorst, G.K., "Use of Oxygen and Carbon Dioxide Detectors to Screen Soil Gas for Subsurface Gasoline Contamination," Ground Water, Vol.. 31, No. 4, 1993. pp. 529-604.

[7] Yanful, E.K., Riley, D.R., Woyshner, M.R. and Duncan, J., "Construction and Monitoring of a Composite Soil Cover on an Experimental Waste-Rock Pile near Newcastle, New Brunswick, Canada," Canadian Geotechnical Journal, Vol. 30, 1993, pp. 588-593.

[8] Hinchee, R.E., Ong, S.K., Miller, R.N., Downey, D.C., and Frandt, R., Test Plan and Technical Protocol for a Field Treatability Test for Bioventing, Environmental Services Office, Air Force Center for Environmental Excellence, May 1992.

[9] Transportation Research Board, Committee on the Comparative Costs of Rock Salt and Calcium Magnesium Acetate (CMA) for Highway Deicing, Highway Deicing: Comparing Salt and Calcium Magnesium Acetate, TRB Special Report No. 235, Washington, D.C., 1991.

[10] Robbins, G.A., Deyo, B.G., Temple, M.R., Stuart, J.D., and Lacy, M.J., "Soil Gas Surveying for Subsurface Gasoline Contamination using Total Organic Vapor Detection Instruments," Groundwater Monitoring Review, Vol. 10, 1990, pp. 122-131.

Russell A. Frishmuth[1], John W. Ratz[1], and John F. Hall[1]

UTILIZATION OF SOIL GAS MONITORING TO DETERMINE FEASIBILITY AND EFFECTIVENESS OF *IN SITU* BIOVENTING IN HYDROCARBON-CONTAMINATED SOILS

REFERENCE: Frishmuth, R. A., Ratz, J. W., and Hall, J. F., **"Utilization of Soil Gas Monitoring to Determine Feasibility and Effectiveness of *In Situ* Bioventing In Hydrocarbon-Contaminated Soils,"** Sampling Environmental Media, ASTM STP 1282, James Howard Morgan, Ed., American Society for Testing and Materials, 1996.

ABSTRACT: To determine the feasibility and effectiveness of *in situ* bioventing, careful monitoring of soil gas chemistry is essential. Prior to design of a bioventing system, initial soil gas surveys should be performed. Concentrations of three constituents, oxygen (O_2), carbon dioxide (CO_2), and total volatile hydrocarbons (TVH), are used in bioventing design. TVH are an indicator of contaminant distribution; O_2 and CO_2 are indicators of biodegradation activity. Analysis of soil gas data collected during pilot-scale testing is the primary design basis for full-scale remediation systems. Biodegradation rates determined from respiration tests are used to estimate the length of time that a system will have to operate to remediate the contamination. Air permeability of the soil, calculated from permeability testing, determines the number and spacing of air injection wells that will be required to ensure adequate oxygen influence through the entire contaminated area.

KEYWORDS: soil gas, *in situ* bioventing, bioremediation, respiration testing, hydrocarbon contamination

[1] Civil engineers, Remediation Department, Parsons Engineering Science, Inc., 1700 Broadway, Suite 900, Denver, CO 80290.

289

DESCRIPTION OF THE *IN SITU* BIOVENTING TREATMENT TECHNOLOGY

When petroleum hydrocarbon products spill or leak from tanks and pipelines, residual hydrocarbon constituents adsorb onto subsurface soil particles and/or become trapped in soil micropores. Without treatment to remove these contaminants, soluble benzene, toluene, ethylbenzene, and xylene (BTEX) compounds and other fuel constituents, such as naphthalene, may migrate through the soil profile to groundwater. Soil contaminated with these compounds could become a continuing source of groundwater contamination through this mechanism. *In situ* bioventing is an effective method of remediating hydrocarbon contamination in vadose zone soils [1] [2]. The purpose of this paper is to present how soil gas data can be used to determine the feasibility and effectiveness of bioventing. This paper should be a companion to existing literature outlining the implementation of bioventing technology.

Bioventing maximizes the *in situ* biodegradation of hydrocarbons by stimulating indigenous microbes with an enhanced supply of oxygen (Figure 1). Although petroleum hydrocarbons can be biodegraded under anaerobic conditions, biodegradation rates are limited by oxygen supply and can be enhanced by oxygenation [3]. Thus, if a site is anaerobic, remediation can be accelerated by providing a continuous source of oxygen to the subsurface. During the remediation process, contaminants are converted into innocuous byproducts (e.g., carbon dioxide and water) and biomass. Hydrocarbon contaminants are biochemically destroyed instead of simply being transferred to another phase or location within the environment as with other common treatment technologies such as soil vapor extraction (SVE) or soil washing.

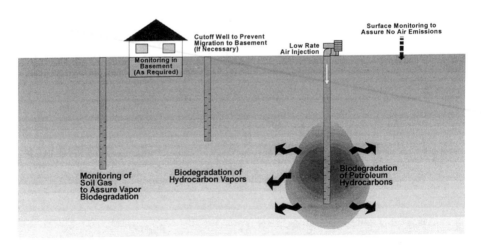

FIGURE 1--Conceptual layout of *in situ* bioventing.

Pilot-Scale Testing

To examine the feasibility of bioventing as a remediation technology, a pilot-scale test must be performed. Pilot testing is generally performed in accordance with procedures such as those described in the "Test Plan and Technical Protocol for a Field Treatability Test for Bioventing" [4]. The objectives of the testing are to assess the potential for supplying oxygen throughout contaminated soil zones, to determine if indigenous soil microorganisms are capable of biodegrading the hydrocarbon residuals, and if so, to quantify the rate at which indigenous microorganisms can biodegrade hydrocarbons when stimulated by oxygen-rich soil gas.

At each pilot test site, a drilling and sampling program is conducted to characterize the geologic and hydrogeologic conditions and the contaminant distribution in the subsurface. The drilling program is usually designed around existing site data from previous investigations or from field investigation techniques such as field screening and analytical analysis of soil gas samples. Four boreholes are drilled at each pilot testing site for installation of a vent well (VW) for air injection and three vapor monitoring points (MPs) for soil gas sampling. Additionally, a background MP is often installed to characterize soil gas conditions in uncontaminated soils. When available, existing background or upgradient groundwater monitoring wells that have screen exposed above the groundwater surface are used as background MPs. Once the VW and all MPs are installed, initial and extended system testing can begin. Tests include respiration testing for the determination of biodegradation rates, permeability testing to determine effective treatment radius, and flux testing to determine the potential for vapor migration. Each of these tests is described in detail in remainder of this document.

Technological Advantages and Disadvantages

In situ bioventing enjoys several advantages over conventional treatments of hydrocarbon-contaminated soils. First and foremost, it is generally much less costly than *ex situ* treatment or even conventional soil vapor extraction [5]. The lower costs result from the limited amount of drilling or excavation required and use of low maintenance equipment. Costs for bioventing average 20 to 33 dollars per cubic meter for sites smaller than 1 000 cubic meters, and less for larger sites. Bioventing is less expensive than soil vapor extraction due to the fact that, generally, no off-gas treatment is required. This also makes permitting of remediation systems much easier in areas where air emissions are a concern. Finally, bioventing is advantageous from a risk standpoint, no large excavations are required and exposure to contaminated soils is minimal during short-term drilling activities. Bioventing is not a rapid process; however, if no regulatory deadlines need to be met, bioventing is often a preferred remedial option.

INITIAL SOIL GAS SURVEY

Soil gas chemistry is one of the primary indicators of the feasibility and effectiveness of bioventing. Initial soil gas surveys can be used to determine the extent of contamination, chemical constituents, and concentrations. Based on these data, a bioventing pilot test can be designed and conducted. Initial soil gas samples taken from MPs installed as part of a pilot test provide baseline data on volatile contaminant concentrations at a site. After system operation, soil gas concentrations can be compared to these initial values to determine system effectiveness. Each of these uses of soil gas will be described in detail below.

<u>Sample Collection</u>

Soil gas surveys are generally conducted with either hand-driven or hydraulically-driven probes. Recently, vans equipped with hydraulic drives and field gas chromatographs (e.g., Geoprobe® systems) have become readily available and are in widespread use. Regardless of the method used, the same type of probe tip is generally used. The tip consists of a section of stainless steel screen that is connected to tubing that runs to the surface. The tip is driven to the desired sampling depth and retracted slightly to expose the open screen section to the soil. The end of the tubing is then connected to a sample pump, a vacuum is applied, and a soil gas sample is collected. After sample collection, the tip and rods are removed, and the hole is backfilled with grout or bentonite. For special applications, tips can be installed permanently or permanent MPs can be constructed out of screened polyvinyl chloride (PVC) piping. A schematic diagram of a typical soil gas probe and sampling system is shown in Figure 2.

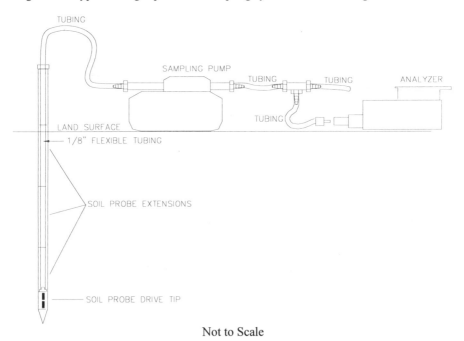

Not to Scale

FIGURE 2--Schematic diagram of soil gas probe and sampling system.

Soil gas samples are commonly collected in airtight sample bags for field screening. Samples to be sent to an off-site analytical laboratory are commonly transferred to a stainless steel vacuum chamber for shipment. For bioventing studies, the majority of the samples are analyzed in the field for O_2, CO_2, and TVH. O_2 and CO_2 are measured as a percentage of the total gas sample volume using a field meter. TVH are measured in parts per million, volume per volume (ppmv) using a photoionization detector (PID) with a lamp voltage appropriate for the expected contaminants at the site. A fraction of the soil gas samples are also commonly sent to an analytical laboratory and analyzed for TVH and the BTEX compounds.

Survey Techniques

Survey techniques vary depending on the type of site and what previous investigation data are available. If a site has not been well characterized then a full grid type of survey should be performed. Samples should be collected at evenly spaced intervals across the site. Sample collection should begin at the center of the area of suspected contamination and move outward. Sample collection should cease once field screening indicates little to no contamination is present. Samples should also be collected at several depth intervals at each sampling point to delineate the vertical extent of contamination. Three-dimensional renderings of probable contamination at a site can be constructed with data from a soil gas survey to aid in pilot test design. Other two-dimensional renderings such as contours maps are also easily developed after a full soil gas survey has been performed.

If a site has already been characterized or the source of contamination is well known, a full soil gas survey may not be required. This is especially true if a soil gas survey has been performed during any previous investigations. In this case, all that is required is a partial survey consisting of a few confirmatory samples in areas of contamination identified previously. Samples are definitely required when no data exists on the oxygen and/or carbon dioxide content of the soil gas. Care should also be taken to assure that the vertical extent of the contamination has been adequately defined. If these data are not available then samples from different depth intervals are required as with a full soil gas survey. Finally, for both types of surveys, a background sample must be collected to determine what typical, non-contaminated, soil gas concentrations are at the site. This sample should be collected as close to the site as possible and from a comparable depth and soil type.

Feasibility Analysis

After data from soil gas surveys have been collected they can be analyzed to determine if the site is suitable for the application of *in situ* bioventing. As mentioned previously, three parameters are examined in determining the feasibility of bioventing. Concentrations of O_2 and CO_2, and TVH are examined and evaluated. Additionally, concentrations of the BTEX compounds may also be evaluated to aid in determination of spill type and age. This analysis, however, is not critical to the determination of bioventing feasibility and may not be performed if the contamination at a site is well characterized.

Oxygen is the primary indicator of whether or not bioventing is an appropriate remedial technology [6]. If soil gas oxygen concentrations are not depleted, oxygen addition to the subsurface will not be beneficial. In most cases, if natural oxygen concentrations are uniformly above 5 percent, bioventing is not necessary [6]. In this type of situation, biodegradation is probably already occurring, and sufficient oxygen is being provided to the microbial population through atmospheric diffusion. This is particularly common at shallow sites with sandy soils and weathered contamination. Background oxygen concentrations should be examined before making a final determination on depleted oxygen concentrations. In isolated instances, naturally occurring processes other than hydrocarbon biodegradation can consume oxygen and create deficiencies in areas that are not contaminated. Natural oxygen sinks include iron-rich soils which consume oxygen to reduce iron, and soils with significant amounts of organic material which consume oxygen as part of the natural decay process. If background levels of oxygen are not depleted then it is apparent that oxygen decline in the source area is due to the biodegradation of hydrocarbons. If oxygen levels are depleted, bioventing is definitely an option and other soil gas constituent concentrations should be examined to further determine how to apply bioventing to the site.

Carbon dioxide concentrations should be examined after it has been determined, based on oxygen concentrations, that bioventing may be applicable. Carbon dioxide is an indicator of complete mineralization of petroleum hydrocarbons. As hydrocarbons are degraded, water and carbon dioxide are released as metabolic by-products. Generally, levels of carbon dioxide above atmospheric concentrations indicate the presence of microbial activity. It should be noted that elevated levels of carbon dioxide can also be found in acidic soils, as carbon dioxide dissolves as a weak acid and could be released from ground water in an acidic environment. Because of this, background data become very important when trying to demonstrate biodegradation on the basis of elevated carbon dioxide levels. Concentrations from samples collected from a contaminated area should be greater than background levels. In some circumstances, carbon dioxide levels may be low, even in the presence of microbial activity. This is common if the soil is alkaline and is buffering carbon dioxide content in the soil gas. It is for these reasons that carbon dioxide concentrations in soil gas are used secondary to oxygen concentrations; carbon dioxide concentrations are not always indicative of subsurface conditions and should be used with care.

TVH and BTEX concentrations are examined to qualitatively assess the type of contamination and how weathered the hydrocarbon is. Because BTEX compounds have greater water solubility than other fuel components, they are more easily biodegraded through natural weathering or enhanced bioventing processes. If samples contain high levels of BTEX and high levels of TVH, it is likely that the source is a recent release of a fuel with a high BTEX/TVH ratio (e.g., unleaded gasoline). If samples contain high levels of TVH and low levels of BTEX, then the source product is either a weathered fuel or the contaminant source had low levels of BTEX (e.g., diesel fuel or fuel oil). This information is particularly significant in light of the recent trend in risk-based remediation. Many cleanups are now targeting BTEX removal to reduce human health risks rather than to meet arbitrary total petroleum hydrocarbon (TPH) standards. ASTM Committee E-50 on Environmental Assessment has created guidelines for these types of cleanups with its publication of the ASTM Guide for Risk-Based Corrective Action. The lower the BTEX content of soil and soil gas, the less source removal may be required. Collection of initial data on TVH and BTEX concentrations also establishes baseline conditions against which the progress of remediation at the site can be measured.

In conclusion, the decision to implement *in situ* bioventing as a remedial technology at a site rests on demonstrating (1) a depleted oxygen environment, (2) elevated carbon dioxide concentrations, and (3) concentrations of TVH and/or BTEX indicating soil contamination levels above regulatory concern. A flow chart outlining the decision process for determining the feasibility of using *in situ* bioventing as a remedial alternative is shown in Figure 3. At most sites with petroleum hydrocarbon contamination in vadose zone soils, the three required conditions are present and readily apparent. If these three conditions are not present, bioventing may not be the best available remedial alternative.

PILOT-SCALE TESTING

Once it has been determined that bioventing is feasible at a site, a pilot-scale test is performed to determine full-scale design parameters. Soil gas data are essential in pilot test design and evaluation. Initial soil gas data will determine VW placement and the depth of the screened interval for air injection. The VW should be placed in the center of the contaminant mass as determined by oxygen and TVH concentrations. The screened interval should extend through the entire depth interval of contaminated soil. Often, VWs are installed so that the screened interval extends one to two feet into ground water to allow for the maximum potential to remediate the capillary fringe and account for seasonal drops in ground water levels [6].

Placement and installation depths of permanent MPs are also determined based on results from an initial soil gas survey. Soil gas MPs serve two functions for a bioventing pilot-scale test: (1) measuring the rate of oxygen utilization to determine *in situ* biodegradation rates, and (2) monitoring the pressure and movement of soil gases in the treatment area. The rate of oxygen utilization is most accurately measured in the soil zones with the lowest oxygen concentrations. As a result, soil gas MPs should be installed in areas shown to have the most oxygen-depleted soil gas during the preliminary survey [6]. A soil gas point should also be placed at the outer limit of contamination as determined by the soil gas survey. This point is used to aid in determining effective treatment radius and vapor migration potential. Monitoring point depths should be determined by examining soil gas data for various depth intervals. Again, whenever possible, points should be located in the intervals of greatest contamination.

Once the VW and all MPs have been installed, soil gas samples are collected from the new permanent locations. Typically, petroleum-contaminated soils containing an acclimated bacterial population will be oxygen depleted and will contain elevated levels

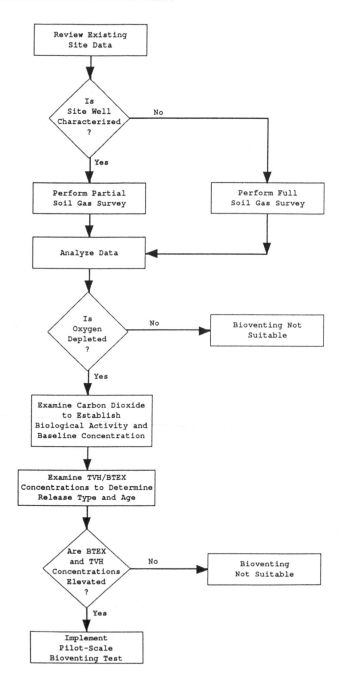

FIGURE 3--Bioventing feasibility decision flow chart.

of carbon dioxide and TVH. The samples are used to confirm an oxygen-depleted environment and to create a baseline for the system. After this initial soil gas sampling has been performed, initial pilot-scale testing can begin. This testing includes respiration testing for determination of biodegradation rates, permeability testing to determine effective treatment area, and flux testing and ambient air monitoring to determine if vapor migration is occurring.

Respiration Testing

In situ respiration testing has been documented as a successful method of measuring aerobic biodegradation of petroleum hydrocarbons [7]. A respiration test is performed during a bioventing pilot test to determine the biodegradation ability of the microbial population at the site. Because this potential varies from site to site, it is important to perform a respiration test as part of every bioventing pilot test. During long-term system operation, respiration testing is performed periodically to assess system performance over time.

To perform a test, air admixed with helium, an inert tracer gas, is injected into contaminated soils, and the rates of oxygen consumption and helium diffusion are monitored periodically (every 1 to 2 hours) after injection ceases. Readings continue until oxygen reaches preinjection concentrations, typically 0 to 5 percent. Helium is an inert, highly diffusive, nonbiodegradable gas, and it is used as a conservative tracer to determine if oxygen diffusion is responsible for a portion of the oxygen loss, or if leakage is occurring during soil gas sampling. At a biologically active hydrocarbon contaminated site, oxygen concentrations decrease, carbon dioxide concentrations generally increase, and helium concentrations remain relatively constant during the course of a respiration test. Because oxygen decreases and helium remains constant, oxygen loss can be attributed to biological uptake rather than diffusion. Increases in carbon dioxide indicate that petroleum hydrocarbon contaminants are being completely mineralized.

After performance of a test, field data are entered into a computer spreadsheet for data analysis. This consists of loading oxygen concentrations versus time into a spreadsheet and plotting the results. The slope of a best-fit line plotted through the linear portion of the resulting curve gives the calculated oxygen utilization rate. This rate is usually calculated in percent decrease per minute. An example of a curve and best-fit line is shown in Figure 4.

After a oxygen utilization rate has been determined, a biodegradation rate can be estimated. The rate calculation is based on the stoichiometry of the chemical formula for the mineralization of hexane:

$$C_6H_{14} + 9.5O_2 \rightarrow 7H_2O + 6CO_2 \tag{1}$$

On a mass basis, this equates to approximately 3.5 grams of oxygen consumed for every gram of hydrocarbon degraded into water and carbon dioxide. Using the air-filled porosity of the soil at the site, a biodegradation rate can be estimated in terms of milligrams of hydrocarbon consumed per kilogram of soil per year. It should be noted that this number is only an estimate, and soil sampling should be performed at the conclusion of the test period to verify biodegradation calculation results.

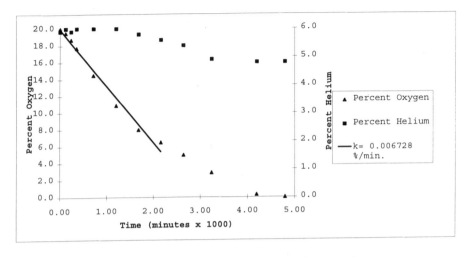

FIGURE 4--Typical respiration test analysis graph.

The primary use of the initial biodegradation rate is determining how long a system may have to operate to reach "clean" standards. If the approximate volume of fuel that was released is known, then it can be easily estimated how long a system may need to operate to remediate a site. This type of analysis can also be done based on laboratory results of soil samples, however, this method is not as accurate. Once an initial biodegradation rate has been determined, testing during system operation can be used to confirm remedial progress. Respiration tests are generally performed at 6-month intervals during the course of system operation. Each time a test is performed, a biodegradation rate is calculated. This rate is then compared to the initial rate to determine if biodegradation is continuing and if so, at what rate. Typically, rates will slow over time as the hydrocarbon content of the soil is reduced and microbial activity declines accordingly. Monitoring the progress of a pilot-scale test in this manner allows for operators to determine progress semi-quantitatively without the expense of soil sampling and analysis.

Permeability and Radius of Influence Testing

Respiration testing is used to determine the level of biological activity and corresponding biodegradation at a site. After the biodegradation rate has been calculated, and it has been determined that *in situ* biodegradation is viable at a site, the areal extent of treatment must be defined. For the purpose of pilot-scale testing, one VW is used to determine this parameter. Once the area of treatment from one well is determined, full-scale treatment systems can be designed quickly and efficiently.

To determine areal extent of treatment it is assumed that every area that receives atmospheric oxygen through air injection will enhance microbial activity and thus promote biodegradation. To determine the extent of air injection influence at a site, two soil gas parameters are measured: soil gas oxygen concentration and pore pressure. To perform a test, air is injected into the VW at a high pressure and flow rate. Corresponding changes in pressure and oxygen concentrations are noted at all MPs during the course of the test. Air injection continues until pressures and soil gas concentrations reach equilibrium. Tests may run from 1 to 2 hours in sandy soils to 1 to 2 days in less permeable clays and silts.

At the conclusion of the test, data are entered into a computer spreadsheet. Using permeability models based on Darcy's Law, a permeability for the soil is calculated in darcys. This number is typically used qualitatively as a method of comparing the permeability of one site with that of another. The primary factor in full-scale design is oxygen concentrations in soil gas at the MPs. During the course of a permeability test, it is possible to see an increase in pressure at a MP without having any influx of oxygen. Oxygen-laden air may push against contaminated soil vapors in the air-filled voids of the soil and create pressure. Fresh air will eventually displace the contaminated vapors; however, in low permeability soils this may take weeks and, as a result, is not evident during shorter permeability tests. A conservative approach to determining the radius of influence uses only increases in oxygen concentration at locations that previously were depleted.

For the purpose of determining the areal extent of treatment, it is considered that any area than can maintain an oxygen concentration of 5 percent or above is within the treatment area. This is easily determined within the area of the monitoring points. However, when the MP furthest from the VW is exhibiting concentrations greater than 5 percent, it is necessary to extrapolate to determine effective treatment radius. This is done by graphing oxygen content of the soil gas versus the linear distance from the injection point. This typically resembles a linear relationship, and a line can be drawn to represent a distance versus oxygen content constant. Using the slope of the line, it is easily determined at what distance from the VW the oxygen content should be 5 percent. In practice, injection wells are designed to overlap by 10 to 20 percent to allow for inherent anomalies in the subsurface. It may also be necessary to interpolate when the MP furthest from the injection point does not receive adequate oxygen to be considered in the treatment area. The same graphical process described above can be used.

Flux Testing

One last soil gas technique that is useful in determining the effectiveness of *in situ* bioventing is the collection of soil gas flux samples from the ground surface. Flux sampling is a method of determining if any vapors are being driven out of the soil and into the ambient air at a site. An air-tight container is placed on the ground surface, ultra-pure air (less than 1 ppmv) is bled into the chamber as sweep air, and then a sample is collected. If any hydrocarbons are migrating out of the ground surface, they will mix with the sweep air and be detected in the sample.

Flux testing is typically performed at a site before a pilot-scale test begins and then again during system startup. Sampling prior to system operation provides background data used for comparison to other samples. Flux testing for bioventing should be performed during initial startup because that is the time at which vapor migration is most likely to occur. After the system has been operational for a period of time, all vapors within the treatment area will reach equilibrium as the pressure in the pore spaces of the soil equalizes.

If contaminant concentrations are apparent in flux samples after system startup, the injection rate of the system should be reduced, and the flux test should be performed again. If reducing the flow rate does not solve the problem, then the system may need to be run for a short period of time in a vapor extraction mode. By running an extraction system and treating the related off-gas, it may be possible to strip out the most volatile compounds over a period of months and then resume injection. This type of operation is well suited for relatively new releases of volatile hydrocarbons such as gasoline spills.

If a bioventing pilot system is installed in the vicinity of occupied buildings, ambient air monitoring should be performed inside the buildings. As with flux testing, measurements should be taken before and after system operation. Ambient air monitoring is especially important if the facility has a basement. This type of monitoring can easily be performed with a field instrument such as a hydrocarbon analyzer or preferably, a PID.

CONCLUSIONS

Soil gas can be used effectively to determine the feasibility and effectiveness of *in situ* bioventing. Proper use of soil gas data allows for efficient design, installation, and monitoring of pilot-scale remedial tests. Respiration testing, the primary factor in calculating biodegradation rates at a site, is one key soil gas method that should be performed methodically at every bioventing site. Once respiration testing has been performed, soil gas data are again used to determine the effective treatment area of the remedial system. Finally, soil gas data in the form of flux testing are utilized to assure that systems are not creating any human health risk while remediating vadose zone soils. As bioventing is utilized more often as a remedial alternative, soil gas data should be used as an inexpensive and efficient tool in aiding system design and performance monitoring.

REFERENCES

[1]   Dupont, R. R., Doucette, W. J., and Hinchee, R. E., "Assessment of *In Situ* Bioremediation Potential and the Application of Bioventing at a Fuel-Contaminated Site," In *In Situ* Bioreclamation Applications and Investigations for Hydrocarbon and Contaminated Site Remediation, R. E. Hinchee and R. F. Olfenbuttel, eds., Butterworth-Heinemann, Boston, 1991.

[2]   Miller, R. N., Vogel, C. C., and Hinchee, R. E., "A Field-Scale Investigation of
       Petroleum Hydrocarbon Biodegradation in the Vadose Zone Enhanced by Soil
       Venting at Tyndall AFB, Florida," In *In Situ* Bioreclamation Applications and
       Investigations for Hydrocarbon and Contaminated Site Remediation, R. E.
       Hinchee and R. F. Olfenbuttel, eds., Butterworth-Heinemann, Boston, MA., 1991.

[3]   Hinchee, R. E. and Miller, R. N., "Bioventing for *In Situ* Remediation of
       Petroleum Hydrocarbons," In Bioventing and Vapor Extraction: Uses and
       Applications in Remediation Operations. Live Satellite Seminar. Sponsored by
       the Air & Waste Management Association and HWAC. April 15, 1992.

[4]   Hinchee, R. E., Ong, S. K., Miller, R. N., Downey, D. C., and Frandt, R., "Test
       Plan and Technical Protocol for a Field Treatability Test for Bioventing,"
       Prepared for the United States Air Force Center for Environmental Excellence,
       Brooks Air Force Base, TX., 1992.

[5]   Reisinger, H. J., Johnstone, E. F., and Hubbard, P., "Cost Effectiveness and
       Feasibility Comparison of Bioventing vs. Conventional Soil Venting," In
       Hydrocarbon Remediation, R. E. Hinchee, B. C. Alleman, R. E. Hoeppel, and R.
       N. Miller, eds., Lewis Publishers, Boca Raton, FL., 1994.

[6]   Hall, J. H., and Downey, D. C., "Addendum One to Test Plan and Technical
       Protocol for a Field Treatability Test for Bioventing - Using Soil Gas Surveys to
       Determine Feasibility and Natural Attenuation Potential," Prepared for the United
       States Air Force Center for Environmental Excellence, Brooks Air Force Base,
       TX., 1994.

[6]   Hall, J. H., and Downey, D. C., "Addendum One to Test Plan and Technical
       Protocol for a Field Treatability Test for Bioventing - Using Soil Gas Surveys to
       Determine Feasibility and Natural Attenuation Potential," Prepared for the United
       States Air Force Center for Environmental Excellence, Brooks Air Force Base,
       TX., 1994.

[6]   Hall, J. H., and Downey, D. C., "Addendum One to Test Plan and Technical
       Protocol for a Field Treatability Test for Bioventing - Using Soil Gas Surveys to
       Determine Feasibility and Natural Attenuation Potential," Prepared for the United
       States Air Force Center for Environmental Excellence, Brooks Air Force Base,
       TX., 1994.

[6]   Hall, J. H., and Downey, D. C., "Addendum One to Test Plan and Technical
       Protocol for a Field Treatability Test for Bioventing - Using Soil Gas Surveys to
       Determine Feasibility and Natural Attenuation Potential," Prepared for the United
       States Air Force Center for Environmental Excellence, Brooks Air Force Base,
       TX., 1994.

[7]   Ong, S. K., Hinchee, R. E., Hoeppel, R., and Scholze, R., "*In Situ* Respirometry for Determining Aerobic Degradation Rates," In <u>*In Situ* Bioreclamation Applications and Investigations for Hydrocarbon and Contaminated Site Remediation</u>, R. E. Hinchee and R. F. Olfenbuttel, eds., Butterworth-Heinemann, Boston, MA., 1991.

Abidin Kaya[1] and Hsai-Yang Fang[2]

CHARACTERIZATION OF DIELECTRIC CONSTANT ON FINE-GRAINED
SOIL BEHAVIOR

REFERENCE: Kaya, A. and Fang, H. Y., "Characterization of Dielectric Constant on Fine-Grained Soil Behavior," Sampling Environmental Media, ASTM STP 1282, James Howard Morgan, Ed., American Society for Testing Materials, 1996.

ABSTRACT: It has been postulated in earlier studies that there is a considerable relationship between some engineering properties of fine-grained soils and the dielectric constant of pore fluid such as with hydraulic conductivity: the lower the dielectric constant, the higher the hydraulic conductivity. However, the mechanisms causing the increase in hydraulic conductivity are not understood well. It is the purpose of this paper to investigate some mechanisms that affect the fine-grained soil behaviors when the dielectric constant of pore fluid is changed. For this reason, the relationships between the dielectric constant of pore fluid and soil-fluid system properties for Atterberg limits, cation exchange capacity, and the zeta potential of kaolinite were determined with different organic solvents. It was observed that as the dielectric constant decreases, kaolinite's liquid limit increases whereas the plastic index remains almost constant. Furthermore, its zeta potential, which is an indication of the thickness of double layer, decreases. It was concluded that when the dielectric constant of pore fluid decreases the surface charge density of soil decreases which causes a higher flow area increasing the hydraulic conductivity.

KEYWORDS: fine-grained soil, hydraulic conductivity, Atterberg limits, cation exchange capacity, zeta potential, surface charge density, dielectric constant

Pore fluid chemistry affects the engineering behavior of fine-grained soils considerably, such as the hydraulic conductivity of clay liners and barrier walls as well as swelling pressure of soils. For example, a three to five order increase in magnitude in hydraulic conductivity of clay barriers has been reported [1]. Although it has not been possible to determine the effect of pore fluid chemistry on soil behavior in a satisfactory manner, several investigators as early as Quincke [2] proposed the use of the dielectric constant of the pore-fluid to quantify the effect of the pore fluid chemistry on fine-grained soils.

[1] Graduate Research Assistant, Civil and Environmental Engineering Department, Lehigh University, Bethlehem, PA 18015
[2] Professor, Civil and Environmental Engineering Department, Lehigh University, Bethlehem, PA 18015

The dielectric constant can be defined as a measure of a material's ability to perform as an insulator. That is, it is a measure of the capacity of a material to reduce conductance of electromagnetic energy. The dielectric constant of the soil is between three and six depending on the orientation of the particles, whereas the dielectric constant of water is about 80 at 20 °C. However, most organic solvents have a dielectric constant less than 80 and as low as two. The dielectric constants of solvents used in this study are presented in Table 1.

CHARACTERIZATION OF DIELECTRIC CONSTANT: HISTORICAL DEVELOPMENT

To investigate the effect of nonpolar fluids on the hydraulic conductivity of fine grained soils, Macy [3] observed that the hydraulic conductivity of soils can increase five to six orders of magnitude compared with the hydraulic conductivity of water on the same soil. He postulated that aggregation and arrangement of particles with emphasis on the effect of the fluid viscosity at the near surface of the clay particles, would cause an increase in hydraulic conductivity.

Waidelich [4] investigated the effect of the dielectric constant on some of the engineering properties of soils such as hydraulic conductivity and compressibility. Although he was able to establish some relationship between dielectric constant and hydraulic conductivity, he was unable to draw solid conclusions between the dielectric constant of the pore fluid and other engineering parameters such as compressibility.

Olsen [5] noted that electroosmotic counter flow, high viscosity and tortuous flow paths failed to account for the hydraulic conductivity characteristics. He observed the effect of the dielectric constant on different types of soils and concluded that unequal pore size is the most important variable that affects the hydraulic conductivity of soils.

Mesri and Olson [6] investigated the mechanisms controlling the hydraulic conductivity of kaolinite, illite, and smectite. They concluded that aggregation of soil particles due to low dielectric constant of pore fluid leads to channelization which causes a higher degree of hydraulic conductivity.

Fernandez and Quigley [7] conducted a study to clarify the contradicting interpretations of the factors controlling the hydraulic conductivity of soils. They determined the hydraulic conductivity of a given soil with different chemical solvents having dielectric constant ranging from 80 (polar) to two (nonpolar). They observed that there was a considerable relationship between hydraulic conductivity and the dielectric constant of the pore fluid which was attributed to a reduction in the thickness of a double layer with a decrease in the dielectric constant because of a decrease in the surface potential. Furthermore, it was speculated that a decrease in surface potential causes flocculation and forms sand size particles or flocks with large interped pores. The results of Fernandez and Quigley [7] along with Waidelich [4] are presented in Figure 1.

Brown and Thomas [8] investigated the effect of the dielectric constant on the swelling of kaolinite, illite, and smectite. They found that there was a significant relationship between the dielectric constant of the solvent and swelling, (see Figure 2). An increase in swelling with a increase in the dielectric constant was attributed to an increase in the flocculation of the clay particles with a increase in the dielectric constant.

MATERIALS AND METHODS

All experiments were performed on commercially available Georgia kaolinite from Georgia Clay^R. Aniline, Methanol, 2-Propanol, and Xylene were used as solvents in addition to water to investigate the effect of dielectric constant of solvent on fine-grained soil behavior. Table 1 presents the properties of materials used in this study.

Table 1. Physico-chemical properties of solvents

Solvent	Density(p) (g/cc)	Viscosity (η) (mPa s)	Dielectric Constant (ϵ)
Aniline	~ 0.8	~ 4.5	~ 7
Methanol	~ 0.8	~ 0.6	~ 33
2-Propanol	~ 0.8	~ 2.4	~ 18
Water	~ 1.0	~ 1.0	~ 80
Xylene	~ 0.88	~ 0.8	~ 2.5

Atterberg Limits

The Atterberg limits are indices of the quantification of clay-size particles, their mineralogical composition and interaction between particles and pore fluid. In general, higher liquid limit and the plasticity index are associated with soils having a greater quantity of clay particles or clay particles having higher surface activity. Casagrande [9] observed that changing the liquid limits at the same plasticity index results in a decrease in both dry strength and toughness, whereas there was an increase in hydraulic conductivity and compressibility of soils.

Cation Exchange Capacity (CEC)

The cation exchange capacity of a negative double layer may be defined as the excess of the counterions that can be exchanged for other cations. A given clay mineral does not have a fixed, single value of exchange capacity, and it depends on various environmental and compositional factors [10]. The values of exchange capacity represent the amount of readily exchangeable cations that can be replaced easily by leaching with a solution containing other dissolved cations of higher replacing power than the adsorbed cation. Sposito [11] describes in great detail the mechanism of cation adsorption on silicates and implies that some of the engineering properties of soils are directly related to their CEC. For example, the higher the CEC, the lower the hydraulic conductivity of the soil. The cation exchange capacity of soil was determined by employing the sodium method (Chapman [12]).

Zeta Potential (ζ)

Zeta potential is the potential drop across the mobile part of the double layer that is responsible for electrokinetic phenomena: for example, electrophoresis. It is assumed that the liquid adhering to the solid surface and mobile liquid are separated by a shear plane. The zeta potential reflects the potential difference between the plane of shear and the bulk phase. The zeta potential calculated from electrophoretic measurements is typically lower than the surface potentials, calculated from the diffuse double layer theory.

The shear plane is an imaginary sphere around a particle inside of which the solvent moves with the particle as the particles and adhering liquid move through the solution. Figure 3 presents the position of zeta potential on the diffuse double layer. Although zeta potential is lower than surface potential, it can be used as a basis to investigate the thickness of the diffuse double layer under different environmental conditions. Low surface potential and/or zeta potential imply that low double layer thicknesses because of low surface charge density and the practical implication of lower zeta potential is flocculation and aggregation of soil particles. Since the effects of the flocculation on engineering behavior of soils are known by geotechnical engineers, they will not be discussed here.

Zeta potential measurements were obtained from the movement of individual clay particles under an applied field as measured by Coulter[R] Delga 440. The zeta potential was then computed based on the Helmholtz Theory [13] which is given as:

$$\zeta = \frac{4 \pi \eta U}{\epsilon}$$

(1)

and

$$U = \frac{v}{E}$$

(2)

where
ζ = zeta potential (mV)
η = viscosity (poise)
ϵ = dielectric constant
U = electrophoretic mobility of a particle (μm . cm / V . s)
v = terminal velocity (μm/s)

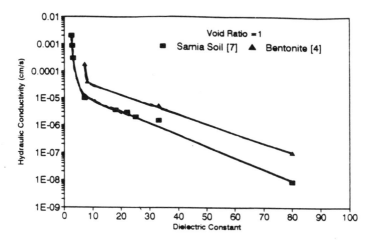

Figure 1. Hydraulic conductivity of soils with different pore fluid as a function
of dielectric constant (data from Waidelich [4] and Fernandez and Quigley [7]).

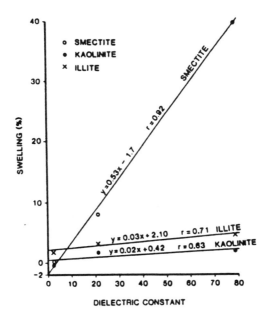

Figure 2. Swelling pressure of soils as a function of dielectric constant (from Brown
and Thomas [8]).

E = applied electric field (V/cm)

It should be noted that Equations 1 and 2 state that zeta potential is proportional to the surface charge density of the particle.

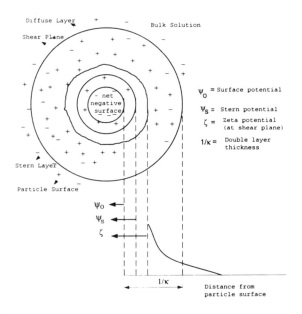

Figure 3. Schematically illustration of diffuse double layer and its zeta potential.

EXPERIMENTAL RESULTS AND DISCUSSION

Figure 4 presents the Atterberg limits of kaolinite as a function of dielectric constant. As can be seen from the figure, as the dielectric constant of the solvent decreases, the liquid limit, LL, of the soil increases whereas plasticity index, PI, remains constant. While mixing kaolinite with the solvents, it was observed that the soil aggregated. It should be mentioned that it was not possible to determine the plastic limit of the soil with Xylene because the soil became so fragile during rolling for plastic limit, in other words, the soil acted very much like silty soil. These observations are compatible with that of Mesri and Olsen [5], and Fernandez and Quigley [7], which is, reduction in dielectric constant of pore fluid causes flocculation and channelization of the soil. Also, the results are very compatible with Casagrande's [9] observation, which is, hydraulic conductivity increases with increase in the liquid limit at the same plasticity index. In other words, decrease in dielectric constant causes aggregation and channelization within the soil pads, thus causing an increase in the flow area. From the data presented in Figure 4, the following relationship is obtained:

$$LL = 77.7 - 17.2 \log \epsilon \qquad (r^2 = 0.99) \qquad (3)$$

where

LL: liquid limit
ϵ : dielectric constant of pore fluid

This empirical relationship should be used with caution because LL is not only related to the type of the pore fluid but also mineralogical composition and size of the particles. The equation should be used in order to gain some understanding of the effect of the organic solvent contamination on the behavior of soils.

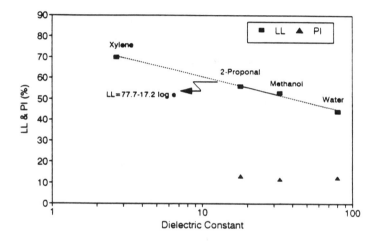

Figure 4. Liquid limit (LL) and plastic index (PI) of kaolinite as a function of dielectric constant of pore fluid.

Cation exchange capacity of the soil was determined by sodium extraction. To completely dissolve the sodium acetate in the solvents, all solvents were prepared with 25 percent water (75% solvent and 25% water). The CEC of kaolinite is presented in Figure 5. From the figure it is obvious that CEC linearly increases with an increase in the dielectric constant of the pore fluid. From regression analysis we obtain the following empirical relationship:

$$CEC = 2.7 + 0.224 \epsilon \qquad (r^2 = 0.96) \qquad (4)$$

where

CEC cation exchange capacity of soils (meq/100gr)

ϵ : dielectric constant of pore fluid

The implications of these results indicate that if the pore fluid is contaminated by organic solvents the soil will not have the same engineering behavior as non-contaminated soil, and some degree of change in certain properties can be estimated by monitoring the changing dielectric constant of the pore fluid.

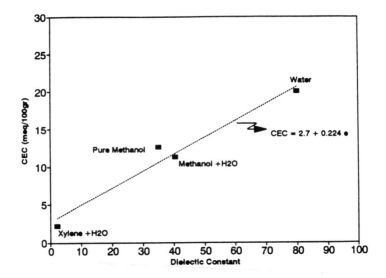

Figure 5. Cation exchange capacity of kaolinite for different solvents.

Zeta potential measurements are presented in Figure 6. From the Figure, it can be seen that the dielectric constant of the pore fluid decreases and reaches zero within the experimental error range. This may be attributed to the fact that surface charge density of the soil particles decreases because of lower proton charge density.

The intrinsic surface charge density, σ_{IN}, is defined by Sposito [14]:

$$\sigma_{IN} = \sigma_o + \sigma_H \tag{5}$$

where

σ_o: the permanent structural charge density, and

σ_H: the net proton charge density

The net proton surface charge density can be written in terms of H^+ and OH^- as following:

$$\sigma_H = F \frac{(q_H - q_{OH})}{S} \tag{6}$$

where

    F = the Faraday constant,

    q_{H_+} = the moles of complexed proton charge, and

    q_{OH^-}= the moles of complexed hydroxyl charge on proton-selective surface functional
        groups per unit mass of the soil

    S = the surface area of average particles

If Eq. 6 is substituted in Eq. 5

$$\sigma_{IN} = \sigma_o + \frac{(q_H - q_{OH})}{S} \tag{7}$$

From Eq. 7, it is clear that as the second part of the right side of the equation (i.e., proton charge) approaches zero, the intrinsic surface charge becomes equal to permanent structural charge and practically speaking the intrinsic charge becomes negligible.

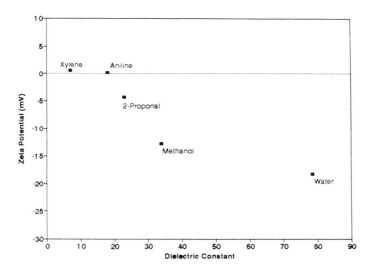

Figure 6 Zeta potential of kaolinite for different solvents.

The decrease in zeta potential is due to a decrease in proton charge density which leads to a reduction in the thickness of double layer. Surface potential of double layer may be written in terms of zeta potential as

$$\psi = c \, \zeta \tag{8}$$

where c is a constant greater than 1 (The unit of both zeta potential and surface potential is the volt). Thus, the surface potential, a distance, x (m), from the surface can be expressed as:

$$\psi_o = c \, \zeta \, e^{\kappa x} \tag{9}$$

and

$$\kappa = (\frac{2 \, n_o \, e^2 \, v^2}{\epsilon \, \epsilon_o \, k \, T})^{1/2} \tag{10}$$

where

κ = Debye-Huckel parameter (m^{-1})
n_o = electrolyte concentration (ions m^{-3})
e = electronic charge (C)
v = valence of cations
ϵ_o = dielectric constant of free space $(C^2 \, J^{-1} \, m^{-1})$
ϵ = dielectric constant of pore fluid
k = Boltzmann's constant $(J \, °K^{-1})$
T = absolute temperature $(°T)$

From Equations 8 and 9, it can be said that the measurement of zeta potential can be used to determine the thickness of double layer and surface potential. In this manner, Figure 6 becomes meaningful and clearly proves that the low dielectric constant of the solvent causes the collapse of the diffuse layer which results in higher surface flow area in the soil-fluid system in the case of the hydraulic conductivity measurements and lower swelling pressure in the case of swelling tests.

SUMMARY AND CONCLUSIONS

Early results of the ongoing study are presented with emphasis on the effect of the dielectric constant on the physico-chemical parameters of fine-grained soils. Atterberg Limits, cation exchange capacity and zeta potential of kaolinite are determined as a function of the dielectric constant to shed some light on understanding of the engineering behavior of soils when contaminated with organic solvents. It was observed that the dielectric constant considerably affects the physico-chemical parameters of the soil as well as its mechanical parameters. From the present study, the following can be concluded:

1. As dielectric constant of the soil decreases, the soil becomes more problematic in an engineering sense, i.e. high liquid limit low plasticity index which means low shear strength and higher hydraulic conductivity which is in good agreement with the early investigators' results.

2. Cation exchange capacity, CEC, of soil decreases due to a decrease in the surface charge density of the particles because of decrease in readily available ions as presented in Figure 5.

3. Zeta potential, ζ, of the soil also decreases, thus the thickness of the double layer decreases as dielectric constant decreases because of the low mobility of the particles under an applied electrical field. A decrease in zeta potential implies collapse of the double layer which leads to an increase in the cross-sectional area available for flow resulting in higher hydraulic conductivity and less swelling as presented by early investigators in Figures 1 and 2.

4. Further research is required in the area of the effect of dielectric constant both on physico-chemical parameters and mechanical parameters.

REFERENCES

[1] Fang, H. Y., and Evans, J. C., " Long Term Permeability Tests using Leachate on a Compacted Clay Liner Material," ASTM Spec. Tech. Publ. STP 986, 1988, 397-404.

[2] Quincke , G., Ann. Phyzik., Vol. 113, 1861, p.513, Quoted in Abramson (1934)

[3] Macy, H. H., " Clay-Water Relationship and the Internal Mechanisms of Drying", Trans. Br. Ceram. Soc. Vol. 41, 1942, pp. 73-121.

[4] Waidelich, W.C., "Influence of Liquid and Clay Mineral Type on Consolidation of Clay-Liquid Systems," Highway Research Board Special Report 40, 1958, pp. 24-42.

[5] Olsen, H. W., Hydraulic Flow through Saturated Clays," Clays and Clay Minerals, Vol. 11, 1960, pp. 131-161.

[6] Mesri, G., and Olsen, R. E., "Mechanisms Controlling the Permeability of Clays," Clays and Clay Minerals, Vol. 10, 1971, pp. 151-158.

[7] Fernandez, F., and Quigley, R. M., "Hydraulic Conductivity of Natural Clays Permeated with Simple Liquid Hydrocarbons," Can. Geotech. J. Vol. 22, 1965, pp. 205-214.

[8] Brown, T. W., and Thomas, J. C., "Mechanism by which Organic Liquids Increase the Hydraulic Conductivity of Compacted Clay Materials," Soil Sci. Soc. Am. J. Vol. 51, 1987, pp. 1451-1459.

[9] Casagrande, A., "Classification and Identification of Soils," Transactions, ASCE, Vol. 113, 1948, pp.901-930.

[10] Mitchell, J.K., Fundamentals of Soil Behavior, 2nd ed., John Willey and Sons, 437 p, 1993.

[11] Sposito, G., The Surface Chemistry of Soils, Oxford University Press, 1989.

[12] Chapman, H.D., "Cation Exchange Capacity" Methods of Soil Analysis, ed. by Black, C.A., Part 2, No:9, Am. Ins. Agronomy, Madison, Wisconsin, pp.891-901.

[13] Helmholtz, H., "Wiedemanns Annalen d. Physik, Vol.7, 1879, p.137.

[14] Sposito, G., The Chemistry of Soils, Oxford University Press, 1984.

Innovative Measurements

Stephen K. Kennedy[1], Gary S. Casuccio[1], Richard J. Lee[1], Gary A. Slifka[2] and Michael V. Ruby[3]

MICROBEAM ANALYSIS OF HEAVY ELEMENT PHASES IN POLISHED SECTIONS OF PARTICULATE MATERIAL - AN IMPROVED INSIGHT INTO ORIGIN AND BIOAVAILABILITY

REFERENCE: Kennedy, S. K., Casuccio, G. S., Lee, R. J., Slifka, G. A., and Ruby, M. V., "Microbeam Analysis of Heavy Element Phases in Polished Sections of Particulate Material — An Improved Insight into Origin and Bioavailability," Sampling Environmental Media, ASTM STP 1282, James Howard Morgan, Ed., American Society for Testing and Materials, 1996.

ABSTRACT: The contamination of environmental materials (soil and dust) by heavy elements is of concern in assessing environmental pollution. Characterization of samples is performed in order to provide information that can be used to assess the potential hazard of the material, to assess remediation strategies, and to determine the source of the heavy element-bearing material. This is difficult because the particles in which the element of interest occur may consist of multiple phases, some potentially hazardous and some benign. Bulk properties of samples are often determined to assess the potential risk related to such a sample. Such an assay may reveal the total content of a contaminant, and so document the maximum hazard potential, but cannot directly address important mitigating factors, including the specific phases present, their size distribution and internal particle structure.

Microbeam analysis using a scanning electron microscope (SEM) or electron microprobe on polished particle mounts allows the size, internal morphology and chemical composition to be determined on a particle-by-particle basis. Computer controlled SEM allows all factors except internal structure to be acquired in an automated fashion.

We present examples of two studies using microbeam techniques. In the first example, we illustrate how cadmium associated with sphalerite (zinc sulfide) in soils can be quantified at levels down to a few parts per million (ppm) using computer controlled SEM techniques. In the second example, we illustrate how lead in soils can be assigned to phases, phase sizes and to internal structure types using manual electron microprobe techniques.

KEYWORDS: heavy elements, lead, cadmium, soil, scanning electron microscope, electron microprobe, mobility, bioavailability

[1] RJ Lee Group, Inc., 350 Hochberg Road, Monroeville, PA 15146

[2] ASARCO Inc., 748 County Road, Leadville, CO 80461

[3] PTI Environmental Services, 2995 Baseline Road, Suite 202 Boulder, CO 80303

INTRODUCTION

The average crustal concentrations of heavy elements are very low, ranging from about 10 to 0.01 ppm [1]. However, the natural processes of ore body emplacement and surface weathering, as well as processes involved in the mining the ores, extracting those elements, and producing products are accompanied by the possibility of increasing heavy element concentrations in the surface and near surface environment. Because ingestion or inhalation of these elements has the potential for adverse health effects it has become desirable to assess soil and dust for both bulk element concentrations and for chemical or morphological characteristics that may influence these health effects. To do this, at least four issues may need to be addressed. The first is to determine the presence and amounts of heavy elements present in the environmental materials. The second, if the concentration is sufficiently high to warrant further investigation, is to assess for potential health hazard of those elements. The third, if a real potential health hazard is deemed to occur at the site, is to determine the remediation strategy based on the site-specific characteristics. The fourth is to identify the source or sources of the heavy elements.

Samples may be analyzed by various techniques and protocols in order to address those issues. The techniques may be selected based on a consideration of both the cost of analysis and the quantity and quality of information yielded by the method. It is the purpose of this paper to summarize some considerations in selecting a method or combination of methods and to present how two microbeam techniques can address these issues.

HEAVY ELEMENT CHARACTERISTICS

Soils and dusts can be described by fully characterizing the composition and the textural relationship among the components. Such characteristics are described below:

Assay

In an elemental assay, the concentration of the element is determined in a bulk sample. Although useful as a screening tool to assist in determining if further analysis is required, this offers no direct information on the source, mobility, or potential health hazard except perhaps in the simplest of settings.

Phase

In the phase analysis, the specific mineral or chemical species in which the heavy element occurs is determined. A heavy element can occur in a variety of phases, not only at a single site but even in a single sample. Each phase has a specific solubility and will be more or less mobile than other phases under the same physical and chemical conditions. Accordingly, a heavy element present at some concentration may be potentially hazardous if it occurs as one phase and relatively benign if it occurs as another. In addition, because specific phases have known stability fields, presence or absence of specific phases may shed light on the chemical and physical conditions in the environment from which the sample was taken.

Size

The size distributions of the heavy element containing particles is important for several reasons. Because chemical reactions occur on surfaces, and because the surface area to volume ratio increases as particle size decreases, particle size influences the potential mobility of the element. Particle size also influences the probability of ingestion and inhalation: pica activity (intentional ingestion of non-food material) may occur with relatively large particles such as paint chips; unintentional ingestion of non-food material occurs with particles typically smaller than a few hundred micrometers which can adhere to surfaces placed in the mouth (e.g., hands and toys); and respirable particles are typically smaller than 10 micrometers. In addition, particle size may shed light on presence or absence of specific sources that may provide the heavy element in particles of a specific size range.

Association

Particles in dust and soils are commonly complex; that is they consist of multiple phases. The physical or spatial relationship among the phases of an individual particle is referred to as the phase association. The association classification used herein is one modified from Link and others [2]. The association relates to heavy element mobility. A heavy element phase may occur as a single phase particle, or exposed at the particle surface in a complex particle. These occurrences are potentially reactive because of the exposed surface. An example of a liberated phase is shown in Figure 1. A heavy element phase may be enclosed within an inert phase and thus present no surface area for chemical reactions. An example of an enclosed phase is shown in Figure 2. In addition, the phase association may indicate a paragenetic sequence such as the alteration of one phase to another as a rim (Figure 3), or the precipitation of a new phase as a rim or cement (Figure 4). These latter two associations result from the phase adjustments to stability in the chemical and physical environment. The occurrence of these associations indicates that a parent phase has been removed and replaced by some alteration product. This process complicates source identification by obscuring the original source signature.

ANALYTICAL METHODS

There are a variety of analytical methods that are used to quantify one or more of the above characteristics. These are described below:

Chemical Analysis

Metal content can be determined by X-ray fluorescence (XRF), atomic absorption (AA) and inductively coupled plasma spectrometry (ICP) techniques. AA and ICP require

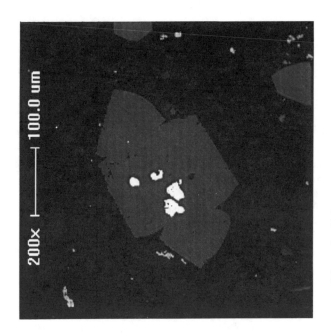

Figure 2 - Galena enclosed in pyrite.

Figure 1 - Liberated lead phosphate.

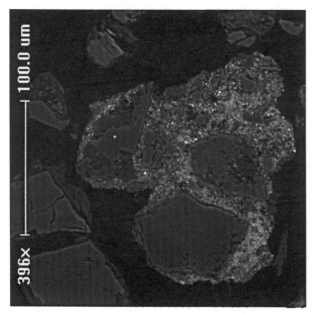

Figure 4 - Quartz cemented by lead-iron-sulfur phase.

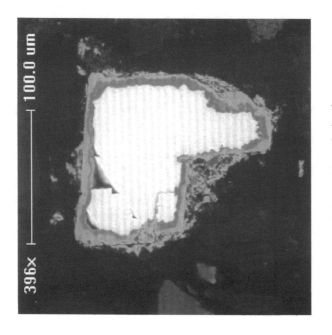

Figure 3 - Galena with anglesite rim.

the sample to be dissolved whereas XRF permits analysis of the solid particulate material. As applied to whole samples, these techniques reveal only the bulk assay. Alternatively, the AA and ICP techniques can be tailored, to some extent, to dissolve only a specific phase or a small group of phases rather than the entire sample. In this approach, referred to as selective sequential extraction, the sample is subjected to a series of dissolving agents of increasing rigor where the phases contributing to the solutions can be inferred.

These methods have problems because dissolution is never as complete or as specific as desired. However, the presence and concentration of multiple phases can be estimated fairly rapidly and at a relatively low cost.

X-ray Powder Diffraction

In XRD analysis, the sample is powdered and bombarded with an angular sweep of x-rays of known wavelength. Interference between the energy and material is related to crystal structure which can be related to specific mineral phases. The limit of detection is fairly high (typically a few percent) but often the sample can be processed to concentrate the components of interest. Although XRD analysis has the same problem of being a bulk technique and no information regarding size or association is obtained, the specific phases present are determined with confidence. Heavy elements occurring as small amounts of impurities in other phases will not be measured.

Electron Beam Instruments

The scanning electron microscope (SEM) and electron microprobe (EMP) are electron beam instruments that combine imaging and spot elemental analysis. The imaging is commonly performed in the backscattered electron mode because brightness is related to the average atomic number of the material being analyzed and so is sensitive to different phases. The beam size allows sub-micrometer features to be imaged. The elemental analysis is performed using energy dispersive spectroscopy (EDS) in the SEM and EMP. Wavelength dispersive spectroscopy (WDS) complements EDS analysis in the EMP. The beam interaction volume allows the composition to be determined for features down to about 5 micrometers in diameter.

Analysis can be performed on samples of mounted particles but epoxy impregnated and polished samples are preferred as this preparation reveals the particle interiors. In the polished samples, multiple phases in a particle can be separated based on backscattered electron intensity and identified using either EDS or WDS. Specific phases are identified, their size determined, and the association is described. In addition, by summing all the metal-bearing occurrences the assay can be determined, although this is more accurately determined by one of the chemical methods discussed above. Thus, these techniques are very powerful and complementary to the chemical analysis methods.

The major deterrent to the electron beam methods is the time involved for analysis. EDS analysis provides data on all elements present at each spot and can be performed in a matter of seconds. However, it has a detection limit in the neighborhood of about one percent and its accuracy is within a few percent. WDS analysis is more time consuming and provides data on a single element at a time but has a much lower detection limit and increased accuracy. In either case, the heavy element phases are usually relatively rare and a significant portion of the analysis time is spent in searching for those phases. This analysis time can be reduced through the use of computer controlled scanning electron microscopy (CCSEM), especially when heavy element phases occur as discrete particles or inclusions of relatively simple shapes.

EXAMPLES

1) CCSEM Analysis

Soils at a site were found to contain cadmium (Cd) at levels up to several hundred ppm. Phases sufficiently rich in Cd to be detected by EDS were very rare, but phases containing Cd at small concentrations were present. One of these observed phases was sphalerite (zinc sulfide) which commonly contains up to 1.6% Cd dispersed in the mineral matrix [3]. The EMP analysis of about 50 sphalerite particles indicated that Cd was present in the sphalerite of these samples at a mean concentration of 0.33%. The amount of sphalerite in the samples was below the detection limit of XRD, but because the sphalerite occurred as discrete phases the association was not of interest and the particles were amenable to CCSEM.

Samples were epoxy impregnated, polished and analyzed in a PERSONAL SEM® using the ZepRun©[1] automated particle analysis program. The program was set for one hour to obtain chemical data on approximately 1000 particles. In this analysis, the sample was observed in the backscattered electron mode where the brightness and contrast of the images were kept constant using copper tape and the aluminum sample holder as standards. An intensity threshold value was set to separate particles darker than sphalerite from the bright particles. This is referred to as a High-Z analysis. During the automated analysis the beam was driven across the field until a bright particle was encountered. The particle was sized, the electron beam was placed in the centroid of the particle, and an EDS spectrum was obtained. The x-ray counts attributable to each element contained in a user defined list were saved for later off-line review. Digital images and the spectra could also be saved if desired. The chemical data so obtained were then related to mineral phases, including sphalerite, by a set of user-defined rules.

Using this procedure, the mineral phases and size (diameter and area) were determined for each bright particle encountered in an analysis. Knowing the total area analyzed (mineral phases plus epoxy) and the mineral fraction of that area, the total mineral area, and the area percent sphalerite were determined. Because the surface analyzed was a random area, the area percent is equivalent to the volume percent [4] and making assumptions about density allows mass percent to be estimated. Finally, the total amount of Cd attributed to the sphalerite was calculated.

---

[1] PERSONAL SEM is a registered trade mark for a scanning electron microscope and ZepRun is a copyrighted automated particle analysis program: RJ Lee Group, Inc., 350 Hochberg Road, Monroeville, PA 15146.

Results

The results of these analyses and calculations are shown in Table 1. This table shows the amount of Cd that can be attributed to the sphalerite in the sample. Cd was detected at levels as low as 1 ppm. Comparing these results of the Cd in sphalerite to the total assay indicates the amount of Cd that must be accounted for by other phases. In this study, Cd was also observed in zinc and iron oxides, in other zinc-sulfur compounds, and in phosphates. The areas of these phases were determined in the CCSEM analysis. Obtaining their Cd concentrations using the EMP would allow estimation of their contributions to the total Cd content.

TABLE 1 -- Cadmium in sphalerite calculations

Samp. Num.	Area Analyzed (um)		Number High-Z Particles	Sphalerite			Cd (ppm)
	Total	Mineral		Area (um)	Area (%)	Mass (%)	
508	53827000	18839450	818	6299	0.033	0.053	1.7
538	28132000	11112140	915	2157	0.019	0.031	1.0
539	29239000	11812556	750	21647	0.183	0.289	9.5
542	18976000	7381664	928	19154	0.259	0.409	13.5
551	42973000	21400554	867	11221	0.052	0.083	2.7
578	7270000	4187520	1040	20964	0.501	0.789	26.1
579	10958000	5610496	1029	2753	0.049	0.077	2.6
580	8275000	3492050	1043	6225	0.178	0.281	9.3
581	26138000	12232584	946	4146	0.034	0.053	1.8
582	20822000	7641674	867	786	0.010	0.016	0.5
613	1701000	729729	1047	0	0.000	0.000	0.0
614	3200000	1641600	961	9	0.001	0.001	0.0
615	66318000	36872808	728	15939	0.043	0.068	2.2
616	21198000	10090248	1051	34813	0.345	0.544	18.0
617	47136000	25877664	841	4185	0.016	0.025	0.8
618	4590000	2400570	1043	0	0.000	0.000	0.0

2) Manual Microprobe Analysis

In this example, the site was the location of over one hundred years of mining and milling and about 75 years of smelting various metals including lead. Across the site, soil samples contained over 1000 ppm lead. A manual microprobe analysis was undertaken to characterize the lead in the soils relative to specific lead phases, as well as phase size, and association distributions. The manual electron microprobe analysis proceeded according to the protocol established for this project [5]. In brief, each sample was sieved and the minus 250 micrometer material was epoxy impregnated and polished. The sample was placed into an electron microprobe and manually scanned in the backscattered electron mode according to a regular grid pattern to identify any of the phases known or thought to occur at this site. (The major lead phases are listed in Table 2.) Several of the phases are stoichiometric minerals, and several are phases containing dissolved lead. Any bright phase was analyzed using the EDS to identify the phase. Sufficient analyses were performed on the variable lead phase occurrences by WDS in order to reasonably estimate the lead content using the EDS. Phases with lead content below ~1 percent (slag and organic particles) were routinely analyzed by WDS to determine the lead content. The analysis continued until 100 lead-bearing particles were documented, or the maximum analysis time of 10 hours elapsed.

TABLE 2 -- Volume and association distributions of the lead phases.

Phase	Est.[2] Vol %	Lead Conc. (%)	Associations (%)			
			Liberated	Enclosed	Cement	Rim
Galena	38.6	86.6	63.3	30.8	4.0	1.8
Slag	31.4	2.2	98.4	1.0	0.5	0.1
MnPb Oxide	6.8	10.0	39.8	21.6	31.0	7.6
FePb Oxide	4.7	2.6	63.0	19.9	6.4	10.7
C(Org)	3.9	2.0	95.0	2.5	2.4	0.2
FePb Sulfate	3.9	9.1	46.2	20.1	20.8	12.9
Anglesite	3.3	68.4	49.5	34.3	12.0	4.2
Cerussite	3.0	77.6	70.2	16.8	8.3	4.8
Pb Phosphate	2.3	28.6	51.0	35.2	9.5	4.3
Other Phases	2.1	-	22.9	66.3	6.7	4.1

When an occurrence was identified as a lead bearing phase, the maximum and minimum diameters were measured, the lead content estimated, and the association determined. All data were entered into a data base along with any comments by the operator and summarized. An example data summary sheet is illustrated in Table 3. The volume was estimated using the maximum diameter.

Results and Discussion

The relative abundance's of the lead phases observed throughout the site are shown in Table 2. Galena was the most abundant phase, about 30% of which is enclosed. The slag is also abundant and liberated, but the lead content is relatively small. Anglesite and cerussite particles were not present in great abundance but their lead content is relatively high. A considerable amount of the anglesite (34%) and cerussite (17%) are enclosed in other phases. More than 10% of MnFe Oxide (31%), FePb Sulfate (21%), and Anglesite (12%) occurred as precipitated cement and so the original phase information is lost.

---

[2] Volume estimate based on long axis.

TABLE 3 -- Example data summary sheet.

DATA SUMMARY SHEET   VALIDATED: YES   DATE: 05/10/93

1. Sample #:   Example   Pb Assay: N/A   Type: Environ   Lab: RJ Lee Group

2. Minerals		Chemical Formula	Freq	T Area um2	Area, %	Frequency Len	Len, %	Length Median (um)
1 Galena	Ga	PbS	97	320.0	0.05	149.0	1.97	1.0
2 Anglesite	Ang	PbSO4						
3 Cerussite	Cer	PbCO3	2	97.0	0.02	13.0	0.17	6.0
4 Pb Oxide	PbO	PbxOy						
5 PbO/carbonate	PbO?		16	16.0	0.00	16.0	0.21	1.0
6 Metalic Pb	Pb	Pb	1	2.0	0.00	2.0	0.03	2.0
7 Pb Phosphates	PbPhos		7	2948.0	0.50	205.0	2.72	20.0
8 Pb-Fe sulphate	PbFeSul	~7.4%	15	3779.5	0.64	314.0	4.16	5.0
9 Pb-arsenate	PbArs		50	464.0	0.08	145.0	1.92	2.0
10 Pb-vanadate	PbVan							
11 Wulfanite	Wulf	PbMoO4						
12 Mn-Pb oxide	MnPbO	~6.5%	8	19042.5	3.25	420.0	5.57	52.0
13 Fe-Pb oxide	FePbO	~1.2%	25	77610.0	13.24	1534.0	20.33	30.0
14 Pb-bearing barite	PbBar							
15 Slag	Slag	~0.9%	32	481500.0	82.15	4670.0	61.88	150.0
16 Pb silicate	PbSi							
17 Pb-bearing C-matter	C(Org)							
18 Paint	Pnt							
19 Solder	Sold							
20 Pb Sulfosalt	PbSfs							
21 Pb Antimonate	PbAnt		3	72.0	0.01	12.0	0.16	2.0
22 Other			16	281.0	0.05	67.0	0.89	5.0

3. Association: 98 particles with lead-bearing phases were analyzed.
All Ga were enclosed, mainly in Slag, as were the Cer, PbO? and Pb. The PbPhos were
liberated, enclosed and rim. The PbFeSul were mostly liberated and enclosed with one rim.
With two liberated exceptions, the PbArs were enlosed in Slag. With the exception of two
enclosed in one particle, MnPbO was liberated. The FePbO were mostly liberated with some
enclosed and rim. The Slag were liberated. The PbAnt were enclosed in Slag.

4. Pb Mineralogical Distribution: The percent length frequency distribution is shown below.
The average lead content of the PbFeSul was 7.4% and no value exceeded 15%. The average lead
content of the MnPbO was 6.5% with only one high value (23%) exceeding 10%. The average lead
content of the FePbO was 1.2% with no value exceeding 5%. The average lead content of the
Slag was 0.9% with no value exceeding 5%. Six Slag analyzed for background content contained
an average of 0.33% lead.

5. Species Bearing Ag __; As __; Bi __; Cd __; Hg __; Se __; Tl __   Observed

   Signature of Analyst: _____   DATE ANALYZED: 04/07/93

Sample: Example

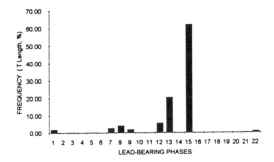

It is clear from the above that there is information that is not contained in a total lead assay. That the chemical form is important with respect to elemental behavior is the basis for sequential extraction (see e.g. [6] and [7]). Further, in vivo, in vitro, and dissolution kinetics studies linked to phase, phase size, and association establish the reasonableness of this link to bioavailability and thus to toxicological effects [8] [9] [10]. The results of this study indicate that there are a considerable number of phases with different solubility's and which are observed in a substantial proportion as being enclosed. Although these data were not used directly in a bioavailability study, the characteristics displayed by the lead occurrences in this site are likely major contributing factors relative to lead bioavailability.

The variability of phase, size, and association makes prediction of lead uptake into the human body very complicated. One such uptake model [11] assumes a default value for the bioavailability of lead in both soil and dust as 30% (Figure 2-11, p. 2-50 in reference [11]) and states "default parameters are recommended unless there is sufficient data to characterize site-specific conditions". Based on the results presented here as well as those presented by other authors, it seems that site-specific characteristics should be routinely obtained when using such predictive models.

CONCLUSIONS

A wide variety of options are available for the analysis of heavy elements in soil and dust. The selection of the "proper" technique depends on the question being asked and the nature of the samples at any particular site. In contrast to the chemical and x-ray diffraction techniques which give bulk results, microbeam techniques yield particle by particle phase, size, and association data. Because these characteristics have been shown to be important in their influence on heavy element mobility their characterization should result in better health risk estimates. Further, computer controlled scanning electron microscopy techniques are suitable for the analysis of heavy elements that occur as discrete phases at levels of a few parts per million.

REFERENCES

[1]   Fergusson, J.E., The Heavy Elements: Chemistry, Environmental Impact and Health Effects, Pergamon Press, New York, 1990.

[2]   Link, T.E., Ruby, M.V., Davis, A. and Nicholson, A.D., "Soil lead mineralogy by microprobe: An interlaboratory comparison", Environ. Sci. Technol., v. 28, No. 5, p. 985-988, 1994

[3]   Barry, L.G. and Mason, B., Mineralogy - Concepts Descriptions Determinations, W.H., Freeman and Company, San Francisco, 1959.

[4]   Chayes, F., Petrographic Modal Analysis, John Wiley and Sons, 1956.

[5]   CDM (Camp Dresser & McKee, Inc., Final Work Plan, Metal Speciation and Source Characterization, California Gulch CERCLA Site, Leadville, Colorado, prepared for Resurrection Mining Company, 1991.

[6]   Tessier, A, Campbell, P.G.C., and Bisson, M., "sequential Extraction Procedure for the Speciation of Particulate Trace Metals", Anal. Chem., v. 51, No. 7, p. 844-851, June, 1997.

[7]   Soto, E.G., Rodriguez, E.A., Rodriguez, D.P., Mahia, P.L., and Lorenzo, S.M., "Extraction and Speciation of Inorganic Arsenic in Marine Sediments", Sci. Tot., Environ., v. 141, p. 87-91, 1994.

[8]   Davis, A., Ruby, M.V., and Bergstrom, P.D., "Bioavailability of arsenic and lead in soils from the Butte, Montana, mining district", Environ. Sci. Technol., v. 28, No. 3, p. 461-468, 1992.

[9]   Davis, A., Drexler, J., Ruby, M.V., and Nicholson, A., "Micromineralogy of Mine Wastes in Relation to Lead Bioavailability, Butte, Montana", Environ. Sci. Technol., v. 27, No. 7, p. 1415-1425, 1993.

[10]   Ruby, M.V., Davis, A., Kempton, J.H., Drexler, J., and Bergstrom, P.D., "Lead Bioavailability: Dissolution Kinetics under Simulated Gastric Conditions", Environ. Sci. Technol., v. 26, No. 6, p. 1242-1248, 1993.

[11]   The Technical Review Workgroup for Lead, "Guidance Manual for the Integrated Exposure Uptake Biokinetic Model for Lead in Children", 1994.

Kathleen K. Fitzgerald[1], Timothy L. Miller[1], Arthur J. Horowitz[2], Charles R. Demas[3], and D.A. Rickert[1]

U.S. GEOLOGICAL SURVEY PROTOCOL FOR MEASURING LOW LEVELS OF INORGANIC CONSTITUENTS, INCLUDING TRACE ELEMENTS, IN SURFACE-WATER SAMPLES

REFERENCE: Fitzgerald, K. K., Miller, T. L., Horowitz, A. J., Demas, C. R., and Rickert, D. A., ''U.S. Geological Survey Protocol for Measuring Low Levels of Inorganic Constituents, Including Trace Elements, in Surface-Water Samples,'' Sampling Environmental Media, ASTM STP 1282, James Howard Morgan, Ed., American Society for Testing and Materials, 1996.

ABSTRACT: The U.S. Geological Survey (USGS) has sponsored a number of studies to evaluate its sampling, processing, and analytical procedures for filtered water samples. The major findings of these studies were that surface-water sampling and processing procedures and equipment used in USGS programs were inadequate to produce contaminant-free trace-element data on a consistent basis. This led to development of a new protocol for collecting, processing, and analyzing filtered water samples for the analysis of trace elements at the 1 microgram-per-liter (ug/L) level using "clean" procedures. The protocol specifies (1) supplies, (2) office and between-site cleaning procedures, (3) samplers, (4) sample splitters, (5) use of disposable capsule filters, (6) quality-assurance measures including use of disposable gloves and shelters for sample processing, and (7) quality-control requirements including blanks, replicates, and split samples.

The USGS is now implementing this protocol for all USGS efforts to produce trace-inorganic data at the microgram-per-liter level on filtered water samples. The protocol will also be used for collecting and processing samples for major ions and nutrients when trace-elements are also being determined.

KEYWORDS: water quality, water sampling, trace elements

Since 1987, the U.S. Geological Survey (USGS) has sponsored a number of studies to evaluate its sampling, processing, and analytical procedures for filtered water samples. The major findings of all these studies were that surface-water sampling and processing procedures and equipment historically used in USGS programs were inadequate to produce contaminant-

---

[1]Hydrologist, Office of Water Quality, U.S. Geological Survey, 412 National Center, Reston, VA 22092
[2]Chemist, U.S. Geological Survey, Peachtree Business Center, Suite 130, 3019 Amwiler Road, Atlanta, GA 30360
[3]Hydrologist, U.S. Geological Survey, 3535 Sherwood Forest Blvd., Suite 120, Baton Rouge, LA 70816-2255

free trace-element data on a consistent basis. This led to the development
of a protocol that ensures contaminant-free (at specific reporting limits)
trace-element samples. This protocol is titled "U.S. Geological Survey
Protocol for the Collection and Processing of Surface-Water Samples for
Subsequent Determination of Inorganic Constituents in Filtered Water" [1].
During development of the protocol, it was also demonstrated to be
applicable to collection and processing of samples for nutrient and major
ion analyses. The protocol represents a significant advance in USGS
guidelines for the collection and processing of water samples for
subsequent chemical analysis. It requires use of "clean hands-dirty hands"
techniques which means that most sampling trips will require at least two
people.

Currently, it appears as if the ambient concentrations of many
dissolved trace elements, at least in large rivers, are in the nanogram-
per-liter (ng/L) range, and analytical capabilities to work at these
ultra-low levels. Therefore, initially, the new protocol is intended for
accurate quantification of trace elements in filtered water at the
microgram-per-liter (µg/L) level. This is the concentration level at which
analytical work is routinely reported and is adequate for the most
stringent Federal drinking-water regulations for trace elements. Field and
laboratory tests carried out during the course of the development of this
protocol show that it probably is usable, in its current form, down to
levels as low, or lower than 200 ng/L.

The tests and evaluations for the new protocol entailed the
examination of surface-water sampling and processing equipment and
procedures. In addition, a series of cleaning techniques for sampling and
processing equipment were designed and evaluated. Finally, a series of
limited field tests, including side-by-side comparisons with non-USGS
personnel, were conducted and evaluated. The list of trace elements
evaluated in these tests is given in (Table 1).

Table 1.--Target trace elements and concentrations for evaluation and
testing of protocol procedures [µg/L=microgram per liter; mg/L= milligram
per liter; ICP-MS=inductively coupled plasma, mass spectrophotometry;
ICP-AES=inductively coupled plasma, atomic emission spectrometry]

Element	Reporting limit and associated analytical error	Analytical Technique
Aluminum	0.5 ± 1.0 µg/L	ICP-MS
Antimony	0.2 ± 0.2 µg/L	ICP-MS
Barium	0.2 ± 0.2 µg/L	ICP-MS
Beryllium	0.2 ± 0.2 µg/L	ICP-MS
Boron	2 µg/L	ICP-AES
Cadmium	0.2 ± 0.2 µg/L	ICP-MS
Calcium	0.02 mg/L	ICP-AES
Chromium	0.2 ± 0.2 µg/L	ICP-MS
Cobalt	0.2 ± 0.2 µg/L	ICP-MS
Copper	0.2 ± 0.2 µg/L	ICP-MS
Iron	3 µg/L	ICP-AES
Lead	0.2 ± 0.2 µg/L	ICP-MS
Magnesium	0.01 mg/L	ICP-AES
Manganese	0.2 ± 0.2 µg/L	ICP-MS
Molybdenum	0.2 ± 0.2 µg/L	ICP-MS
Nickel	0.2 ± 0.2 µg/L	ICP-MS
Silica	0.01mg/L	ICP-AES
Silver	0.2 ± 0.2 µg/L	ICP-MS
Sodium	0.2 mg/L	ICP-AES
Strontium	0.5 µg/L	ICP-AES
Thallium	0.2 ± 0.2 µg/L	ICP-MS
Zinc	0.5 ± 1.0 µg/L)	ICP-MS

Note that the target concentrations are substantially below the microgram-per-liter reporting level that the protocol was intended to meet and wherever feasible were set at levels at least 50 percent below the proposed reporting limits (Table 1). This was done intentionally to ensure that, even when summed, the potential contamination from the equipment, sampling and processing procedures, and cleaning techniques would be negligible compared to the designated reporting limit.

The USGS is now implementing this protocol as the standard operating procedure for all USGS efforts that produce trace-inorganic data at the microgram-per-liter level on filtered water samples. The protocol will also be used for collecting and processing samples for major ions and nutrients, when trace elements are also being determined. For the few situations where trace-element concentrations are high, such as acid mine drainage areas, protocol modifications may be permitted, provided sufficient quality-control data are generated to prove that any contamination will not affect the data at the designated laboratory reporting limits.

Development of the protocol has shown that the two most important factors in avoiding or reducing sample contamination are:

1. An awareness of potential contaminant sources, and

2. Strict attention to the work being done.

These two factors should be viewed as being equally as important as actually following the protocol itself.

Other than carelessness during sample collection and processing, the two biggest sources of contamination of samples for inorganic analysis are:

1. Improperly cleaned and maintained equipment, and

2. A variety of atmospheric inputs (dirt and dust from the field vehicle, from the sampling platform, and from the local environment).

SUPPLIES

Only those devices that contain no metal parts that might come in contact with the sample will be used to collect samples. All parts of samplers to be used for the collection of materials for subsequent inorganic analysis, and actually come in contact with the sample, must be composed of uncolored or white polypropylene, polyethylene, Teflon, or other suitable non-metallic material.

OFFICE AND BETWEEN-SITE CLEANING PROCEDURES

A number of the procedures and precautions described in this protocol are intended to ensure that all sampling and processing equipment are adequately cleaned prior to taking them into the field. Precautions also are detailed to maintain the cleanliness of the equipment prior to actual use. Because cross-contamination is always a possibility, and many sampling trips entail the collection of more than one sample at more that one site, detailed procedures are provided for cleaning and preparing equipment between sites. Finally, because atmospheric inputs appear to be a major source of potential contamination, some existing equipment will be modified and all sample processing and preservation will be carried out within a controlled environment designed to reduce atmospheric inputs and (or) cross contamination (for example, the introduction of processing and preservation chambers).

The following procedures make up the protocol:

    Procedure 1. Office Preparations and Cleaning of Equipment
    Procedure 2. Field Rinsing of Equipment Prior to Sampling
    Procedure 3. Sample Collection, Processing, and Preservation
    Procedure 4. Field Cleaning to Prevent Cross-Contamination
                 Between Sites Where Other Inorganic Samples are
                 to be Collected at Additional Sites.

SAMPLERS

On the basis of comparisons done on the Mississippi River, as well as other experiments, it has been shown that a number of samplers in current use can contaminate water samples at the microgram-per-liter level. Data have shown that, although some samplers are cleaner than others, no sampling devices currently in use can be cleaned sufficiently with either native water or deionized water (DIW) alone to completely eliminate contamination at the microgram-per-liter level. The following samplers, DH48, DH59, DH76, D74, P61, P63, and P72, produce unacceptably high levels of trace-element contamination and their use for collecting trace-element samples has been discontinued. Some of these samplers may be acceptable for other inorganic constituents such as major ions and nutrients. If they are used for non-trace-element sampling, the user must collect appropriate quality-control samples to prove that they do not add contaminants for the constituent(s) of interest and must document this information.

The D77 sampler with a standard plastic bottle, the D77 with a Teflon sampler bottle, the D77 with a collapsible bag, and the frame-type (3- and 8-liter) sampler with either a rigid bottle or a collapsible bag appear to be adequate for collection of samples for filtered trace-element analyses. Extended experiments using the cleaning procedures (Procedure 1) documented in this protocol were done to evaluate the listed samplers. The evaluations showed that all the listed samplers were acceptable for use, at the microgram-per-liter level, after they had been cleaned following the procedures outlined in the protocol. By inference, since the same sampling bottles, caps, and nozzles are used in the DH81 as in the D77 sampler, the DH81 is also an acceptable sampler at the microgram-per-liter level provided the metal rod is covered with plastic shrink-wrap. The samplers that are acceptable for collecting samples for determination of trace elements at the microgram-per-liter level, major ions, and nutrients are listed in (Table 2).

Table 2.--Acceptable samplers for the collection of inorganic
          (trace-element, nutrient, and major ion) samples

D77 with plastic bottle, cap, and nozzle
D77 with Teflon bottle, cap, and nozzle
Frame-type sampler with plastic or Teflon bottle, cap, and nozzle
D77 or Frame-type sampler with Teflon or Reynolds Oven collapsible bag,
    Teflon or plastic cap and nozzle
DH81 (with "shrink-wrapped" handle) with Teflon or plastic bottle,
    cap and nozzle

SAMPLE SPLITTERS

The protocol, as currently written, uses the churn splitter to composite and split samples. To decrease the possibility of contamination from atmospheric sources, two modifications are required. The first is to place a cappable funnel in the lid of the splitter to limit exposure while sample water is being poured into the splitter. The funnel that is being used is made by cutting the bottom off of a 1 L sample bottle and inserting it into a 1 inch hole in the cover of the churn splitter. A cap for the funnel is made by cutting the top portion off of a 1 L Nalgene sampler bottle. The second modification is to double bag the splitter in plastic bags and then place it in a covered plastic carrier (a plastic trash can or similar container). The churn splitter remains in the carrier throughout the sample collection process and the carrier is opened only to pour sample water into the splitter. The splitter is left in its plastic bags and removed from the carrier when placed in the vehicle for sample processing.

USE OF DISPOSABLE CAPSULE FILTERS

The almost universally accepted, as well as the traditional USGS operational definition of a dissolved constituent is what is contained in a water sample after it has passed through a 0.45 μm pore-size membrane filter. Recent evidence indicates that this definition does not ensure that the conditions for a consistent operational definition are met; it is too non-specific and filtered sample concentrations (especially for such constituents as iron, aluminum, manganese, copper, zinc, and lead) may be affected by many other factors such as filter type, filter diameter, filtration method, volume of sample processed, suspended-sediment concentration, suspended-sediment grain-size distribution, concentration of colloids and colloid-associated trace elements, and concentration of organic matter [2]. These factors may be as important as the filter pore size itself because during the course of sample processing they act to continuously reduce the nominal pore size of the membrane. Some of the factors cited above are beyond the control of field personnel, whereas others can be controlled. In addition, experiments were conducted to evaluate the contamination potential of various membrane filters. For these reasons, the USGS is recommending the use of 0.45-μm pore size capsule filters for filtering trace-element samples in order to eliminate the clogging problems associated with filtering samples containing large concentrations of suspended sediment.

Traditionally, the USGS has used plate filters to process whole-water samples to obtain filtrates for the determination of dissolved inorganic constituents. This usually means the use of a 142-mm, 0.45-μm pore size, tortuous path filter in a non-metallic backflushing system using a peristaltic pump for positive pressure. For purposes of a "clean" protocol, these systems are usable. However, the use of plate filters involves a good deal of handling (to seat the filters, process the samples, clean the system between samples). As a result, they need to be opened and closed frequently. Because of the increased handling and cleaning requirements associated with the use of large plate filters and the difficulties of contaminant-free field cleaning between uses, a decision was made to allow plate filters to be used only to obtain filtered samples for the subsequent determination of nutrients, some major ions and (or) radiochemicals, but not for trace elements.

Testing of the capsule filter by the USGS has shown that certain capsule filters can be used for processing samples for the subsequent determination of microgram-per-liter level trace elements, nutrients, and major ions [3]. Capsule filters have a number of advantages relative to the more traditional plate filters: (1) they require minimum precleaning (at the microgram-per-liter level, a 1-liter deionized water rinse), (2) because they are sealed units, their use will entail much less handling; hence there is much less likelihood for contamination, (3) because they have a large surface area, they are much less subject to clogging than plate filters, and (4) they will not require post-cleaning prior to processing additional samples because they are "throwaway" units, only intended for a single use. The only major disadvantage associated with capsule filters is that they are about six to eight times more expensive than membrane filters. This higher cost should be offset by the savings associated with the labor savings attributable to reduced field handling, no need for between-sample cleaning procedures potentially required for plate filters, and the need to process fewer field blanks because of the decreased potential for contamination.

QUALITY-ASSURANCE MEASURES

Use of Disposable Gloves

This protocol requires the use of non-powdered disposable vinyl gloves. Experience has shown that use of this protocol will lead to the use of large numbers of these gloves. As time goes on, familiarity with the requirements of this protocol will reduce glove usage. Regardless, when in doubt about the cleanliness of the gloves, it is better to err on the side of caution, and change them. Changing gloves in the middle of a procedure, or out in the open during sampling, processing, or preservation can be difficult, especially with wet hands. If it is known ahead of time that a particular procedure will/may require several glove changes, it is more convenient to put on several pairs at once before beginning the procedure than to change them during the procedure. Then, if a glove change is necessary, all that is required is the removal of the outer pair. Both "clean" hands and "dirty" hands wear gloves. That way, if "clean" hands needs help, all "dirty" hands has to do is remove the outer pair of gloves and thus, at least temporarily, become a second pair of "clean" hands.

Sample Processing and Preservation Chambers

All sample processing must be done in an enclosure or chamber in order to keep atmospheric inputs from contaminating the samples. The chambers that are being used are either portable ones made out of plastic pipe with plastic bags suspended inside or permanently-installed ones made out of wood and plexiglass.

Bridge and Cableway Sampling - Special Precautions

For obvious reasons, given the option, sampling from metal bridges and (or) cableways should be avoided, if at all possible. However desirable this might be, in many cases this is not feasible. Therefore, special precautions need to be taken when sampling from and (or) around metallic structures. Some or all of the following suggestions may be useful; several are applicable to bridge sampling in general:

1. Whether or not an individual is designated as "clean" hands or "dirty" hands, pay particular attention to where hands are placed; if a metallic object is touched, the gloves should be changed.

2. Lay a large plastic sheet over the bridge rail, or cableway side, to provide a barrier between the sampler and the metal structure; use of the sheeting may also limit the chances for touching metal, which will require a glove change. In the case of bridge sampling, move the sheet along the bridge rail as sampling proceeds along a cross section. In the case of cableway sampling, plastic sheeting may represent a potential hazard because of slipping. In such cases, a potential substitute for plastic sheeting is plastic strips held in place by tape.

3. The use of a carrier for the churn splitter, along with the presence of two individuals, on a cableway may not be feasible because of space limitations. Under such circumstances, the churn carrier may have to be eliminated; however, the churn should still be double-bagged in plastic. In any event, it is incumbent on the sampling crew to pay particular attention to potential atmospheric contamination from either the cableway

itself and (or) atmospheric sources.

4. As the sampler is raised or lowered at the start and completion of a vertical, try to control the ascent and descent to eliminate contact with the bridge; this is particularly important in regard to the sampler cap and nozzle.

5. If the traffic patterns at a particular bridge sampling site are known, try to arrive when the traffic is as light as possible. One of the biggest problems with bridge sampling is the potential contaminants from vehicular exhausts and dust that the vehicles throw up from the roadway.

6. Even if the number of vehicles using the bridge during sampling is limited, as the sampler is raised or lowered, pay attention to the traffic patterns on the bridge and try to time sampler recovery to lulls. This is especially important when emptying the sampler into the churn splitter. Use the lid of the churn carrier as an additional windbreak and (or) as a barrier to material thrown up by vehicular traffic during the transfer of sample from the sample container to the churn splitter. Always open the lid so that its back faces the roadway, not the bridge rail.

7. Even though it can substantially increase the time it takes to obtain a sample, if traffic is particularly heavy, consider sealing the sampler nozzle after recovery (with a disposable glove) and walking off the bridge to a more protected spot before pouring the sample into the churn splitter.

QUALITY-CONTROL REQUIREMENTS

Field-collected quality-control samples are generated during the course of collecting and processing actual environmental samples. The data generated from these quality-control samples are required for evaluating the quality of the sampling and processing techniques, as well as for the environmental data. Without quality-control data, sample data cannot be adequately interpreted because the errors associated with the sample data are unknown. The various types of field-collected quality-control samples are: (1) Field blanks which are designed to assess potential sample contamination levels that could occur during field sampling and sample processing; (2) Split field samples which are designed to determine the analytical precision (reproducibility) for various constituents in an environmental sample matrix; (3) Concurrent field samples which are two samples taken as closely together, in time and space, as possible and are intended to provide the user with a measure of sampling precision or reproducibility and (or) to indicate inhomogeneities (spatial or temporal) in the system being sampled; and (4) standards and reference samples which are used to evaluate bias and accuracy associated with the analytical procedures.

PROTOCOL PROCEDURES

Procedure 1: Office Preparations And Cleaning Of Equipment

Rationale--New and (or) previously used and stored equipment is likely to contain, or have adhering to it, a wide variety of potential contaminants. The purpose of the universal cleaning procedure is to ensure that these are removed prior to using the equipment. This procedure is applicable for samplers that will be used to obtain material for subsequent inorganic analysis. This is a four-step procedure using detergent, tap water, dilute acid, and deionized water. The detergent is basic, and is used in conjunction with brushes to remove any adhering material such as sediment or algae. Most of the detergent residue will be removed with tap water.

Any remaining organic films and detergent residues will be removed with
the acid. Finally, any acid residues will be removed with deionized water.
This procedure must be used for all sampler bottles/containers with the
exception of Reynolds oven bags. In this case, omit the detergent step as
it is impossible to scrub the bag with a brush without tearing it or to
completely remove detergent residues.

Requisite Supplies--

1. Deionized Water.
2. Concentrated, trace-element free hydrochloric (HCl) acid which is
   diluted with deionized water to 5 percent by volume (50 mL acid into
   1 L water) and stored in a non-contaminating container.
3. Assorted safety-labeled wash bottles for deionized water and dilute
   acid (do not use the ones with colored caps).
4. Liquid detergent that does not contain either phosphates or NTA.
5. Disposable, non-powdered vinyl gloves.
6. Non-contaminating, non-metallic clear/uncolored polypropylene/high-
   density polyethylene basins (minimum of four) sufficiently large to
   immerse all parts of the sampling and processing equipment (sampler
   nozzles, sampler caps, sampler bottles, plate filters, pump tubing),
   with the exception of the churn splitter.
7. Various non-contaminating (non-metallic, uncolored) brushes.
8. Assorted sealable plastic bags for storage and transport after
   cleaning.
9. White or clear polyethylene or polypropylene container (carboy or
   jerrican), 25 to 30 liter, for use as neutralization container.
10. Marble chips, 1 to 2 centimeter, 25 to 50 pound bag.

Procedure--

1. Before cleaning the equipment, clean the basins and wash bottles; label
   each basin and bottle with a waterproof marker. Each item must be
   cleaned with (a) detergent, (b) tap water, (c) dilute acid, and (d)
   deionized water. Read through this procedure and follow the
   appropriate steps for the items, as if they were part of the sampling/
   processing equipment, before beginning to clean the equipment itself.
   It is not necessary to clean the storage bags.

2. Clean the processing chamber in the same way as the basins, following
   the four-step procedure. This step isn't necessary if the chamber
   covers are clipped to the inside of the chamber frame.

3. Disassemble all equipment (sampling and processing), including any
   pump tubing that will be used, and immerse all parts in the detergent
   solution. Make sure that the pump tubing is filled with the detergent
   solution.

4. Allow the equipment to soak in the detergent for at least 30 minutes.

5. Put on a pair of disposable gloves and, using the appropriate brushes,
   thoroughly scrub all the equipment with the detergent.

6. Once scrubbed, place the cleaned items in a second pre-cleaned, non-
   contaminating basin.

7. Partially fill the churn splitter with detergent solution and
   thoroughly scrub it. Pay particular attention to the paddle and the
   area around the spigot. Make sure that the spigot and cappable funnel
   are cleaned as well.

8. Change gloves.

9. Thoroughly rinse the scrubbed items with warm tap water until there

is no sign of any detergent residue (until the soap bubbles all disappear). Fill the churn about one-third full through the cappable funnel with the tap water and swirl it around to remove any detergent residues. Make sure to allow some of the water to pass through the spigot. Force the tap water through any tubing that has been cleaned with the detergent. If necessary, use a wash bottle filled with tap water to clean out any hard-to-reach places.

10. Change gloves.

11. Place all the tap-water-rinsed items in a pre-cleaned, non-contaminating basin. Immerse the equipment in the dilute (5 percent) acid and soak for at least 30 minutes. Fill the churn splitter with the dilute acid and allow it to soak for the same amount of time. Alternatively, the churn splitter may be thoroughly rinsed three times by swirling a small volume of acid solution within it, discarding the solution into the neutralization container between each of the rinses.

12. At the end of the soak, remove the equipment and place it in a pre-cleaned, non-contaminating basin. Drain the acid from the churn splitter through the spigot into the neutralization container.

13. Change gloves.

14. Fill the basin and the churn splitter with deionized water. Using either a deionized water faucet and (or) a wash bottle, thoroughly rinse all the equipment with deionized water. Swirl the deionized water in the churn splitter and drain it through the spigot into the neutralization container.

15. Repeat step 14 two more times. If rinse water is neutral to litmus paper at this point, it does not have to be emptied into the neutralization container.

16. All the parts, except the churn splitter, should be placed inside two sealable plastic bags. The churn splitter with cappable funnel should be double-bagged and placed inside the churn carrier. Filtration gear should be reassembled and double-bagged. All pump tubing required for sample filtration/processing should be sealed in double plastic bags.

Procedure 2: Field Rinsing Of Equipment Prior To Sampling

Rationale--Field rinsing is required to ensure that all cleaning solution residues are removed and to equilibrate the sampling equipment to the sampling environment. Before starting this procedure, one member of the sampling team is designated as "clean" hands and the other member as "dirty" hands. Both individuals should wear gloves.

Requisite Supplies--

1. Disposable, non-powdered vinyl gloves.

Procedure--

1. Put on a pair of disposable gloves.

2. Collect sufficient quantities of native water from areas of low sediment concentrations and surface contamination with the sampler to completely fill the sampler bottle; shake, then empty the bottle by pouring the water through the nozzle.

3. Collect aliquots of native water with the sampler and pour into the churn through the cappable funnel until the volume in the churn is 2

to 4 liters.

4. Remove the churn splitter, still contained within its inner plastic bag, from the churn carrier; leave the outer plastic bag inside the carrier. Move the churn paddle up and down several times to ensure that the inside is thoroughly wetted, and then swirl the water in the churn so that the entire system has been rinsed.

5. Force the churn spigot through the inner plastic bag and drain all of the rinse water through the spigot holding the churn at an angle to maximize discharge of any sediment.

6. Once draining is complete, pull the inner plastic bag back over the spigot, rotate the churn so the spigot is no longer near the hole in the plastic bag, and replace the churn inside its inner plastic bag back inside the outer plastic bag and the churn carrier.

Procedure 3: Sample Collection, Processing, And Preservation

Rationale--Trace-element samples will be processed with a capsule filter. Samples for other inorganic determinations may be processed either with a plate filter system or a capsule filter system. Before starting this procedure, one member of the sampling team is designated as "clean" hands and the other member as "dirty" hands. Both individuals should wear gloves.

Requisite Supplies--

1. Deionized Water.
2. Disposable, non-powdered vinyl gloves.
3. 0.45-am pore-size capsule filters.
4. Non-metallic, 142 mm plate-type filtration system (if capsule filters are not being used).
5. Appropriate peristaltic pumps and pump tubing (silicon, C-flex, or Teflon) for field use.
6. Non-metallic (ceramic, Teflon, plastic), uncolored tweezers for handling the 142-mm filters (coated metallic tweezers are unacceptable because it is too easy to abrade the coating).
7. Pre-cleaned sampler bottle, cap, and nozzle or bottle for weighted bottle sampler.
8. Appropriate sampler and required suspension equipment.
9. Pre-cleaned churn splitter with cappable funnel.
10. Processing chamber.
11. Preservation chamber.

Procedure--

1. Park the field vehicle as far away from any nearby road(s) as possible and turn off the motor. Road dust and emissions from vehicles and (or) the field van can contaminate trace-element samples for microgram-per-liter analysis. The door to the sample processing area should face away from the road.

2. Put on a pair of disposable gloves.

3. Collect the whole-water sample, using an appropriate sampler and following whatever acceptable procedure is appropriate to the site and the flow conditions. Even though one individual has been designated as "clean hands" and another as "dirty hands" it is still extremely important to pay attention while the sampling operation is in progress to avoid, as much as possible, contact with any potential source(s) of contamination (for example, don't touch metal bridge parts; try not to touch the sounding weights). When operating from metallic structures, it may be useful to spread a large plastic sheet

over the area where sampling is to take place. If contact is made with a potential contaminant, dispose of the gloves and put on a new pair before transferring any sample water to the churn splitter.

4. Fill the churn splitter with each collected aliquot by opening the churn carrier lid and the plastic bags and pouring the water through the cappable funnel in the lid. Remember, only remove the cap when filling the churn splitter. After adding the sample aliquot to the churn splitter, re-seal the plastic bags. In a high traffic area, try to time the recovery of the sampler, and the filling of the churn splitter to periods when there is little or no traffic on the bridge. Further, place the open-lidded side of the churn carrier in such a way that it can serve as a barrier to the prevailing wind and (or) to turbulence caused by moving vehicles.

5. When sampling is complete, move the churn splitter inside its carrier and plastic bags, and the sampling equipment back to the field vehicle.

6. Remove the churn splitter from the carrier leaving it in the inner plastic bag. Leave the churn carrier and outer plastic bag outside the vehicle.

7. Attach the pump tubing through a hole in the side of the processing chamber. Keep the pump tubing as short as practical.

8. Pass 1 L of deionized water through the pump tubing and through the capsule filter. After passage of the 1 L of deionized water, remove the tubing from the deionized water reservoir, and continue to run the pump until all air is expelled from the capsule to drain as much of the deionized water remaining in the system as possible. Protect the intake of the pump tubing from dust by leaving it inside the deionized water bottle while pumping air through the filter. Removal of entrained water can be facilitated by shaking the capsule filter. Discard all the deionized water.

9. Transfer the pump tubing to the churn splitter through the cappable funnel and reseal the plastic bag around the tubing.

10. Remove the pump tubing from the filtration system, start the peristaltic pump, and pump sufficient sample to fill all the pump tubing; place the end of the tubing in the disposal funnel or "toss" bottle to prevent spillage in the processing chamber.

11. Open a bottle for a filtered trace-element sample, place the outlet of the capsule filter over the opening, and filter 25 mL of water (fill the bottle to the top of the bottom lip). Cap the bottle, shake, and discard the water.

12. Process the filtered trace-element sample by filling the rinsed 250 mL sample bottle to the top of the upper lip of the bottle (about 200 mL).

13. Follow the same procedure for a mercury sample, if one is being processed.

14. Process sufficient water to permit adequate rinsing of any remaining sample bottles, but no more than 100 mL.

15. Complete any other requisite filtrations for any remaining water-quality determinations. If other inorganic constituents are to be determined, the order of collection must be (a) nutrients, (b) major ions, and (c) radiochemicals.

16. Once all the filtrations are complete, remove each sample bottle from the processing chamber, one at a time, and place in the preservation chamber if preservation is required. Add the correct preservative to each bottle, making sure that preservation is done in an order that will prevent contamination of other samples (nitric acid to trace element samples first; then potassium dichromate/nitric acid to mercury samples; change gloves and preservation chamber covers; then sulfuric acid to nutrient samples, if required; change gloves and preservation chamber covers; then any other samples). Once the preservative has been added, tightly cap the bottle and chill if necessary.

Procedure 4: Field Cleaning To Prevent Cross Contamination Between Sites

Rationale--Field experiments [4] have shown that cross contamination between sites can occur if the sampling and processing equipment are not adequately cleaned before they are reused. The field-cleaning procedures to eliminate between-site cross contamination are usually less rigorous than the office cleaning procedures because the equipment has not had a chance to dry out; thus, material has not had a chance to strongly attach to the various components. The procedure described herein was used effectively to eliminate cross contamination between an acid mine drainage site with high trace-element concentrations (Iron = 50,000 μg/L, Manganese = 5,000 μg/L, Zinc = 5,000 μg/L, Copper = 400 μg/L, Cobalt = 125 μg/L) and a nearly pristine rural/agricultural site where the trace-element concentrations were at or near current (microgram per liter) reporting limits with no apparent cross contamination. In situations under which equipment shows signs of recalcitrant contamination, spare, precleaned equipment or site-dedicated equipment needs to be used and the contaminated equipment returned to the office for rigorous cleaning. This procedure should be carried out at the first sampling site when the equipment is still wet, and before driving to the second site. The procedure involves two steps using a combination of dilute acid and deionized water.

Requisite Supplies--

1. Deionized Water.
2. 5 percent (by volume) hydrochloric acid.
3. Assorted wash bottles for deionized water and dilute acid, one filled deionized water bottle should have been stored inside the glove box/ processing chamber, whereas the others (acid, deionized water) are kept outside.
4. Disposable, non-powdered vinyl gloves.
5. Sealable plastic bags.
6. White or clear polyethylene or polypropylene container, 25 to 30 liter, for use as neutralization container.
7. Marble chips, 1 to 2 centimeter, 25 to 50 pound bag.

Procedure--

1. Put on a fresh pair of disposable gloves.

2. Disassemble the sampler bottle, cap, and nozzle so that all of the pieces can be thoroughly wetted with the various rinses. Where appropriate, vigorously agitate the cleaning fluid inside the container (sampler bottle, churn splitter) to facilitate cleaning and rinsing.

3. Thoroughly rinse the sampler bottle, cap, and nozzle with deionized water; use a stream of deionized water from the appropriate wash bottle, if required.

4. Thoroughly rinse the sampler bottle, cap, and nozzle with dilute acid;

use a stream of dilute acid from the appropriate wash bottle, if required. Discard rinse into the neutralization container.

5. Thoroughly rerinse the sampler bottle, cap, and nozzle with deionized water; use a stream of deionized water from the appropriate wash bottle, if required. Discard rinse into the neutralization container.

6. Repeat step 5 two times. If rinse water is neutral to litmus paper at this point, it does not have to be emptied into the neutralization container. Reassemble the sampler bottle, cap, and nozzle and place this unit in double plastic bags.

7. Remove the churn splitter from its plastic bags and discard the bags. Thoroughly rinse the churn splitter with deionized water. Pour the water through the cappable funnel until the churn splitter is about half full; swirl the deionized water in the churn splitter, and drain some of the rinse through the spigot prior to discarding the remaining rinse water.

8. Thoroughly rinse the churn splitter with dilute acid (2 to 3 liters). Add dilute acid through the cappable funnel; swirl the acid in the churn splitter, and drain some of the acid rinse through the spigot into the neutralization container. Discard the remaining dilute acid rinse into the neutralization container.

9. Thoroughly rerinse the churn splitter with deionized water. Pour the water through the cappable funnel until the churn splitter is about half full; swirl the DIW in the churn splitter and drain some of the rinse through the spigot prior to discarding the remaining rinse water into the neutralization container.

10. Repeat step 9 two times. If rinse water is neutral to litmus paper at this point, it does not have to be emptied into the neutralization container.

11. Repackage the churn splitter in two new plastic bags, seal, and place the entire unit back inside the churn carrier.

12. Place the end of the pump tubing, which normally connects to the capsule filter, inside the disposal funnel or "toss" bottle in the bottom of processing chamber.

13. Pass 1 liter of dilute acid through the system using the same pump and pump tubing used to filter the sample. Discard rinse into the neutralization container.

14. Pass 2 liters of deionized water through the system using the same pump and pump tubing used to filter the sample. Discard rinse into the neutralization container.

15. Remove the pump tubing from the hole in the processing chamber and repackage it in double plastic bags.

16. Discard and replace the disposable cover if using a portable chamber or wash down the inside of the processing chamber with deionized water to remove any spilled native water, suspended solids, or wash solutions spilled/dropped during filtration.

17. Discard the last preservation chamber cover. Do not replace it until ready to preserve additional samples at the next sampling site.

18. Proceed to the next sampling site and conduct Procedures 2 and 3.

SUMMARY

The USGS has, over the past several years, developed and tested a new protocol for collecting and processing surface-water samples for subsequent determination of inorganic constituents in filtered water. The protocol has been shown to work well at and, in some cases below, the microgram-per-liter level. Experience in the field has shown the protocol to be easy to use with some training and practice.

REFERENCES

[1]   Horowitz, A.J., Demas, C.R., Fitzgerald, K.K., Miller, T.L., and Rickert, D.A., "U.S. Geological Survey Protocol For The Collection And Processing Of Surface-water Samples For Subsequent Determination Of Inorganic Constituents In Filtered Water," U.S. Geological Survey Open-File Report 94-539, 1994.

[2]   Horowitz, A., Elrick, K., and Colberg, M., "The Effect Of Membrane Filtration Artifacts On Dissolved Trace Element Concentrations," Water Research, Vol. 26, 1992, pp 753-763.

[3]   U.S. Geological Survey, Office of Water Quality, "Evaluation of Trace-Element and Nutrient Contamination Associated with Capsule Filters," unpublished Office of Water Quality Technical Memorandum 93.05, 1993.

[4]   U.S. Geological Survey, Office of Water Quality, "Trace-Element Contamination: Findings of Studies on the Cleaning of Membrane Filters and Filtration Systems," unpublished Office of Water Quality Technical Memorandum 92.13, 1992.

Sampling Quality Assurance/Quality Control

Robert Bellandi[1], James T. Mickam[1], Vernon C. Burrows[2]

DECONTAMINATION OF FIELD EQUIPMENT: AN INSTITUTIONAL STATUS
REPORT

REFERENCE: Bellandi, R., Mickam, J. T., and Burrows, V. C., "Decontamination of
Field Equipment: An Institutional Status Report," Sampling Environmental Media,
ASTM STP 1282, James Howard Morgan, Ed., American Society for Testing and
Materials, 1996.

ABSTRACT: Advances in hazardous waste remediation have forced the development
and refinement of correlative activities such as the decontamination of field equipment.
Specific decontamination procedures are routinely included in remediation requirements
such as the quality assurance project plan (QAPP). The authors surveyed the decontamination practices of several states and major federal agencies in 1988 before ASTM
promulgated D 5088-90, "Decontamination of field equipment used at nonradioactive
sites." They found a diversity of practices with little consistency of detail or attention.
    The authors revisited the practices of several states and the USEPA regions to
assess the effect of D 5088-90. Their presumption was that the ASTM standard would
make inroads on practices. They found that D 5088-90 has had less than the expected
effect because too few practitioners are aware of it.

KEYWORDS: decontamination, ground water sampling, sampling equipment, quality
assurance, quality control

[1]O'Brien & Gere Engineers, Inc., Syracuse, NY
[2]O'Brien & Gere Engineers, Inc., Edison, NJ

This research was supported by O'Brien & Gere Engineers, Inc.

OBJECTIVE

In 1989, the authors published their review of decontamination procedures used by several of the states and federal agencies [1]. That review showed that equipment decontamination techniques suffered from a lack of standardization. This circumstance was viewed as a shortcoming since equipment decontamination is a foundation of quality assurance and quality control for ground water and vadose zone sampling.

At that time, too, ASTM was developing a standard for equipment decontamination. That standard has since been promulgated [2]. The authors decided to undertake another review of the procedures that government agencies recommend for decontamination of equipment that contacts samples to see if the procedures had become more standardized since publication of the ASTM standard. The objective of the current review was two-fold:

- To see if agencies had begun to standardize with ASTM D 5088-90

- To examine whether the procedures of ASTM D 5088-90 were reflected, explicitly or not, in procedures that regulatory agencies recommended.

MATERIALS AND METHODS

Officials responsible for quality assurance or ground water sampling of several state agencies with responsibility over hazardous waste site remediation were contacted. Similar officials in the ten USEPA regions were contacted. Each was asked to describe or to cite the guidance the agency required or recommended for decontamination of field sampling equipment that is used at hazardous waste sites under its jurisdiction.

Most government officials noted that they prefer the use of disposal equipment, but that is not always possible, they observed. Their next preference is for field sampling equipment to be decontaminated under the controlled conditions of the laboratory. If even that is impossible, field decontamination techniques may be used. This study concentrates on field decontamination techniques.

The standard of practice for decontamination was assumed to be ASTM D 5088-90. The essential components of ASTM D 5088-90 with regard to sample-contacting equipment are presented below [3] (Figure 1).

Responses were forthcoming from each of the ten USEPA regions. The decontamination procedures of the following states were also examined:

Alabama	Arkansas
California	Connecticut
Delaware	Florida
Georgia	Hawaii
New Jersey	New York

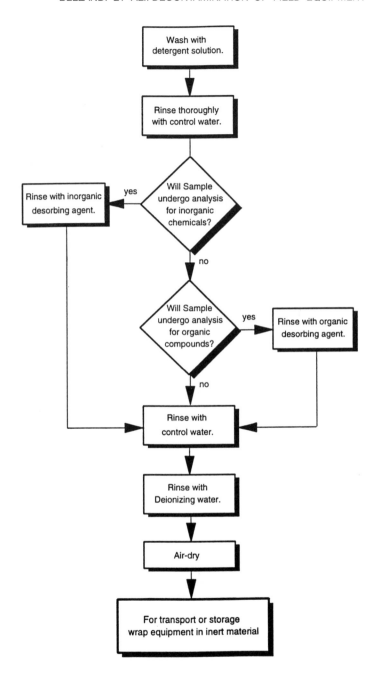

Figure 1--Simplified method for decontamination of sample-contacting equipment according to ASTM D 5088-90

RESULTS

The various decontamination procedures that were examined are summarized here. Officials of the USEPA regions were readily able to identify and specify documents that they follow and recommend. The responses of the states varied, however, with one respondent stating that it had a procedure but it was not written down.

Procedures of USEPA Regions

The various USEPA regions use different decontamination procedures. The procedures were developed by the region itself, or they were taken from USEPA guidance developed on the national level. Officials from only two regions, IX and X, stated that they use the same guidance, USEPA's RCRA guidance for decontamination [4].

Region I, Boston--The approach of USEPA Region I is conventional and similar to ASTM D 5088-90:

1. Rinse with Alconox.
2. Rinse with tap water.
3. Rinse with isopropanol (a change; most approaches recommend methanol).
4. Rinse with deionized (DI) water.
5. Air dry.

With oily material, hexane may be substituted for the isopropanol. For samples that may contain metals, many procedures recommend the use of an acid solution for the rinse. Region I recommends that no acid be used on carbon steel equipment.

Region II, New York--Region II follows its own CERCLA quality assurance manual [5]. The official contacted believes that the method needs updating.
The Region II manual includes the following information:

- Decontamination before beginning field work
- Decontamination during field work
- Avoiding cross-contamination
- Decontamination process.

The decontamination procedure that it recommends consists of the following steps:

1. Wash and scrub with low phosphate detergent.
2. Rinse with tap water.
3. If the sample may contain metals, rinse with 10% nitric acid (HNO_3), ultrapure. When it is necessary to use split spoon sampling devices of carbon steel rather than stainless steel, the nitric acid rinse may be lowered to 1% to reduce the possibility

of leaching metals from the spoon itself.

4. Rinse with tap water.
5. If the sample may contain organic compounds, rinse with acetone only or methanol then rinse with hexane. (Solvents must be pesticide grade; Region II is considering recommending isopropanol in place of methanol.)
6. Rinse with DI water that has been demonstrated to be analyte-free. The volume of water used during this rinse must be at least five times the volume of solvent used.
7. Air dry.
8. Wrap in aluminum foil for transport.

Region III, Philadelphia--The procedure used by USEPA Region III consists of the following steps:

1. Wash with soap and water.
2. Rinse with DI water.
3. If the equipment is to be used for organic compounds, a drying solvent such as methanol or acetone is used. Hexane is also used as a drying solvent, but it might not be needed if the matrix contains no oil.
4. If the equipment will be used for metals, a nitric acid rinse is used before the drying solvent.
5. Provide a final rinse, and air dry.

Region IV, Atlanta--The USEPA of Region IV distributes its decontamination procedure with its quality assurance manual on computer disk [6]. The standard cleaning procedures are in Appendix B of the manual and include equipment decontamination procedures. The manual states:

> Equipment that is used to collect samples of hazardous
> materials or toxic wastes or materials from hazardous
> waste sites, RCRA facilities, or in-process waste
> streams shall be decontaminated before it is returned
> from the field. At a minimum, this decontamination
> procedures shall consist of washing with laboratory
> detergent and rinsing with tap water. More stringent
> decontamination procedures may be required, depending
> on the waste sampled.

The manual specifies cleaning procedures for the following types of equipment used for the collection of samples for trace organic compounds or metal analysis or both:

- Teflon or glass
- Stainless steel or metal equipment

Specific procedures for automatic waste water sampling equipment are also presented, sample tubing, field tapes, pumps, hoses, and so forth.

Region IV requires that the effectiveness of field-cleaning procedures be monitored. The cleaned equipment is required to be rinsed with organic-free water, and the rinsate is analyzed for residue. The analysis is to detect low levels of extractable organic compounds including pesticides and also a standard ICP scan. [7].

The manual specifies the following field decontamination procedure [8]:

1. Clean with tap water and laboratory detergent using a brush if necessary to remove particulate matter and surface films.
2. Rinse thoroughly with tap water.
3. Rinse thoroughly with DI water.
4. Rinse twice with solvent.
5. Rinse thoroughly with organic-free water and allow to air dry as long as possible.
6. If organic-free water is unavailable, allow equipment to air dry as long as possible. Do not rinse with DI or distilled water.

Region V, Chicago--The USEPA office of Region V follows essentially the same procedure specified for Region IV and the ASTM standard. Quality assurance personnel, however, recommend against using a phosphate detergent - as do most procedures - and against using solvents.

Region VI, Dallas--The quality assurance personnel of USEPA Region VI follow rigorous procedures in part because of their affiliation with USEPA's National Enforcement Investigation Center (NEIC).

Region VI uses only teflon bailers, no pumps and no metal bailers, when it takes ground water samples. For its own sample work, it uses a careful procedure. The bailers are cleaned in the field only sufficiently to provide for safe handling. They are then cleaned in the Region's laboratory in anticipation of the next use.

It uses the following laboratory procedure for decontamination:

1. Thorough wash with Alconox.
2. Tap water rinse
3. DI water rinse
4. Nitric acid rinse (0.1 N)
5. Tap water rinse
6. Hexane rinse
7. DI water rinse

8. Air-drying and storage.

This entire procedure is used for all the bailers that are decontaminated. Because USEPA personnel cannot anticipate whether the bailer will be used the next time for organic or inorganic or mixed samples, both the nitric acid and the solvent rinses are used.

Region VI personnel also take blank samples and other quality control samples associated with the decontamination procedure. These measures are used to establish the adequacy of the decontamination procedure.

Region VII, Kansas City--USEPA Region VII recommends the decontamination procedure specified in its SOP 2006 [9]. This document treats decontamination procedures broadly and includes the topics of interference and potential problems, reagents, quality assurance and quality control, and health and safety. The guidance itself comprises the following elements:

Abrasive cleaning methods

- Mechanical cleaning methods use brushes of metal or nylon.
- Air blasting is used to clean large equipment such as bulldozers, drilling rigs, or auger bits.
- Wet blast cleaning is used on large equipment also.

Non-abrasive cleaning methods

- High pressure water
- Ultra-high pressure water.

Disinfection and rinse methods

- Disinfection, to inactivate infectious agents
- Sterilization, impractical for large equipment
- Rinsing, to remove contaminants through dilution, physical attraction, and solubilization.

The cleaning procedure for field sampling equipment begins with one of the removal procedures listed above. It continues with the following steps:

1. Wash equipment with non-phosphate detergent solution.
2. Rinse with tap water.
3. Rinse with distilled or DI water.
4. Rinse with 10% nitric acid if the sample will be analyzed for trace organic compounds [sic].

5. Rinse with distilled or DI water.
6. Use a solvent rinse, pesticide-grade, if the sample will be analyzed for organic compounds.
7. Air-dry the equipment completely.
8. Rinse again with distilled or DI water.

SOP 2006 specifies that a rinsate blank is to be taken from the decontamination fluids. It is analyzed in conjunction with field blanks and trip blanks to evaluate the effectiveness of the field decontamination procedure.

Region VIII, Denver--USEPA Region VIII has recently revised its field sampling procedures [10]. Its guidance discusses decontamination of both small and large field sampling equipment. For the small equipment, it recommends the following steps:

1. Scrape or wipe to remove all visible contamination.
2. Scrub with a brush and nonphosphate detergent.
3. Rinse three times with potable tap water; collect a tap water sample before use for analysis for chemical constituents that could bias the analytical results.
4. Rinse three times with DI or distilled water.
5. If significant concentrations of inorganic compounds are expected, rinse with dilute (10%) nitric acid.
6. If significant concentrations of organic compounds are expected, rinse with acetone or pesticide-quality hexane.
7. Allow equipment to air dry, and wrap in plastic before transporting to the next sampling location. If the equipment will be used to sample for the analysis of volatile compounds, it should be wrapped in metal foil rather than plastic.

Region IX, San Francisco, and Region X, Seattle--Both Region IX and Region X of USEPA recommend the use of USEPA's RCRA guidance for decontamination [11]. This document presents the following steps for decontamination: [12]:

1. Wash the equipment with a nonphosphate detergent.
2. Rinse the equipment with tap water.
3. If the samples may have organic compounds in them, rinse the equipment with pesticide-grade hexane or methanol (methyl alcohol). If inorganic constituents may be present, rinse with dilute (0.1 N) hydrochloric or nitric acid. Hydrochloric acid is preferred over nitric acid when cleaning stainless steel because nitric acid may oxidize the steel.
4. Rinse the equipment with reagent grade acetone. (This step is not used for samples that may contain inorganic constituents.)
5. Rinse the equipment with organic-free water.

6. Allow equipment to air dry in a dust-free environment.

7. The equipment should be wrapped in aluminum foil, labeled, and stored.

The procedure notes that if acetone, hexane, or methanol are analytes of interest, a different solvent, such as isopropanol, should be chosen. Officials of Region X also mentioned that some samplers are reluctant to take acetone or methylene chloride to the field because of concern over flammability and other hazards.

USEPA Regions and the ASTM procedure--Region I sometimes sees project plans that reference the ASTM procedure in place of its own. Typically, though, it suggests the USEPA national plan with the few changes that the Region has made [13]. Its approach is essentially the same as the ASTM procedure. The official of Region II who was contacted was unaware of the ASTM procedure. In Region III, the ASTM standard is referenced in a field decontamination method, and the USEPA official was aware of it. Officials of Region VI are aware of the ASTM standard but do not use it. The official of Region VII was unaware of the ASTM standard.

Responses of the states

The subject of decontamination of field sampling equipment is arcane. As this survey was conducted, it appeared that only a few specialized individuals in the state agencies were aware of the requirement or need for decontamination. Finding these individuals in the state offices, therefore, was often difficult.

States with few requirements--Some of the states that responded to this study have limited or no formal requirements. Alabama, for example, had no specific guidelines or procedures but follow the procedures stated in the National Contingency Plan (NCP) [14]. Similarly, the official contacted in Hawaii said that the state has no written procedures, but encourages the use of disposable materials. Its general procedure is to wash with Alconox, double-rinse, and air-dry.

The official of the state of Connecticut said that nothing more is required than the USEPA prerequisites. A Quality assurance project plan (QAPP) or sampling plan must, therefore, follow USEPA guidance.

States with prevalent requirements--Some of the states require the basic steps of decontamination. Arkansas, for example, recommends a wash with nonphosphate detergent such as Alconox or Liquinox, a rinse with tap water, and a rinse with DI water. Delaware uses the RCRA ground water monitoring guidance discussed above for USEPA Regions IX and X.

States with detailed requirements--The state of California, does not, according to the geologist contacted, specify a decontamination procedure for use by others. It has, however, developed its own procedures [15]. Its guidance recommends the following steps for decontamination:

• Pumps should be cleaned in the field by pumping a solution of nonphosphate detergent through the pump and associated tubing. This solution should be followed by tap water and then three times by purified water.

- Bailers should be disassembled and cleaned by washing in nonphosphate detergent followed by rinses with tap water, purified spectroscopic-grade solvent, and DI water. They should then be air-dried, reassembled using powderless surgical gloves, and wrapped in aluminum foil.

- Filtering glassware should be cleaned in a solution of a nonphosphate detergent. It should then be rinsed with tap water and a 5 to 10% nitric acid solution. A final rinse is with purified water.

The California procedure recommends that blank samples be collected from the equipment cleaned in the field and reused. In this way, contamination not removed or introduced through the decontamination procedure can be detected.

The state of Florida also specifies an extensive decontamination procedure [16]. The state emphasizes that the use of pre-cleaned equipment is preferred. The following steps are required to clean teflon and stainless steel sampling equipment in the field for all extractables (organic compounds, metals, nutrients, and so forth) [17]:

1. Clean with tap water and lab grade soap (Liquinox or equivalent) using a brush, if necessary, to remove particulate matter or surface film.
2. Rinse thoroughly with tap water.
3. If trace metals are to be sampled, rinse with 10 to 15% reagent grade nitric acid (HNO_3). The acid rinse should not be used on steel sampling equipment.
4. Rinse thoroughly with DI water. Enough water is to be used to ensure that all equipment surfaces are flushed with water.
5. Rinse twice with isopropanol. One rinse may be used as long as all equipment surfaces are thoroughly wetted with free-flowing solvent.
6. Rinse thoroughly with analyte-free water, and allow to air dry as long as possible.
7. Wrap clean sampling equipment, if appropriate, in aluminum foil or in untreated butcher paper to prevent contamination during storage or transport to the field.

The Florida procedures include many qualifications depending on the analytes and circumstances of the sampling. Additional procedures are set out for automatic waste water samplers, sampling trains, bottles, filtration equipment, sample tubing, pumps, analyte-free water containers, and ice chests and shipping containers.

The state of Georgia has extensive and detailed procedures that it has developed for field sampling. [18]. Decontamination requirements are detailed in the Georgia procedures as follows [19]:

Bailer cleaning   The best procedure is one bailer for one well. However, when this is not possible a single bailer may be cleaned between wells as follows:

1. The sampler, without removing gloves, will untie the rope and will

open the bailer to allow the helper to pour distilled water into and around the bailer. This will be shaken and poured out.

2. The helper will then pour spectro-grade isopropanol into and around the bailer. It is again shaken and poured out.

3. A final rinse is now performed with distilled water in copious amounts into and around the bailer. This should be done more than once.

4. A fresh piece of aluminum foil is placed on the plastic sheet and the bailer is placed in it. The foil is folded around the bailer for carrying.

5. It is important to sample the upgradient wells first and then proceed to the more contaminated wells.

6. The bailer is then returned to the laboratory for a thorough cleaning with Alconox and distilled water rinse, and foil wrapping.

The procedure notes that for wells that are contaminated with insoluble wastes field cleaning is not recommended. Additionally, if isopropanol appears in the test, a resampling will have to be done.

The state of New Jersey has more sites listed on the National Priorities List (NPL) than any other state. It has developed careful decontamination procedures.

New Jersey recommends the following decontamination procedure for field use:

1. Laboratory grade glassware detergent plus tap water wash

2. Generous tap water rinse

3. Distilled and DI water (ASTM Type II) water rinse

4. If the sample is to be analyzed for metals, 10% nitric acid rinse (trace metal or higher grade HNO_3)

5. If the sample is to be analyzed for organic compounds:

   a. Pesticide-grade acetone rinse
   b. Total air dry or pure nitrogen blow out
   c. Distilled and DI water (ASTM Type II) rinse.

Auxiliary requirements are noted in the procedure. Additional procedures describe the decontamination of heavy equipment, backhoes and drill rigs, and pumps.

New York State uses its own procedures. They are similar to those of New Jersey in both content and coverage. Like New Jersey, too, New York has a large number of hazardous waste sites under its jurisdiction. New York officials also had the following comments:

- For large pieces of equipment, steam cleaning is sufficient.
- The state is trying to reduce the use of solvents in the field.
- For small pieces of sampling equipment, the preferred procedure is cleaning with a brush and non-sudsing detergent.
- For water samples, dedicated equipment is preferred.

• Samplers are encouraged to take sufficient supplies of laboratory-decontaminated equipment into the field to accommodate the full day of sampling.

The states and the ASTM standard--Most of the individuals contacted in the various states were unaware of the ASTM standard. The New York official was familiar with it. The state of New Jersey references the ASTM standard in its field sampling procedures manual.

Conclusion

Decontamination procedures for field equipment must be flexible because of the highly variable conditions encountered at hazardous waste sites both with regard to contaminants of concern and the physical circumstances of the site. This variability itself may pose an obstacle to the adoption of D 5088-90. Differences from D 5088-90 in decontamination practices found in state and federal guidance appear to be the result of circumstances that specific regulators have encountered in the field. The global procedures promoted by USEPA and many of the states are, however, the same as ASTM D 5088-90 in essential respects.

Although many USEPA officials were aware of the promulgation of ASTM D 5088-90, the regional offices were not using them and in most cases were not referencing them. The officials at the state level were generally unaware of the ASTM standard. If the goal of more standardized decontamination procedures is desirable, ASTM should promote the use of the standard by publicizing its content and intent. States that have few or no procedures themselves would probably be the most likely to adopt ASTM D 5088-90 or to find it a useful standard around which to develop their own procedures.

How has the institutional status of decontamination changed since the authors conducted their first survey in the late 1980s? These surveys are not directly comparable because different agencies responded to each survey, and the current survey decided to concentrate on the procedures of the USEPA regions rather than consider the USEPA a monolithic agency as the first survey did. A conclusion of the authors from the first survey is still relevant, however [20]:

> Both the approach to and the documentation of
> equipment decontamination practices and
> procedures are highly variable and incomplete
> across the United States.

Although the diversity of hazardous waste sites would counsel against USEPA's hegemony in the area of decontamination, the agency should perhaps raise the awareness of the place of decontamination in quality assurance and quality control among professionals. It is the geologists and engineers rather than only the individuals charged with quality assurance and quality control who should be made cognizant of the steps that constitute adequate decontamination.

The USEPA could join with ASTM to provide increased education about decontamination. They could present ASTM D 5088-90 as the basis for the states to develop a decontamination procedure suited for the states' individual circumstances. The wealth of USEPA reference procedures including those developed by the various regions, could give the states additional options with which they could customize the decontamination procedures. Given the dissimilarity of decontamination procedures and even the dissimilarity in the attention paid to decontamination, such an effort would be profitable to achieve more consistent quality assurance at hazardous waste sites.

ACKNOWLEDGEMENT

The authors are grateful to the many state and federal officials who contributed to this paper by supplying useful and timely information on decontamination techniques. Any benefit that the field of hazardous waste remediation derives from the continued discussion of decontamination procedures fostered by this paper is a result of the generosity of these professionals. This paper is dedicated to them:

Greg Allen, USEPA Region III; Linda Baker, Georgia Department of Natural Resources, Environmental Protection Division; Kathy Baylor, USEPA Region IX; Bill Bokey, USEPA Region IV; Steven Caleo, USEPA Region VIII; Dan Cooper, State of Alabama; Robert Eisner, Connecticut Department of Environmental Protection; John Evenson, New Jersey Department of Environmental Protection and Energy, Environmental Measurements Bureau; Lynette Gandle, USEPA Region VIII; Walt Helmich, USEPA Region VI; Charlie Hopper, USEPA Region VI; Jerry Marcott, State of California; Sandy McClain, USEPA Region IV; Dave Parson, California Department of Toxic Substances Control; Charles Perfert, USEPA Region I; Laura Scalise, USEPA Region II, Edison, New Jersey; George Schupp, USEPA Region V; Betty Sealy, New York Department of Environmental Conservation; Sharon Smith, USEPA Region X; Barry Townes, USEPA Region X; Eric Trinkle, Delaware Department of Natural Resources and Environmental Control; Alana Trusket, Georgia Department of Natural Resources; Jeff Wankee, USEPA Region VII; Penny Wilson, Arkansas Department of Pollution Control and Ecology; Bruce Woods, USEPA Region X; and the following personnel of O'Brien & Gere Engineers: Laura J. Wilgan in Tampa and Cathleen Miller in Syracuse.

REFERENCE

[1] Mickam, J.T., R. Bellandi, and E.C. Tifft, Jr. 1989. "Equipment decontamination procedures for ground water and vadose zone monitoring programs: status and prospects," Ground water monitoring review. pp. 100-121.
[2] ASTM. "Standard practice for decontamination of field equipment used at non-radioactive waste sites." Philadelphia, PA. D 5088-90. Approved June 29, 1990; published September 1990.
[3] ASTM D 5088-90. 7.2.
[4] USEPA. 1992. RCRA ground-water monitoring: draft technical guidance. EPA/530-R-93-001.
[5] USEPA Region II. 1989. CERCLA quality assurance manual. Final copy.
[6] USEPA Region IV. 1991. Environmental Compliance Branch standard operating procedures and quality assurance manual.
[7] Ibid., Section B.2
[8] Section B.8

[9] USEPA. 1991. Compendium of ERT waste sampling procedures. EPA/540/P-91/007.

[10] USEPA Region VIII. 1994. Standard operating procedures for field sampling activities. Version 2.

[11] USEPA. 1992. RCRA ground-water monitoring: draft technical guidance. EPA/530-r-93-001.

[12] Section 7.3.4, pp. 7-17 and 7-18.

[13] USEPA. 1992. RCRA ground-water monitoring: draft technical guidance. EPA/530-r-93-001.

[14] 40 CFR 300-430.

[15] Hazardous Materials Laboratory (HML), California Department of Toxic Substances Control. 1993. HML user's manual.

[16] Florida Department of Environmental Regulation. 1992. DER SOP 4.1.

[17] DER SOP 4.1.4.1.

[18] Georgia Environmental Protection Division, Hazardous Waste Management Branch. 1994. Guidance for field sampling, sample handling, and lab report evaluation.

[19] Georgia Department of Natural Resources, Environmental Protection Division. 1991. Hazardous waste management program, groundwater testing, appendix IX, "Georgia modified standard method," p. 39.

[20] Mickam, Bellandi, Tifft. 1989. p. 113.

Lazowski, Deborah D. [1]

USE OF PHASED APPROACH TO SAMPLING SOIL AND SEDIMENT FOR SITE
CHARACTERIZATION AT THE PANTEX PLANT IN AMARILLO, TEXAS

REFERENCE: Lazowski, D. D., ''**Use of Phased Approach to Sampling
Soil and Sediment for Site Characterization at the Pantex Plant
in Amarillo, TX,**'' Sampling Environmental Media, ASTM STP 1282,
James Howard Morgan, Ed., American Society for Testing and Materials, 1996.

ABSTRACT: In this study, the phased approach is used to
expedite collection of data and information about the site.
The approach consists of sampling soils and sediments at a
variety of locations, using applicable sampling devices. The
validated analytical measurements acquired from sampling is
entered into a database. This data then can be applied in a
geographical information system (GIS). Further analysis of
material can be reduced and saved in the database. The chain-
of-custody records throughout the phased approach ensures
careful and accurate documentation.

KEYWORDS: sampling, soil, sediment, phased approach, sampling
devices, geographical information system.

Pantex Plant is located in the Texas panhandle in Carson
County, near Amarillo. The site consists of approximately
16,000 acres, of which 10,080 acres are owned by the
Department of Energy with an additional 6,000 acres leased
from Texas Tech University for a safety and security buffer
zone (figure 1).

The Pantex Plant site is flat; situated on the Southern
High Plains at an elevation of approximately 3,540 ft. The
Plant site is characterized as mixed prairie of native and
introduced grasses. Five natural playas are fed by rainwater
and snow melt. In January 1989, the U.S. Environmental
Protection Agency (EPA) conducted a Resource Conservation and

---

[1]Project Scientist, Environmental Protection Department,
Battelle Memorial Institute, Amarillo, TX 79177.

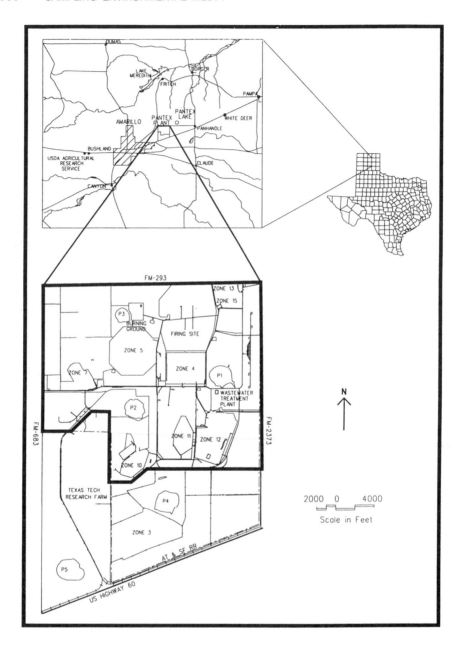

FIG 1--Location of Pantex Plant.

Recovery Act (RCRA) Facility Assessment to identify Solid
Waste Management Units (SWMUs) which may require investigation
and/or corrective action under the 1984 Hazardous and Solid
Waste Amendments to the RCRA.

In September 1989, a Draft Administrative Order on Consent, pursuant to Section 3008 (h) of the RCRA, was issued identifying 143 solid waste management units (SWMUs). The SWMUs were organized into groups that are to be carried through the corrective action process on independent schedules.

A RCRA Facility Investigation (RFI) was conducted on ten SWMUs;

SWMU #1   Building 12-17 Drainage Ditch
SWMU #2   Building 12-43 Drainage Ditch
SWMU #3   Building 11-44 Drainage Ditch
SWMU #4   Building 11-50 Drainage Ditch
SWMU #5   All other remaining drainage ditches
SWMU #6   Playa #1
SWMU #7   Playa #2
SWMU #8   Playa #3
SWMU #9   Playa #4
SWMU #10   Pantex Lake

A Work Plan [1] was developed that specified field methods for soil and sediment sampling, locations of sampling sites, laboratory analyses to be performed, and worker health and safety precautions to be followed during the investigation, as well as defined Quality Assurance and Waste Management Plans.

The site investigation requires sediment and soil samples. The exact parameters per sample site are chosen based on the history of potential releases that may have reached the sampling site.

The data needed regarding the characteristics of the sediment, soil, and vadose zone and how it will be determined are listed in Table 1.

A sample numbering system has been established for identification and type of sample. All locations will be numbered sequentially. This will consist of a two-part identifier; (PTX) the facility code and (08) the project code. The type of sampling; either boreholes (2000), sediment sampling in ditches (3000), sediment sampling in playas (4000), will have one of these numbers after the two-part identifier. This is to easily identify which sampling activity took place.

A data search was performed prior to writing of the work plan. Data regarding past reports, spills/releases at the various SWMUs within Zones 11 and 12, discharges into the playas, and results from past wastewater discharge, surface water, sediment, and soil sampling were used to prepare the work plan. Process knowledge as well written records, and operations performed, were useful in this task. A site visit was conducted in order to locate various ditches of interest. This accumulated information became very useful in the project decision process.

Table 1 SEDIMENT/SOIL/VADOSE ZONE CHARACTERISTICS

Data Needs	How Data Will be Determined
Surface soil distribution	Field surveys from Soil Conservation Service
Soil moisture content	Laboratory measurements
Thickness of vadose zone	Measurements of depth to groundwater in nearby wells
Pore water velocity	Calculation using soil porosity and permeability
Percolation	Field measurements
Volumetric water content	Calculation of volume of water in bulk volume of soil
Depth of contamination	Chemical analysis of the soil samples
Extent of contamination	Sampling and analysis of soil samples
Contamination characterization	Published information/data regarding constituents
Areal extent of contamination within	Composite sediment samples taken across transect of the ditches
Seepage rates of ditch discharge waters to underlying groundwater	Use soil permeability
Vertical and lateral hydraulic conductivity of vadose zone migration	Use packer test

A phased approach based on decision trees was used to determine the actual type, location, and number of samples taken during the RFI. This phased approach will enable the project to proceed in a timely and cost-effective manner. By prior knowledge of site conditions enable investigators to effectively determine the number of samples to be taken. The Texas Natural Resource Conservation Commission approved the decision-tree based approach. Cooperation of the federal and state organizations when using this approach, the more efficient field sampling progression resulted and the time required to perform the RFI, while meeting data collection objectives, will be minimized.

The overall objective is to obtain quality data as necessary to meet the following Data Quality Objectives:

   1) Define the nature and extent of contaminated sediment and soil;
   2) Define the local hydrogeological system;
   3) Perform a corrective measures study;
   4) Analyze the risk to human health and the environment;
   5) Perform the corrective measures study engineering design;
   6) Construct the corrective measures study adopted.

SEDIMENT

Sediment investigations of the ditch and playa group of SWMUs were designed to identify and delineate areas of contamination due to discharge and transport of industrial wastewaters. Initial field investigations were conducted and the resulting analytical data evaluated to determine whether all areas of apparent contamination were adequately defined. From this investigation further sampling was deemed necessary to locate the contamination.

Ditch sediment sampling was performed at SWMUs 1, 2, 3, 4, and 5. Grab samples were collected from the center line bottom of the ditch, as shown in Figure 2, at each sample location. This provided the most conservative characterization of any contamination present within the ditches, since the highest concentrations of contaminants were likely to be present at that point. Field composite samples were taken across the ditch at sample locations for all requested parameters except for volatiles, samples for which composited in the laboratory. Composite samples were taken from a minimum of 3 and a maximum of 5 subsamples collected at locations across the ditch transect (Figure 2). The locations at which composite samples were collected were determined in the field. The sample location was based on prior knowledge of the site conditions and probable discharges. The results of the analyses of the composite samples were correlated to the results of the analyses of the grab samples collected at the same locations to characterize contamination across the ditch.

Sediment samples were collected to a minimum depth of 12 inches as shown in Figure 2. If the sediments were present to

main zone perimeter ditch. Sample locations are chosen within selected 500-foot reaches of the ditches into which the discharges directly flow.

The perimeter ditches transport the wastewater from the confluence point to the delta of a playa. Sample locations chosen within all of the 500-foot reach of the main zone perimeter ditches. These sample locations were based on prior knowledge of the site and prior site investigations. The 500-foot reach was determined to be the maximum length that a possible contaminate may have traveled. This was determined from previous information regarding contamination. found in downstream areas from outfalls of buildings.

The main ditches are stratified into 1,000-foot reaches. The sample locations were chosen according to prior information and data records. These sampling locations were then identified within the 1,000-foot reaches of the main ditches flowing to the playa.

DITCH GRAB SAMPLING SCHEME

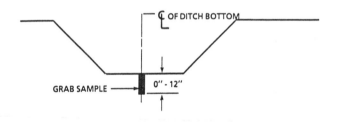

DITCH COMPOSITE SAMPLING SCHEME

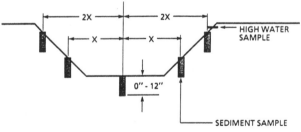

X - DISTANCE TO SAMPLE LOCATION

FIG 2--Ditch sampling scheme

a depth below 12 inches, samples were collected to the depth at which the sediment/soil interface was found. The sediment/soil interface was visually determined in the field, based on apparent changes in soil characteristics. This is a standard practice for this type of sampling.

Because the ditches do not usually contain water except after a rain event, the ditch sediment sampling was conducted in a dry ditch basin. Samples were collected using a 2-inch stainless steel sediment coring tube for those soils that can be easily removed from the tube. A 3-inch stainless steel bucket auger was used collection of those sediments that can not easily be removed from a coring tube.

Triplicate samples were collected at a portion of the sample locations for quality assurance/quality control purposes. Side-by-side triplicate samples were collected, because significant variations within distances of a few inches were not expected in the bottoms of the ditches, and side-by-side samples required less handling than split samples. Split samples were collected for analysis of other analytical parameters.

Playa sediment samples were collected from 2 distinct depth horizons in order to determine extent of vertical migration of contaminants in playa sediment. Grab samples were collected from the sediment surface to approximately 12 inches deep. Additional sediments were collected to a depth below 30 inches in order to determine the depth of the Randall clay layer. The characteristic soil types for these landscape positions are Pullman clay loam in the interplaya uplands and Randall clay in playa bottoms.

Stratified random sampling of the ditch sediments was accomplished by subdividing the lengths of the ditches into non-overlapping reaches, of specific lengths and then randomly selecting sediment samples form each reach, except those reaches that contain discharge, confluence, and standing water [2,3].

Ditches that have been stratified for differential sampling intensity, were based on sedimentation differences. These sedimentation differences were known by prior investigations and documented knowledge of the area. Ditches of the facility have been separated into three designations for sampling: discharge ditches, main zone perimeter ditches, and main ditches. Sampling differed for each type of strata. The lengths of the stratified reaches and the frequency of sampling within the stratified reaches reflect the assumption that contaminates are most likely present in the sediments immediately downgradient of the discharge point. The discharge ditches are immediately downgradient of a discharge point and are therefore stratified into 100-foot reaches from the point of discharge to the point of confluence within the

Sediment investigations are designed to identify and delineate areas of contamination due to discharge and transport of industrial wastewaters [4,5]. A flow chart that identifies the logic for gathering and evaluating analytical data resulting from collection and analysis of sediment samples is presented in Figure 3.

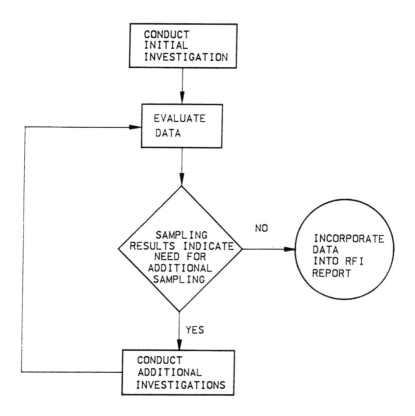

FIG 3--Sediment decision tree.

SOIL

The soil investigation approach at Pantex Plant was also conducted in a phased approach. Initial investigations included drilling and sampling at 48 locations. Subsequent phases of investigation were based on the results from sediment and soil sampling in the first phase. Whether additional data was needed was determined based on the criteria listed in Figure 3 and utilizing the decision tree shown in Figure 4. Soil borings, locations, depth, number of samples, and parameters measured are summarized in Table 2.

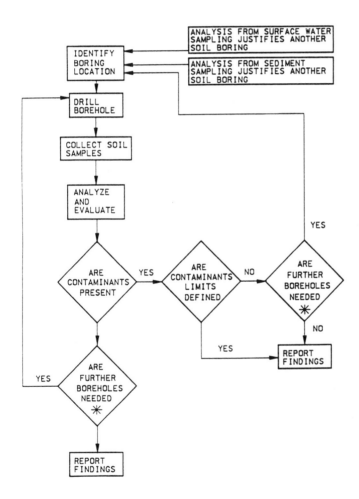

FIG 4--Soil boring decision tree

Table 2 Proposed soil borings

Location	ID Number PTX08	Minimum Depth (feet)	Minimum Number of Samples
SWMU 1	2006	30	5
SWMU 2	2010	30	5
SWMU 3	2018	30	5
SWMU 4	2016	70	7
SWMU 5	2001	30	5
12-51	2002	5	2
12-67	2003	5	2
12-68	2004	30	5
12-9	2005	30	5
12-19	2007	30	5
12-73	2008	30	5
12-21	2009	30	5
12-81	2011	10	3
12-41	2012	30	5
12-36	2013	30	5
	2014	30	5
	2015	30	5
11-51	2019	30	5
11-20	2020	30	5
SWMU 6	2021	30	5
	2022	30	5
	2023	30	5
	2024	30	5
	2025	30	5
	2026	30	5

Soil borings were drilled to various depths ranging from 5-feet to 70-feet to obtain samples for chemical and physical analysis. Field screening was conducted at each borehole, by performing a head space analysis of samples using a photoionization detector, Draegar tubes, or other appropriate devices.

Boreholes were 5 to 30 feet in depth according to location. Drilling methods employed were those applicable to site conditions and thus allowed samples to be analyzed for chemical and physical parameters with little or no changes to the soil matrix samples themselves. Methods used included ASTM standard methods, Pantex Plant Standards and Pantex Plant Operating & Instruction Methods for sampling and safety (O&Is).

All borings were lithologically logged and the soil samples collected and analyzed according to the unified Soil Classification System using ASTM approved methods. Drilling was interrupted at designed 5-foot intervals to collect samples for chemical and physical analysis. Samples for chemical analysis were taken with sampling tools (split-spoon, hollow stem auger) appropriate to the subsurface geology and site conditions. Undisturbed soil samples were analyzed for physical parameters as necessary and collected with appropriate sampling tools (Dennison tin, Shelby Tube) dependent upon subsurface geology and site conditions.

Following drilling, the samples were submitted for chemical and physical analysis, and the information obtained was used to determine the potential presence and mobility of contaminants in the soil. The drilling log books, chain-of-custody papers, and laboratory data records provided the documentation and audit trail information necessary for this type of effort.

These analyses then were placed into a database that can be used by a geographic information system (GIS) to create maps for a visual display of the chemical constituents. In addition, we can now carry out spatial analysis using the powerful tools of a geographic information system. A GIS allows us to carry-out sophisticated analyses of spatial variations. This could include analyzing the spatial and time dependent variations that are inherent with various chemical parameters.

CONCLUSION

The phased approach at Pantex Plant was used to expedite collection of data and information. The rigorous process of sampling different environmental matrices at a variety of locations, using applicable sampling devices, was achieved. The validated data that was acquired form the sampling process was collected and placed into a database. This data was then copied into a GIS for further exploratory design of the contamination plumes. Further analysis of the material was reduced and saved into the database. The chain-of-custody records throughout the phased approach ensured careful and accurate documentation. This sampling event has achieved its goals of project timeliness and cost effectiveness. It has also produced data of good quality, to be referenced for other projects and give an overall feel for the contamination at the site. This type of sampling will be used for future projects on-site.

REFERENCES

[1]   U.S. Army Corps of Engineers, <u>Final Work Plan for RCRA Facility Investigations at Ditches and Playas ADS 1216,</u> U.S. Army Corps of Engineers, Tulsa District, Tulsa Oklahoma, 1991

[2]   Ott, L. 1977. <u>An Introduction to Statistical Methods and Data Analysis,</u> Wadsworth Publishing Company, Inc., Belmont, California, 1977

[3]   Snedecor, G.W., and Cochran W.G., Statistical Methods, Iowa State University Press, Ames, Iowa, 1967.

[4]   Purtyman, W.D., and Becker, M.N., Supplementary Documentation for an Environmental Impact Statement Regarding Pantex Plant: Geohydrology, Report LA-9445-PNTX-I, Los Alamos National Laboratory, Los Alamos, New Mexico 1982.

[5]   Purtyman, W.D., and Becker, M.N., Supplementary Documentation for an Environmental Impact Statement Regarding Pantex Plant: Geohydrologic Investigations, Report LA-9445-PNTX-H, Los Alamos National Laboratory, Los Alamos, New Mexico 1982.

Christopher C. Calkins[1], Craig A. Gabriel[2], and Jeffrey E. Banikowski[3]

SOIL GAS SAMPLE ANALYSIS METHOD EVALUATION AND COMPARISON

REFERENCE: Calkins, C. C., Gabriel, C. A., and Banikowski, J. E., ''Soil Gas Sample Analysis Method Evaluation and Comparison,'' Sampling Environmental Media, ASTM STP 1282, James Howard Morgan, Ed., American Society for Testing and Materials, 1996.

ABSTRACT: The quality of soil gas survey data is a function of many variables, particularly the analytical method used to analyze the collected samples. Several analytical methods are commonly employed, including direct reading photoionization detectors, portable gas chromatographs (GCs), and mobile laboratories equipped with benchtop GCs. Two analytical methods employing Hewlett-Packard (HP) 5890 Series II GCs were compared relative to their sensitivity, precision, accuracy, practicality, and productivity. The first method utilized direct injection as the method of sample introduction, and the second, a Tekmar LSC 2000/ALS 2016 purge and trap autosampler. The direct injection method provided greater accuracy, precision, and ease of use, while the purge and trap autosampler method provided increased sensitivity and higher productivity.

KEYWORDS: vadose zone, direct injection, purge and trap, autosampler, gas chromatograph, accuracy, precision, percent recovery, standard deviation, relative standard deviation

---

[1]Project Scientist, Environmental Toxicology/Industrial Hygiene, O'Brien & Gere Engineers, Inc., Syracuse, New York 13221.

[2]Scientist, Environmental Toxicology/Industrial Hygiene, O'Brien & Gere Engineers, Inc., Syracuse, New York 13221.

[3]Managing Scientist, Environmental Toxicology/Industrial Hygiene, O'Brien & Gere Engineers, Inc., Syracuse, New York 13221.

INTRODUCTION

Selection of analytical instrumentation and the method for analysis of soil gas samples depends on the data quality objectives, budget and schedule of the program. Over the past two years O'Brien & Gere Engineers, Inc. has performed several large soil gas survey's at United States Air Force (USAF) installations in the United States. The surveys were designed to evaluate the concentration, chemical composition, and areal extent of volatile organic compounds (VOCs) in the vadose zone. Data obtained from the soil gas surveys were used to guide subsequent placement of soil borings and monitoring wells. As a result, emphasis was placed on obtaining high quality analytical data. To accomplish this objective and provide the quality of data necessary, a Hewlett-Packard (HP) 5890 Series II GC was selected for analysis of the samples. Analytical methodologies employed for two such soil gas surveys are discussed here. The first survey, consisting of approximately 400 sampling points, utilized direct injection as the method of sample introduction, and the second survey, consisting of approximately 2500 sampling points, utilized a Tekmar LSC 2000/ALS 2016, 16-port purge and trap autosampler [1].

EXPERIMENT

Sample Collection

For both surveys, samples of soil gas were collected using slotted aluminum shield points attached to a length of teflon® tubing. The teflon® tubing-shield point assemblies were driven to a depth of 1 meter below grade using hardened steel probes and a manual slide hammer. Prior to collection of a sample, the hardened steel probe was retracted approximately 10 cm to expose the vapor intake slots of the shield point. A sample was collected by attaching a vacuum box to the exposed end of the teflon® tubing. A 1 L tedlar® bag equipped with a polypropylene fitting, was placed inside the vacuum box, and the bag was filled with sample using a manual vacuum pump. Prior to collection of a sample, five teflon® tubing volumes of soil gas was purged from the system to minimize ambient air contributions to analytical results. After filling, sample bags were labelled and placed in a cooler without ice and transported to the on-site laboratory for analysis.

Sample Analysis

For the first survey, samples were analyzed by direct injection of 100 μL of soil gas into a HP 5890 Series II GC equipped with a purged-packed inlet, 30 m by 0.53 mm ID SPB-1 wide bore capillary column, and electron capture (ECD) and flame ionization (FID) detectors. GC method programming, data acquisition, data reduction, and report preparation was performed using the HP 3365 Chemstation software. This package eliminates the need for manual number crunching, thereby saving time and money and minimizing errors in sample quantitation. The software was loaded on a 386 computer equipped with a 20 MHz microprocessor, 2-MB RAM, 80 MB hard drive, and VGA color monitor. The samples were analyzed using a column oven temperature of 35°C and a carrier gas (ultra high purity nitrogen) flow rate of 5 mL/min, allowing for an analytical run time of 20 min. Based on historical analytical data, the following target analyte list was selected for GC calibration: ECD analytes of 1,1-dichloroethene (1,1-DCE), 1,1-dichloroethane (1,1-DCA), chloroform, trichloroethene (TCE), and tetrachloroethene (PCE), and FID analytes of benzene, toluene, chlorobenzene, ethylbenzene, m-xylene, and o-xylene.

The second method employed a HP 5890 Series II GC equipped with a 30 m by 0.53 mm ID SPB-1 wide bore capillary column, ECD and photoionization (PID) detectors, and the HP 3365 Chemstation software package. The GC was interfaced with a LSC 2000/ALS 2016 16-port purge and trap autosampler with 5 ml fritted spargers, which served as the method of sample introduction. A soil gas sample of 30 mL was transferred from a tedlar® bag into a sparger containing 5 mL of HPLC grade water, using a 30 mL syringe with a luer® tip. The samples were purged for 10 min with helium at a rate of 40 mL/min. The sample was preconcentrated on a Tenax®/Silica Gel/Charcoal Trap, desorbed for 3 min at 240°C, and passed to the analytical column for analysis. A carrier gas (ultra high purity helium) flow rate of 6 mL/min was utilized, and the column oven temperature started at 35°C for 5 min, and was ramped to 160°C at a rate of 5°C/min. The GC was calibrated to the following target analytes: ECD analytes of 1,1,1-trichloroethane (1,1,1-TCA), carbon tetrachloride, TCE, and PCE, and PID analytes of trans-1,2-dichloroethene (t-1,2-DCE), cis-1,2-dichloroethene (c-1,2-DCE), benzene, toluene, ethylbenzene, m-xylene, p-xylene, o-xylene, and naphthalene.

RESULTS AND DISCUSSION

Analysis Method Comparison

The detection limit, accuracy, and precision of the direct injection method and the purge and trap method, using both 5 mL and 25 mL spargers, were calculated to evaluate the performance of the methods. The analyses were performed for TCE and PCE, the two target analytes used in both surveys. Detection limits were estimated by placing the minimum detectable area into the respective calibration equations calculated from a four point calibration. Accuracy and precision were calculated by performing a statistical analysis on a set of ten replicate analyses. The statistical analysis involved calculation of percent recoveries for ten standard runs for each method, followed by calculation of the relative standard deviation (RSD). Percent recovery numbers provide a measure of accuracy for the

methods and the RSD provides a measure of precision. Results are
summarized in (Table 1).

TABLE 1 -- Results of analyses.

Method	TCE M%R[a]	TCE SD[b]	TCE %RSD[c]	PCE M%R	PCE SD	PCE %RSD
DI[d]	108	2	2	92	7	7
P&T(5 mL)[e]	121	9	7	110	6	15
P&T(25 mL)[f]	121	8	6	132	10	7

a.   M%R = Mean percent recovery.
b.   SD = Standard Deviation.
c.   %RSD = Percent relative standard deviation.
d.   DI - Direct injection method.
e.   P&T(5 ml) - Purge and trap method with 5 ml spargers.
f.   P&T(25 ml) - Purge and trap method with 25 ml spargers.

The detection limit of the direct injection method was
estimated to be 200 parts per billion by volume (ppb/v) using an
injection volume of 100 μl. Detection limits of 10 and 25 ppb/v
were obtained for the purge and trap method for the 25 mL and 5
mL spargers, respectively. The higher sensitivity of the purge
and trap method resulted from the larger sample volume and sample
preconcentration.

Mean percent recoveries (M%R) of 108 and 92 were obtained
for TCE and PCE, respectively, for the direct injection method.
For TCE, percent recoveries ranged from 104 to 112, while PCE
percent recoveries ranged from 80 to 103. For the purge and trap
method, mean percent recoveries of 121 and 110 were observed for
TCE and PCE, respectively, using the 5 mL spargers and 121 and
132 using the 25 mL spargers. For the 5 mL spargers, percent
recoveries ranged from 108 to 137 for TCE and from 79 to 131 for
PCE. For the 25 mL spargers, percent recoveries ranged from 108
to 135 for TCE and from 113 to 141 for PCE. These results show
the direct injection method provides better accuracy than the
purge and trap method.

For the direct injection method, RSD values of 2 and 7 were
obtained for TCE and PCE, respectively. For the purge and trap
method, RSD values of 7 and 15 were obtained for TCE and PCE,
respectively, using the 5 mL spargers, and 6 and 7 for the 25 mL
spargers. The precision was slightly better for the direct
injection method but was good for both methods.

Finally, the practicality and productivity of the two methods were compared. From the standpoint of practicality, the direct injection method required less steps, as the purge and trap method required that the spargers be drained, rinsed, and refilled with HPLC grade water. However, the purge and trap method allows for automated analysis of up to sixteen samples, without the analyst being present. This capability can be especially important when the project schedule is aggressive and a high number of analyses per day is required. Based on past experience, the direct injection method allowed for the analysis of approximately 15 to 20 samples per unit, for each 10 hr work day, and the purge and trap method 30 to 35 samples per unit per day.

CONCLUSION

It has been shown that the use of a HP 5890 Series II GC in conjunction with direct sample injection or purge and trap samplers provided quality analytical data for soil gas surveys. Greater analytical accuracy and precision were achieved via direct sample injection, while increased sensitivity and higher productivity were achieved through the use of a Tekmar LSC 2000/ALS 2016 purge and trap autosampler.

REFERENCES

[1]   Engineering Science, Inc., Standard Operating Procedure for GC Analysis of Selected Volatile Organic Compounds in Soil Gas, October 1993.

M. A. Gabr[1] and M. H. Akram[2]

SAMPLE PREPARATION TECHNIQUES FOR FILTRATION TESTING OF FLY ASH WITH
NONWOVEN GEOTEXTILES

REFERENCE: Gabr, M. A. and Akram, M. H., ''Sample Preparation Techniques
for Filtration Testing of Fly Ash with Nonwoven Geotextiles,'' Sampling
Environmental Media, ASTM STP 1282, James Howard Morgan, Ed., American
Society for Testing and Materials, 1996.

ABSTRACT: A laboratory investigation was conducted to evaluate the
applicability of five preparation techniques of fly ash/geotextile
specimens for the gradient ratio (GR) test (ASTM D 5101-90). These
techniques included the current ASTM method, a modified ASTM method, a
vibro-prep technique, a dry preparation, and slurry method. Using the ASTM
method for preparation, measured specific discharge and GR across the fly
ash-geotextile systems were $8.0x10^{-4}$ cm/sec and 1.25, respectively.
However, during the placement of the fly ash in the permeameter some
intrusion of the particles into the geotextile openings took place. A
technique for placing fly ash in the permeameter was modified after ASTM
D5101. Using this modified method, the specific discharge and GR were
measured to be $6.8x10^{-4}$ cm/sec and 0.80, respectively. The specimens
prepared using the slurry method exhibited a relatively stable behavior.
The specific discharge had an average value of $2.5x10^{-4}$ cm/sec after 600
hours of testing. The GR decreased from 1.52 to 1.37 during this period.
The extent of piping under different testing conditions was quantified in
this investigation. The specimens prepared using dry method exhibited an
unsteady behavior and piping was observed to be dominant. Although piping
of fine particles through the geotextile system was less than the
suggested limit of 0.25 g/cm^2 for soils, it is proposed that in the case
of fly ash specimens with geotextile, stable systems should be categorized
as those with piping rate less than 0.03 g/cm^2.

KEYWORDS: filtration, fly ash, geotextiles, gradient ratio, laboratory,
piping, sampling, specific discharge

---

[1] Assistant Professor, Department of Civil and Environmental Engineering,
Box 6103, West Virginia University, Morgantown, WV 26505-6103.

[2] Graduate Research Assistant, Department of Civil and Environmental
Engineering, West Virginia University, Morgantown, WV 26505.

INTRODUCTION

Fly ash is an industrial by-product from coal-burning electric utilities. The U.S. Environmental Protection Agency (USEPA) estimated that annual production of fly ash will reach 120 million tons by year 2000. At present, approximately 20% of the produced fly ash is utilized in various applications with the remainder being landfilled or deposited in surface impoundments [1]. Utilization of fly ash as a construction material is gaining a wide acceptance specially with a recent USEPA ruling that fly ash from coal burning facilities is not to be classified as hazardous waste. Fly ash can be commercially used in construction activities including landfill liners, landfill covers, highway embankments and road bases, grout mixes, and structural backfills. Such utilization is considered as means of recycling by which disposal cost is reduced while savings in the cost of natural materials are realized.

Successful utilization of fly ash in earth structures requires adequate filtration and separation of the fly ash from the surrounding environment. Geotextiles are the emerging material for filtration and drainage applications. The geotextile material acts as a filter through which drainage is allowed while ash particles are retained from migrating off site. The selection of a proper geotextile for a given application can be accomplished through the use of the gradient ratio test for soils (ASTM D 5101-90). Using the results of this test, an apparent opening size (AOS) of the geotextile can be selected to accommodate a given soil's particle size distribution.

A laboratory investigation was conducted to determine the applicability of various specimen preparation techniques for filtration testing of fly ash material with geotextile. Specimens preparation techniques investigated in this study included a Dry method, the ASTM D 5101-90 method, a Modified ASTM method, a Slurry method, and a Vibro-Prep method. A modified specimen placement procedure was devised to overcome the initial intrusion of the fly ash particles into the geotextile openings and to allow for achieving the desired 101.6 mm height of the specimen after back saturation. This placement procedure was modified after ASTM D5101-90 [2].

A nonwoven geotextile with an apparent opening size ranging from 0.15-0.22 mm was used. Approximately 80% of the fly ash specimens used in the testing passed through sieve #200 (0.074 mm). The specimen preparation techniques included a procedure to quantify the loss of fine particles during placement and during the filtration tests. Results of the gradient ratio tests using specimens prepared according to the various techniques are presented and discussed.

PREVIOUS STUDIES

Filtration using geotextile material involves the movement of liquid through the geotextile with the retention of the media particles on its upstream face. A successful filtration requires simultaneous soil retention and adequate permeability. The accumulation of fine particles at the upstream face of the geotextile has been referred to as filter cake formation. Formation of filter cakes depends on the geotextiles pore size distribution, percent opening area (porosity), and the thickness of the fabric. Lawson [3] indicated that geotextiles in filtration applications act as a catalyst for the formation of the filter cake. After this

formation, geotextiles serve only as a separation medium. The selection of
an appropriate opening size of the geotextile was considered critical to
the formation of stable and effective filter cake.

Rollin et al. [4] observed three distinct types of behavior during
long-term filtration tests, as shown in Figure 1. These included (1)
normal behavior where particles moving towards the geotextile were blocked
and soil density increased with time resulting in reduced flow; (2) piping
phase that was characterized by loss of particles through the geotextile;
and (3) combined behavior in which loss of soil particles was followed by

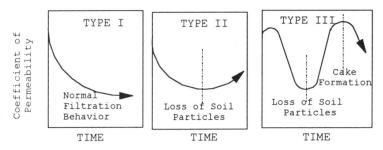

Figure 1. Long -Term Filtration Behaviors (Rollin et. al.   1985)

the formation of filter cake at the soil/geotextile interface.

Rao et al. [5] discussed the effect of specimens density and fine
contents on the long-term filtration behavior of soil-geotextile systems.
Rao et al. reported that time required to achieve a stable condition
depended on specimen density, soil fines content, and the type of
geotextile. Siva and Bhatia [6] presented the results of long-term
filtration tests performed on different geotextiles and soils using dry,
dry-saturated, slurry paste, and slurry suspension samples. Results from
their study indicated that the reduction in the specific discharge was the
highest when flow was initiated in a dry sample and was the lowest for the
slurry paste samples. Fanin et al. [7] discussed a water pluviation
technique for preparation of medium and coarse sand specimens for gradient
ratio testing. They reported that, using this technique, homogeneous and
saturated specimens can be prepared at any target density.

Nonetheless, fly ash is rheologically different from natural and
synthetic soils. For example, the majority of fly ash particles is
spherical in shape and includes crystalline matter and carbon with a
considerable percentage of its particles passes # 200 sieve. The study
presented in this paper examines different methods for preparing fly ash
specimens with geotextiles for gradient ratio testing.

MATERIAL AND METHODS

Geotextile. A nonwoven geotextile with an apparent opening size (AOS) of
70-100 sieve equivalent was used in this investigation. The properties
of the geotextile are presented in Table 1.

Fly Ash. Fly ash samples were obtained from Fort Martin Power Plant in

West Virginia. The grain size distribution of the fly ash is presented in Figure 2.

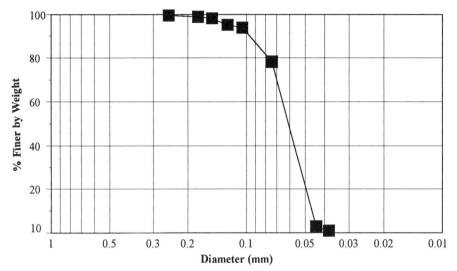

Figure 2. Sieve Analysis of Fort Martin Fly Ash

Table 1 - Characteristics of Nonwoven Geotextile

Brand	Fabric Weight (g/m^2)	Flow Rate $(l/s)/m^2$	Permittivity (sec^{-1})	AOS (Sieve size)
A	156	128.8	2.54	70-100

Permeameter for Gradient Ratio and Filtration Test. The gradient ratio test is used for evaluating the performance of the soil/geotextile system under controlled conditions. The gradient ratio is defined as the ratio of the hydraulic gradient through the soil-geotextile system to the hydraulic gradient through the soil alone. Although the test has been standardized, no standard equipment was readily available. After survey of the literature, a design was selected and permeameters were fabricated as shown in Figure 3.

Specimen Preparation Techniques. The fly ash particles are generally spherical in shape with a considerable percentage passing # 200 sieve (0.074 mm). Although the specimen preparation technique for testing of soils in a Gradient Ratio device has been stipulated in ASTM D 5101-90, no prior research exists for testing fly ash in a gradient ratio device. Five specimen preparation techniques were used in this study as follows:

Dry Preparation-No Back Saturation. The preparation technique as presented in ASTM D 5101-90 was generally followed. The geotextile sample was placed on the top of a geonet and secured between the funnel and the center section of the permeameter (Figure 3). A layer of silicone gel was applied around the geotextile to minimize in-plane flow. An end cap made

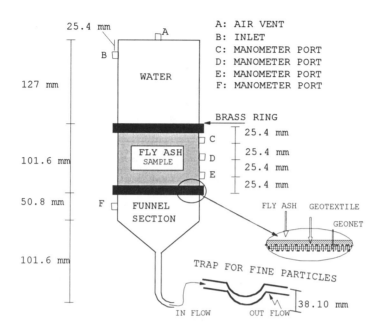

Figure 3. Schematic of a Modified Permeameter for Testing Fly Ash with Nonwoven Geotextiles

from PVC pipe was installed at the bottom of funnel stem (Figure 3)

to retain fine particles of fly ash escaping through the geotextile during placement. The fly ash specimen was placed on the top of the geotextile in layers that were 25 mm in thickness. Compaction of the specimen was achieved by tapping the sides of the permeameter six times with a wooden rod 19 mm in diameter. Carbon dioxide gas was not used for expelling air from the specimens because it resulted in pumping out fly ash particles from the manometer ports and caused specimens disturbance. The end cap was removed and the retained fine fraction was weighed. Flow was initiated on the dry specimens and monitored for filtration behavior.

Dry Preparation-ASTM. In addition to the procedure followed for dry preparation, specimens were back-saturated with degassed water for a 24 hour period under low hydraulic heads ($\approx$101.6 mm). After back-saturation was completed, fine particles retained inside the S-trap (shown in Figure 3) were weighed. The specimens were then tested for filtration and gradient ratio under hydraulic gradients of up to 6.0.

Modified ASTM. Fly ash was placed in 12.7 mm layers and the sides of the permeameter were tapped at locations oriented 60° with respect to each other. Fly ash was placed to an approximate height of 127 mm using an extension collar that was attached to the middle section of the permeameter. The specimens were back-saturated with degassed water for 24 hours under low hydraulic heads ($\approx$101.6 mm). After back-saturation was completed, fine particles retained inside the S-trap were weighed. The extension collar was removed, the top of the fly ash specimens was leveled with the excess fly ash used to evaluate the specimen's moisture content. The density of the specimen was back-calculated using specific gravity (Gs) and water content (w) data. The specimens were tested for filtration and gradient ratio under hydraulic gradients of up to 6.0.

Vibro-Prep. In order to prepare the fly ash specimens at higher densities without breaking down the particles, specimens were prepared using a vibratory table. Syntron Vibrating Table-Model VP181C conforming to the ASTM D4253-83 specifications was used in this preparation technique [8]. The table operates with a straight line vibratory motion(0-3600 vibrations per minute). The amplitude of the vibrations can be changed by regulating the rheostat knob.

The specimen of the geotextile was secured between the funnel and the center section of the permeameter (Figure 3). A Proctor mold was placed in the center of the vibrating table and the permeameter assembly was placed such that the funnel section was inside the Proctor mold. The control unit was set at the desired amplitude and the vibration was turned on. The fly ash was then placed in the permeameter using a long-stem (228 mm) table spoon in concentric circles, ensuring that the height of drop did not exceed 25 mm. The vibrations table was switched 'off' when the permeameter was filled. The top surface of the permeameter was leveled with a straight edge ruler and the specimen was weighed to determine its density. Specimens were then back-saturated with degassed water for 24 hours under low hydraulic heads (≈101.6 mm) and then tested for filtration and gradient ratio under hydraulic gradients of up to 6.0.

Slurry Preparation. Fly ash was mixed with 35-40% water by weight. Mixing was performed using a rubber spatula until a homogeneous slurry was achieved. The slurry was then poured on the top of the geotextile. The specimens were back-saturated with degassed water for 24 hours and then tested for gradient ratio and filtration characteristics.

Specimens Compression. It was noticed that during back-saturation with degassed water, the height of the specimens was reduced to less than the 101.6 mm recommended by ASTM D5101. Specimens prepared using the ASTM method were compressed approximately 12.5 mm. Specimens prepared using the modified ASTM technique compressed 8.5 to 10 mm. The specimens prepared using the vibro-preparation technique compressed 1 to 1.5 mm. In order to overcome this problem, an extension collar, 50.8 mm in height, was designed whereby the initial specimen height was increased to 127 mm. After the completion of the back saturation process (24 hours), the top of the specimen was trimmed down to the recommended 101.6 mm. The schematic of the extension-collar is presented in Figure 4.

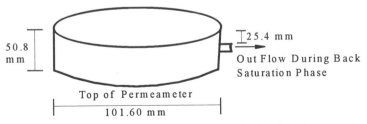

Figure 4. Extension Collar Used With the Permeameter

RESULTS AND DISCUSSION

Filtration behavior of non-woven geotextile with specimens prepared using the techniques described above was monitored as a function of time. In addition, the weight of particles passing through the geotextiles was also quantified.

ASTM Method. The filtration behavior of specimens prepared using ASTM method is presented in Figure 5 (a,b). The variation of specific discharge, defined as flow rate per unit area [6], and gradient ratio across the fly ash-geotextile system can be described by two stages. During the first stage, labeled as AB, specific discharge across the fly ash-geotextile increased from $3.9x10^{-4}$ to $9.0x10^{-4}$ cm/sec and was accompanied by a gradient ratio (GR) decrease from 1.75 to 0.95. Piping was observed across the fly ash-geotextile interface and fine particles were washed out through the geotextile. During the second stage, BC, the specific discharge decreased from $9.0x10^{-4}$ to $8.0x10^{4}$ cm/sec and was accompanied by an increase in the GR from 0.95 to 1.25. During this stage,

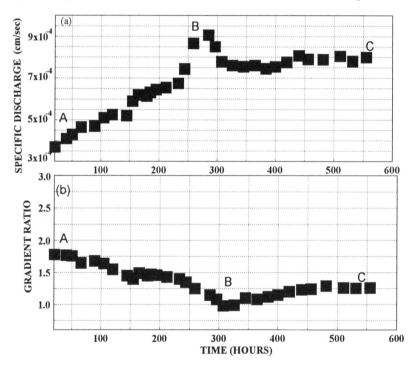

Figure 5. Filtration Behavior of Specimens Prepared Using ASTM Method: (a) Specific Discharge, and (b) Gradient Ratio

the particles of fly ash were retained at the interface which resulted in a decrease in the specific discharge and an increase in GR. A steady state flow pattern was achieved after approximately 450 hours.

Modified ASTM Method. Variation of the specific discharge and GR for specimens prepared using the modified ASTM D 5101-90 procedure is presented in Figure 6 (a,b). The pattern of flow regime across the fly ash-geotextile system can be divided into three stages. During the first stage, AB, the specific discharge increased from $2.0x10^{-4}$ to $7.8x10^{-4}$ cm/sec and was accompanied by a decrease in the GR from 1.5 to 0.95. Piping was the dominant phenomenon. During the second stage, BC, the specific discharge decreased from $7.8x10^{-4}$ to $5.8x10^{-4}$ cm/sec. It was accompanied by an initial increase in GR from 0.95 to 1.10 which later decreased to 0.90. During the initial phase of the second stage, blinding and blocking mechanisms were prevalent. During the third stage, CD, the specific discharge increased from $5.8x10^{-4}$ to $7.0x10^{-4}$ cm/sec and then attained a stable behavior after 450 hours. The GR decreased from 0.90 to 0.8 during this stage.

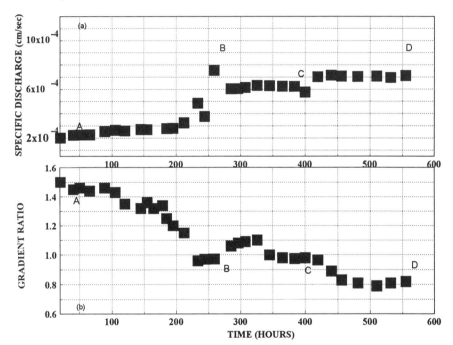

Figure 6. Filtration Behavior of Specimens Prepared using Modified ASTM
Procedure: (a) Specific Discharge, and (b) Gradient Ratio

Vibro-Prep. The density of the specimens prepared using this method increased with increasing the vibration amplitude. The density of specimens at different amplitudes is presented in Table 2. It was noticed that rheostat setting beyond 50 resulted in the segregation of particles.
    The variation of specific discharge and GR for specimens prepared using the vibrating table with rheostat position at 12 is presented in Figure 7(a,b). This density is similar to that of the specimens prepared using ASTM method and modified ASTM technique.

Table 2. Effect of Amplitude on Specimen Density

Rheostat Position	Density pcf(kN/m^3)	Standard Deviation pcf(kN/m^3)
0	66.02(10.56)	0.271(0.04)
2	76.59(12.25)	0.271(0.04)
6	77.62(12.42)	0.136(0.02)
8	80.12(12.82)	0.241(0.04)
12	80.44(12.87)	0.241(0.04)
14	83.42(13.35)	0.138(0.02)
25	88.59(14.17)	0.200(0.03)
40	88.47(14.16)	0.190(0.03)
50	89.17(14.27)	0.198(0.03)

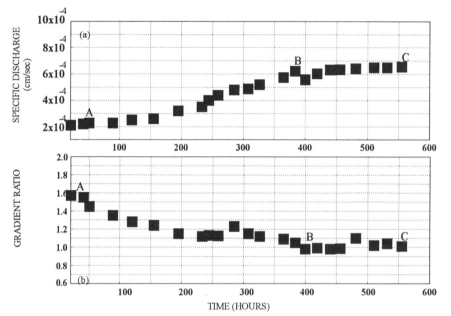

Figure 7. Filtration Behavior of Specimens Prepared using Vibrating Table: (a) Specific Discharge, and (b) Gradient Ratio

Test results for these specimens indicated that the pattern of specific discharge and GR can be divided into two stages. During the first stage, AB, the specific discharge increased from 2.0×10^{-4} to 6.0×10^{-6} cm/sec and was accompanied by a decrease in GR from 1.6 to 1.0. Piping was observed to be the dominant mechanism. During the second stage, BC, the system appeared to be stabilizing after approximately 450 hours. The specific discharge attained a value of 6.5×10^{-4} cm/sec with a GR of 1.0.

Dry Method. The filtration behavior of specimens prepared using the dry method is presented in Figure 8. The flow regime across the fly ash-geotextile system was dominated by piping mechanisms. The specific discharge increased from 1.8×10^{-4} to 1.5×10^{-3} cm/sec during approximately

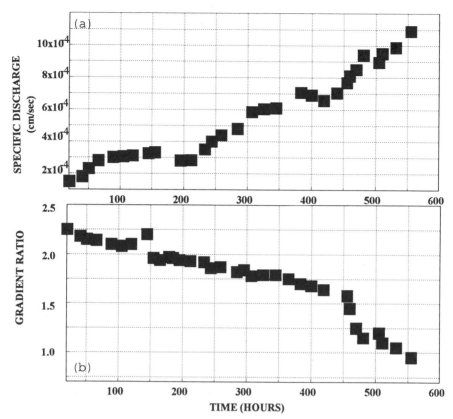

Figure 8. Filtration Behavior of Specimens Prepared Using Dry
Method: a) Specific Discharge, and (b) Gradient Ratio

550 hours of testing. This increase was accompanied by a decrease in GR
from 2.5 to 0.95. No steady state behavior was observed after the 560
hours of testing.

Slurry Method. The variation of specific discharge and GR for the
specimens prepared using the slurry method is presented in Figure 9(a,b).
The specific discharge exhibited a relatively stable behavior with an
average value of 2.5×10^{-4} cm/sec. The GR decreased from 1.52 to 1.37 after
600 hours of testing. It appears that sorting of particles was taking
place within the specimens which resulted in a slight unsteady behavior
and GR variation.

Comparison of Filtration Behavior. The comparison of filtration behavior
of specimens prepared using the ASTM method, the Modified ASTM method, the
Vibro-Prep method, the Dry method and the Slurry method is presented in
Figure 10. Maximum variation in specific discharge was observed for
specimens prepared using the dry method; 1.8×10^{-4} to 1.5×10^{-3} cm/sec. The
least was measured for the slurry specimens, which exhibited a generally
stable behavior with an average specific discharge of 2.5×10^{-4} cm/sec. The

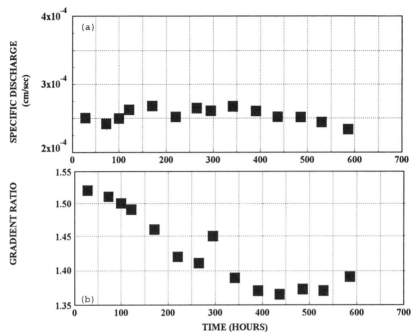

Figure 9. Filtration Behavior of Specimens Prepared Using the Slurry
Method:(a) Specific Discharge, and (b) Gradient Ratio

average specific discharge for all methods was estimated to be
approximately 6.5×10^{-4} cm/sec. The specific discharge for specimens
prepared using vibro-prep method was 5.8×10^{-4} cm/sec. This method had the
advantage of preparing specimens at different densities with a better
quality control.

Fly Ash-Geotextile System Stability. Lafleur [9] reported that stable
soil-geotextile systems can be categorized as those in which the amount of
soil passing through the filter was less than 0.25 g/cm^2. In order to
evaluate the piping potential, the fly ash passing through the geotextile
during specimens preparation and after the termination of the filtration
tests was collected and weighed. The weight of fly ash passed through
the geotextile is presented in Table 3.

Table 3.   Weight of Soil Passed for Different   Specimens

Specimen Preparation Technique	Weight of Soil Passed		
	Preparation	Filtration	Total Rate
ASTM Method	0.45 grams	1.25 grams	0.020 g/cm^2
Modified ASTM	0.45 ,,	1.00 ,,	0.018 ,,
Vibro-Prep	1.55 ,,	1.15 ,,	0.033 ,,
Dry Method	0.50 ,,	2.00 ,,	0.031 ,,
Slurry Method	0.10 ,,	1.00 ,,	0.014 ,,

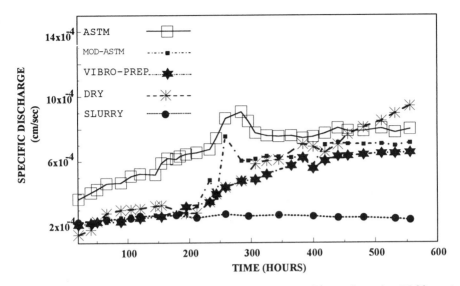

Figure 10. Comparison of Testing Results Using the Different
Specimen Preparation Techniques

Results suggest that in the case of fly ash specimens with geotextile, stable systems should be categorized as those in which the rate of piping is less than 0.03 g/cm^2 instead of the 0.25g/cm^2 specified for soils.

CONCLUSIONS

The filtration behavior of fly ash specimens prepared using five methods was investigated. Based on the results of this investigation, the following conclusions can be drawn:

1. The filtration behavior across the fly ash-geotextile specimens is affected by the method of specimen preparation.

2. Results indicated that specific discharge and GR across the fly ash-geotextile systems were 8.0x10^{-4} cm/sec and 1.25, respectively, using the ASTM method, and 7.0x10^{-4} cm/sec and 0.80, respectively using the modified ASTM method.

3. Specimens at different target densities were successfully prepared using vibro-prep technique. The target densities (10 to 14 kN/m^3) were achieved with a standard deviation varying from 0.02 to 0.04 kN/m^3. This indicated that filtration behavior of specimens with simulated field densities can be tested in the laboratory.

4. The filtration behavior of specimens prepared using vibro-prep technique exhibited a relatively stable behavior as compared to specimens prepared using the dry method, the ASTM method and the modified ASTM method. The specific discharge attained a value of 6.5x10^{-4} cm/sec with a GR of 1.0 after 600 hours of testing.

5. The specimens prepared using the dry method exhibited an unsteady behavior and piping was observed to be dominant. These specimens exhibited continuous piping for the test duration. The specimens prepared using the slurry method exhibited a relatively stable behavior.

6. All systems can be categorized as stable except specimens prepared using the dry method. The piping of fine particles through the geotextile system was less than the suggested limit for soils of 0.25 g/cm^2. This may suggest that the criterion established for soils is inadeqate when fly ash material is used.

ACKNOWLEDGMENTS

This study was supported by the National Research Center for Coal and Energy. The authors would like to thank Dr. Carl Irwin, the project manager, for his interest in the work. In addition the support provided by Monongahela Power Company and a number of geotextile manufacturing companies is greatly appreciated.

REFERENCES

[1] USEPA, "Wastes from the Combustion of Coal by Electric Utility Power Plants," USEPA Report 530-sw-88-002, 1988.

[2] ASTM, "Standard Test Method for Measuring the Soil-Geotextile System Clogging Potential by the Gradient Ratio Method," ASTM Standards on Geosynthetics, 1993, pp. 88-94

[3] Lawson, C. R, "Filter Criteria for Geotextiles : Relevance and Use," J. of Geotech. Engg.Div., ASCE Vol. 108(GT10), 1982, pp. 1300-1317

[4] Rollin, A. L., Broughton, R. S. and Bolduc, G. " Synthetic Envelopment Materials for Subsurface Drainage Tubes," Paper Presented at CPTA Annual Meeting, 1985, Fort Lauderdale, FL.

[5] Rao, G. V., Gupta, K. K. and Pradhan, M. P. S. "Long-Term Filtration behavior of Soil-Geotextile Systems," Geotech. Testing J., GTJODJ, Vol. 15(3), pp. 238-247.

[6] Siva, U. and Bhatia, S. K. "Filtration Performance of Geotextile with Fine-Grained Soils," Geosynthetics'93-Vancouver, Canada, pp. 483-499.

[7]. Fanin, R. J., Vaid, Y. P. and Shi, Y. "A Critical evaluation of the Gradient Ratio Test," Geotech. Testing Journal, GTJODJ, Vol. 17(1), 1994, pp. 35-42.

[8] ASTM ,"Standard Test Methods for Maximum Index density of Soils using a Vibratory Table," ASTM Annual Book of Standards, Section 4.08 Soil and Rock, 1993, D4253-83, ASTM, Philadelphia, PA, pp. 554-565.

[9] Lafleur, J., Mlynarek, J. and Rollin, A. L. "Filtration of Broadly Graded Cohesionless Soils," J. of Geotech. Engg. Vol. 115(12), 1984, pp. 1747-1768.

Author Index

Subject Index